Signal and System Fundamentals in MATLAB and SIMULINK

Mohammad Nuruzzaman

Electrical Engineering Department
King Fahd University of Petroleum & Minerals
Dhahran, Saudi Arabia

BookSurge Publishing
7290 B Investment Drive
Charleston, SC 29418

Dr. Mohammad Nuruzzaman
Electrical Engineering Department
King Fahd University of Petroleum and Minerals
KFUPM BOX 1286
Dhahran 31261, Saudi Arabia
Email: nzaman@kfupm.edu.sa
Web Link: http://faculty.kfupm.edu.sa/EE/NZAMAN/

ISBN-10: 1-4196-9934-2
ISBN-13: 978-1419699344

Printed in the United States of America

This book is printed on acid-free paper.

ISBN 978-1-4196-9934-4
51899 >
9 781419 699344

To my parents
Mohammad Shamsul Haque & Nurbanu Begum

Preface

The text *Signal and System Fundamentals in MATLAB and SIMULINK* is intended to serve as a basic reference for signal and system course taught in undergraduate curriculum of electrical engineering around the world. The book stresses concept-to-implementation approach of signal and system terminologies in a facilitating fashion. Core understanding of signal and system is extremely essential in the sense that these elementary topics pave the way for understanding of special branch like power system, control system, communication, signal processing whether analog or digital, and many others. Today's signal and system based education and research are intertwined with the software and hardware activities. Concentration on hardware for the most part goes to industry and research in the field. Academic community in the field has a slight bias towards the software activity. It is this motivation which drives us write the text.

Twenty first century students do not just read and pass the examinations. Superb advances in computer technology specially the easy accessibility to the personal computer based window system quite factually make the software tools available for study and analysis of signal associated problems almost to every student's reach. MATLAB which is the short for matrix laboratory is a great companion software. It offers so many advantages to solving electrical engineering problems that we find one toolbox (a family of MATLAB built-in functions) for all majors of the engineering.

Since MATLAB is rich in possessing function-files, virtual electric systems can quickly be formed in MATLAB without rigorous programming. Few words from keyboard implement so much of the signal and system that one appreciates using the computational capability of the software. SIMULINK whose elaboration is SIMUlation and LINK is devised to operate in the platform of MATLAB and can be regarded as an additional part of MATLAB. So to activate SIMULINK, one needs to activate MATLAB first. If MATLAB is a world of matrices, then SIMULINK is the world of blocks. Most programming aspects in SIMULINK happen through dialog window or graphical user interface (GUI) instead of typing in MATLAB Command Window or entering through an M file (source codes of MATLAB are written in an M file). Pre-designed blocks of SIMULINK added another enhancing feature for signal and system study.

Signal and system classroom problems are solved in a systematic procedure and in a tutorial fashion throughout the text. Chapter 1 presents a brief introduction to MATLAB and SIMULINK's getting-started features. Signal sample generation is addressed in chapter 2 mainly in MATLAB context whereas chapter 3 separately demonstrates modeling of signals in SIMULINK. Usage of linear signal operations is widespread in practical applications. Basic linear signal operation techniques are found in chapter 4. At the heart of signal processing, transform domain based analysis plays a vital role. Signal and system study is never complete without the signal transforms on whose account chapters 5, 6, and 7 address the problems and issues of Fourier, Laplace, and Z transforms. Chapter 8 develops the background of system formation. A parallel treatment is drawn wherever possible to focus the code writing or modeling approach to form a system mostly in continuous time. Ultimate goal of the subject is to make the reader able to analyze a system exposed to a variety of inputs and to identify the responsive behavioral difference on these inputs. Keeping this in center of attention, chapter 9 pedagogically demonstrates neoteric tools for the analysis. Nevertheless chapter 10 highlights miscellaneous topics related to the subject which we believe are important for the course material study. Finally chapter 11 randomly presents the exercise problems under the heading mini project ranging from the simplest to toughest with the solutions whatsoever.

I wish to express my acknowledgement to the King Fahd University of Petroleum and Minerals (KFUPM). I earnestly appreciate the library facilities that I received from the King Fahd University. All signal and system problems covered in the text have been solved or simulated by a Pentium IV Laptop operated on Microsoft Windows.

Mohammad Nuruzzaman

Table of Contents

Chapter 1
Introduction to MATLAB and SIMULINK

Chapter 2
Signal Sample Generation

Chapter 3
Signal Modeling

Chapter 4
Signal Operations

Chapter 5
Fourier Series and Transform

Chapter 6
Laplace Transform

Chapter 7
Z Transform

Chapter 8
System Implementation

Chapter 9
System Analysis

Chapter 10

Miscellaneous Signal Topics

Chapter 11

Mini Problems on Computation/Simulation

Appendices

Chapter 1

Introduction to MATLAB and SIMULINK

MATLAB is a computing software, which provides the quickest and easiest way to compute scientific and technical problems and visualize the solutions. As worldly standard for simulation and analysis, engineers, scientists, and researchers are becoming more and more affiliated with MATLAB and SIMULINK. The general questionnaires about MATLAB or SIMULINK platform before one gets started with are the contents of this chapter. SIMULINK is designed to function in the platform of MATLAB. Much of the MATLAB computational approach presupposes that the element to be handled is a vector or matrix. Whereas, SIMULINK quickly and accurately maps a technical problem into computer model through elementary blocks. Our highlight covers the following:

- MATLAB and SIMULINK features found in the MATLAB command window
- The easiest and quickest way to get started in MATLAB/SIMULINK beginning from scratch
- Frequently encountered questions while working in MATLAB/SIMULINK environment
- Relevant introductory topics and forms of assistance about MATLAB/SIMULINK

1.1 What is MATLAB?

MATLAB is mainly a scientific and technical computing software whose elaboration is matrix laboratory. The command prompt of MATLAB (>>) provides an interactive system. In the workspace of MATLAB, most data element is dealt as a matrix without dimensioning. The package is incredibly advantageous for the matrix-oriented computations. MATLAB's easy-to-use platform enables us to compute and manipulate matrices, perform numerical analysis, and visualize different variety of one/two/three dimensional graphics in a matter of second or seconds without conventional programming in FORTRAN, PASCAL, or C.

1.1.1 MATLAB's opening window features

If you do not have MATLAB installed in your personal computer, contact MathWorks (owner and developer, www.mathworks.com) for the installation CD. If you know how to get in MATLAB and its basics, you can skip the chapter. Assuming the package is installed in your system, run MATLAB from the Start of the Microsoft Windows. Let us get familiarized with MATLAB's opening window features. Figure 1.1(a) shows a typical firstly opened MATLAB window. Depending on the desktop setting or MATLAB version, your MATLAB window may not look like the figure 1.1(a) but the descriptions of the features by and large are appropriate.

❖❖ Command prompt of MATLAB

Command prompt means that you tell MATLAB to do something from here. As an interactive system, MATLAB responds to user through this prompt. MATLAB cursor will be blinking after >> prompt once you open MATLAB that says MATLAB is ready to take your commands. To enter any command, type executable MATLAB statements from keyboard and to execute that, press the Enter key (the symbol ↵ for the '**Hit the Enter Key**' operation).

❖❖ MATLAB Menu

MATLAB is accompanied with six submenus namely File, Edit, Debug, Desktop, Window, and Help. Each submenu has its own features. Use the mouse to click different

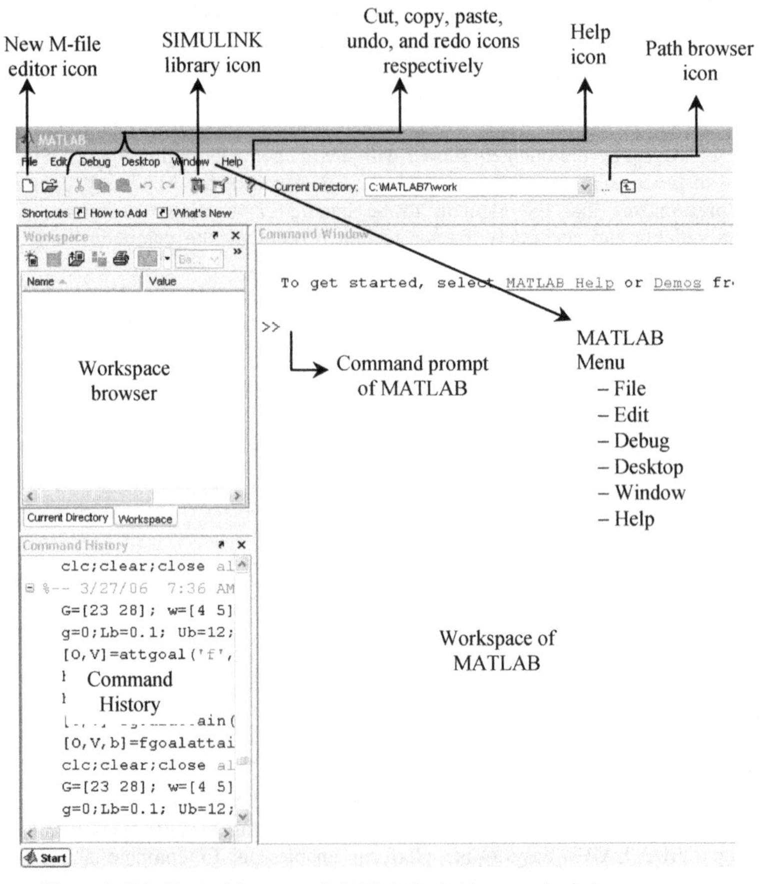

Figure 1.1(a) Typical features of MATLAB's firstly opened window

submenus and their brief descriptions are as follows:

Submenu File: It (figure 1.1(b)) opens a new M-file, figure, model, or Graphical User Interface (GUI) layout maker, opens a file which was saved before, loads a saved workspace, imports data from a file, saves the workspace variables, sets the required path to execute a file, prints the workspace, and keeps the provision for changing the command window property.

Figure 1.1(b) Submenu File

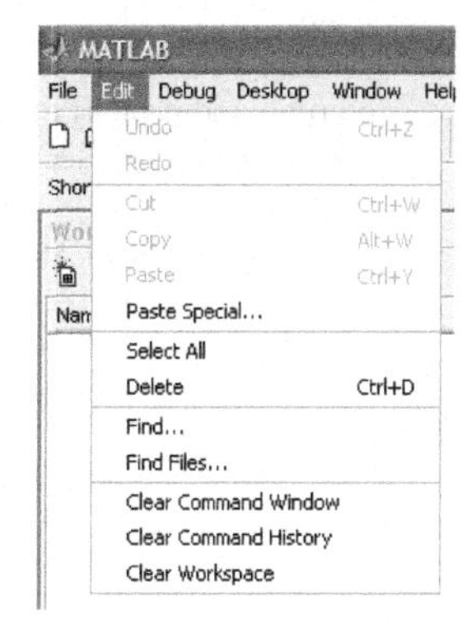

Figure 1.1(c) Submenu Edit

Submenu Edit: The second submenu Edit (figure 1.1(c)) includes cutting, copying, pasting, undoing, and clearing operations. These operations are useful when you frequently work at the command prompt.

Submenu Debug: The submenu Debug (figure 1.1(d)) is mainly related with the text mode or M-file programming.

Submenu Desktop: The fourth submenu Desktop (figure 1.1(e)) is accompanied with MATLAB command window viewing functions such as displaying the

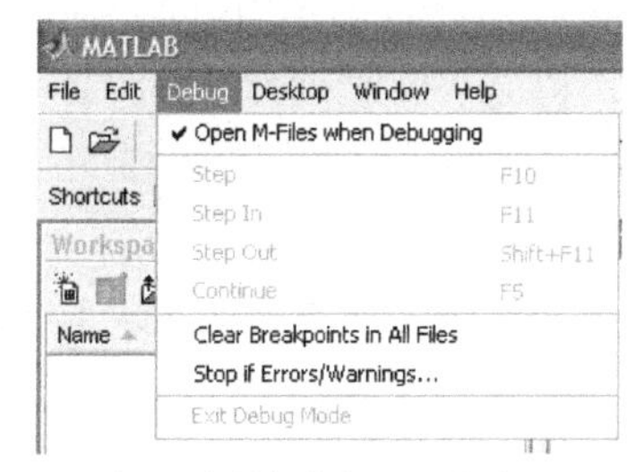

Figure 1.1(d) Submenu Debug

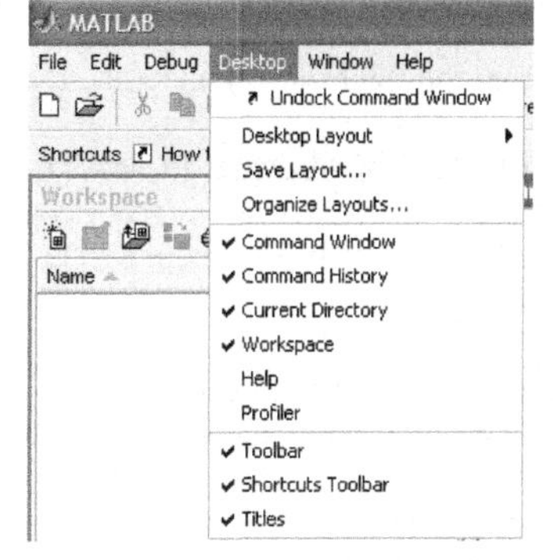

Figure 1.1(e) Submenu Desktop

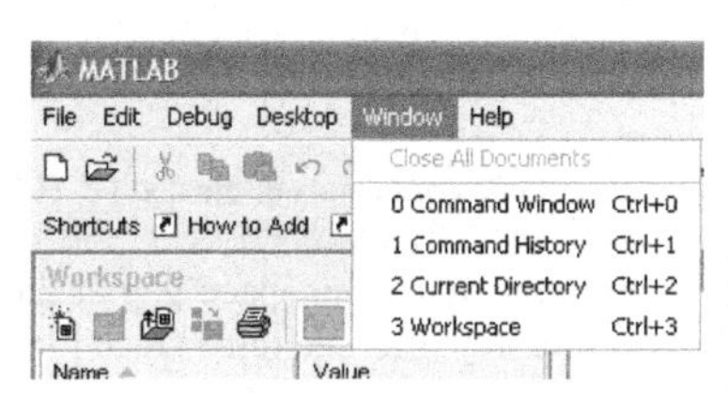

Figure 1.1(f) Submenu Window

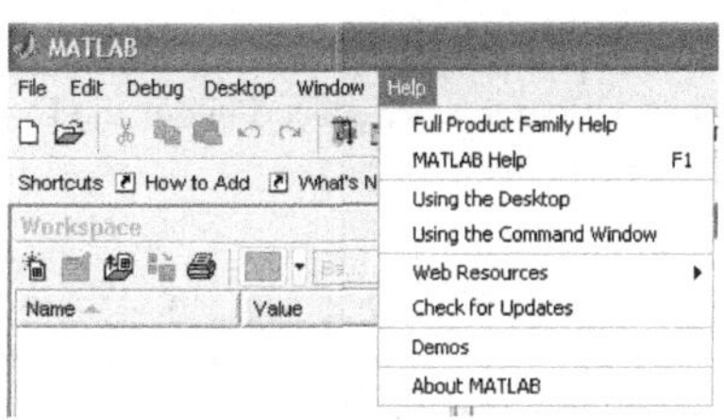

Figure 1.1(g) Submenu Help

workspace variable information, current directory information, command history, etc.

Submenu Window: You may open some graphics window from MATLAB command prompt or running some M-files. From the fifth submenu Window (figure 1.1(f)), one can see how many graphics window under MATLAB are open and can switch from one window to other by clicking the mouse to the required window.

Submenu Help: MATLAB holds abundant help facilities. The last submenu shows Help (figure 1.1(g)) in different ways. Latter in this chapter, we mention how one gets specific help. The submenu also provides the easiness to get connected with the MathWorks Website provided that your system is connected to Internet.

❖❖ Icons

Available icons are shown in the icon bar (down the menu bar) of the figure 1.1(a). Frequently used operations such as opening a new file, opening an existing file, getting help, etc are found in the icon bar so that the user does not have to go through the menu bar over and over.

❖❖ MATLAB workspace

Workspace (figure 1.1(a)) is the platform of MATLAB where one executes MATLAB commands. During execution of commands, one may have to deal with some input and output variables. These variables can be one-dimensional array, multi-dimensional array, characters, symbolic objects, etc. Again to deal with graphics window, we have texts, graphics, or object handles. Workspace holds all those variables or handles for you. As a subwindow of the figure 1.1(a), its browser exhibits the types or properties of those variables or handles. If the browser is not seen in the opening window of MATLAB, click the Desktop down Workspace in the menu bar to bring the subwindow (figure 1.1(e)).

❖❖ MATLAB command history

There is a subwindow in the figure 1.1(a) called Command History which holds all previously used commands at the command prompt. Depending on the desktop setting, it may or may not appear during the opening of MATLAB. If it is not seen, click the Command History from the figure 1.1(e) under the Desktop.

1.1.2 How to get started in MATLAB?

New MATLAB users face a common question how one gets started in MATLAB. This tutorial is for the beginners in MATLAB. Here we address the terms under the following bold headings.

❖❖ How one can enter a vector/matrix

The first step is the user has to be in the command window of MATLAB. Look for the command prompt >> in the command window. One can type anything from the keyboard at the command prompt. Row or column matrices are termed as vectors. We intend to enter the row matrix R=[2 3 4 −2 0] into the workspace of MATLAB. Type the following from the keyboard at the command prompt:

>>R=[2 3 4 -2 0] ← Arial font set is used for executable commands in the text i.e. R⇔R

There is one space gap between two elements of the matrix R but no space gap at the edge elements. All elements are placed under the [ ]. Press Enter key after the third brace] from the keyboard and we see

```
R =
        2   3   4  -2   0
>>      ← command prompt is ready again
```

It means we assigned the row matrix to the workspace variable R. Whenever we call R, MATLAB understands the whole row matrix. Matrix R is having five elements. Even if R had 100 elements, it would understand the whole matrix that is one of many appreciative features of MATLAB. Next we wish to enter the column matrix $C=\begin{bmatrix} 7 \\ 8 \\ 10 \\ -11 \end{bmatrix}$. Again type the following from the keyboard at the blinking cursor:

```
>>C=[7;8;10;-11] ↵        you will see (↵ means 'Press the Enter Key'),

C =
        7
        8
       10
      -11
>>                         ← command prompt is ready again
```

This time we also assigned the column matrix to the workspace variable C. For the column matrix, there is one semicolon ; between two consecutive elements of the matrix C but no space gap is necessary. As another option, the matrix C could have been entered by writing C=[7 8 10 -11]'. The operator ' of the keyboard is the matrix transposition operator in MATLAB. As if you entered a row matrix but at the end just the transpose operator ' is attached. After that the rectangular matrix $A=\begin{bmatrix} 20 & 6 & 7 \\ 5 & 12 & -3 \\ 1 & -1 & 0 \\ 19 & 3 & 2 \end{bmatrix}$ is to be entered:

```
>>A=[20 6 7;5 12 -3;1 -1 0;19 3 2] ↵     you will see,

A =
         20     6     7
          5    12    -3
          1    -1     0
         19     3     2
```

Two consecutive rows of A are separated by semicolon ; and consecutive elements in a row are separated by one space gap. Instead of typing all elements in a row, one can type the first row, press Enter key, the cursor blinks in the next line, type the second row, and so on.

◆◆ How one can use the colon and semicolon operators

The operators semicolon ; and colon : have special significance in MATLAB. Most MATLAB statements and M-file programming use these two operators almost in every line. Generation of vectors can easily be performed by the colon operator no matter how many elements we need. Let us carry out the following at the command prompt to see the importance of the colon operator:

```
>>A=1:4 ↵                 you will see,

A =
       1   2   3   4      ← We created a vector A or row matrix where A=[1   2   3   4]
```

Let us interact with MATLAB by the following commands:

```
>>R=1:3:10 ↵              you will see,

R =
       1   4   7   10     ← We created a vector or row matrix R whose elements form an
                            arithmetic progression with first element 1, last element 10,
                            and common difference or increment 3
```

Vector with decrement can also be generated:

```
>>C=[0:-2:-10]' ↵         you will see,

C =
```

```
          0
         -2
         -4      ← We created a vector or column matrix C whose
         -6        consecutive elements have the decrement 2 with the
         -8        first element 0 and the last element −10
        -10
```

MATLAB is also capable of producing vectors whose elements are decimal numbers. Let us form a row matrix R whose first element is 3, last element is 6, and increment is 0.5 and which we accomplish as follows:

```
>>R=3:0.5:6 ↵        you will see,

R =
      3.0000   3.5000   4.0000   4.5000   5.0000   5.5000   6.0000
```

Then, what is the use of the semicolon operator? Append a semicolon at the end in the last command and execute that:

```
>>R=3:0.5:6; ↵       you will see,
>>                                   ← Assignment is not shown
```

Type R at the command prompt and press Enter:

```
>>R ↵

R =
      3.0000   3.5000   4.0000   4.5000   5.0000   5.5000   6.0000
```

It indicates that the semicolon operator prevents MATLAB from displaying the contents of the workspace variable R.

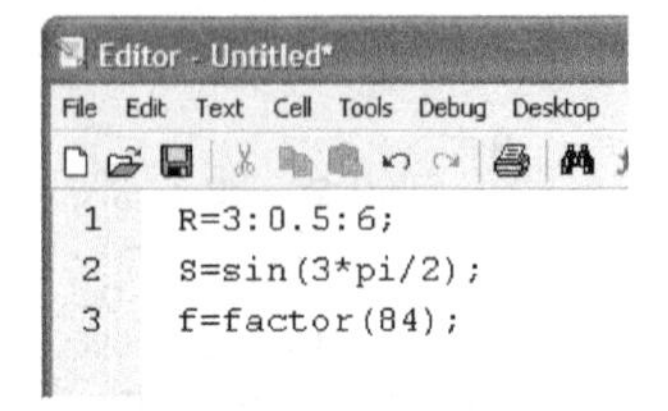

Figure 1.2(a) Last three executed statements are typed in the M-file editor of MATLAB

♦♦ How one can call a built-in MATLAB function

In MATLAB, thousands of M-files or built-in function files are executed. Knowing the descriptions of the function, the numbers of input and output arguments, and the nature of the arguments is mandatory in order to execute a built-in function file at the command prompt. Let us start with a simplest example. We intend to find $\sin x$ for $x = \frac{3\pi}{2}$ which should be −1. The MATLAB counterpart (appendix B) of $\sin x$ is sin(x) where x can be any real or complex number in radians and can be a matrix too. The angle $\frac{3\pi}{2}$ is written as 3*pi/2(π is coded by pi) and let us perform it as follows:

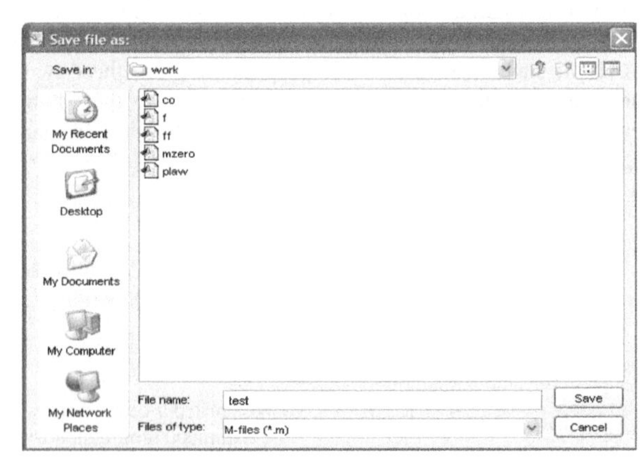

Figure 1.2(b) Save dialog window for naming the M-file

```
>>sin(3*pi/2) ↵

ans =
     -1
```

By default the return from any function is assigned to workspace ans. If you wanted to assign the return to S, you would write S=sin(3*pi/2);. As another example, let us factorize the integer 84 ($84=2\times2\times3\times7$). The MATLAB built-in function **factor** finds the factors of an integer and the implementation is as follows:

```
>>f=factor(84) ↵

f =
        2   2   3   7
```

The output of the **factor** is a row matrix which we assigned to workspace **f** in fact the **f** can be any user-given name. Thus you can call any other built-in function from the command prompt provided that you have the knowledge about the calling of inputs to and outputs from the function.

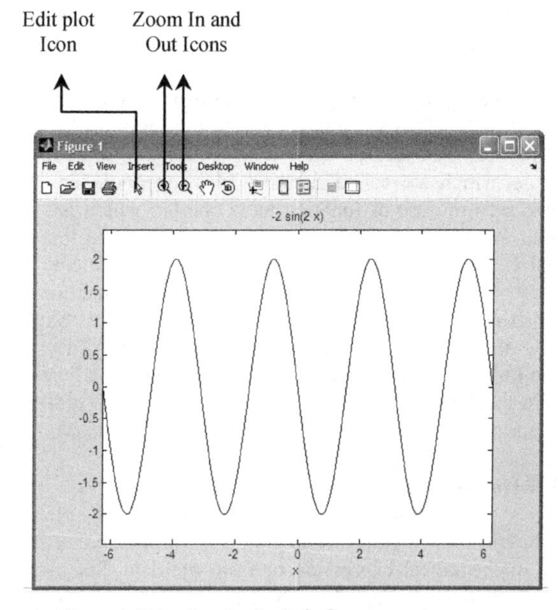

Figure 1.2(c) Graph of $-2\sin 2x$ versus x

♦♦How one can open and execute an M-file

This is the most important start up for the beginners. An M-file can be regarded as a text or script file. A collection of executable MATLAB statements are the contents of an M-file. Ongoing discussion made you familiarize with entering a matrix, computing a sine value, and factorizing an integer. These three executions took place at the command prompt. They can be executed from an M-file as well. This necessitates opening the M-file editor. Referring to the figure 1.1(b), you find the link for the M-file editor as **File → New → M-file** and click it to see the new untitled M-file editor. Another option is click the New M-file editor icon of the figure 1.1(a). However after opening the new M-file editor, we type the last three executable statements in the untitled file as shown in the figure 1.2(a). The next step is to save the untitled file by clicking the Save icon or from the File Menu of the M-file editor window. Figure 1.2(b) presents the File Save dialog window. We type the file name as the **test** (can be any name of your choice) in the slot of the File name in the window. The M-file has the file extension **.m** but we do not type **.m** only the file name is enough. After saving the file, let us move on to the MATLAB command prompt and conduct the following:

```
>>test ↵
>>          ← command prompt is ready again
```

It indicates that MATLAB executed the M-file by the name **test** and is ready for the next command. We can check calling the assignees whether the previously performed executions occurred exactly as follows:

```
>>R ↵

R =
      3.0000   3.5000   4.0000   4.5000   5.0000   5.5000   6.0000
>>S ↵                                   >>f ↵

S =                                     f =
      -1                                       2   2   3   7
```

This is what we found before. Thus one can run any executable statements in the M-file. The reader might ask in which folder or path the file **test** was saved. Figure 1.1(a) shows one slot for the Current Directory in the upper middle portion of the window. That is the location of your file. If you want to save the M-file in other folder or directory, change your path by clicking the path browser icon (figure 1.1(a)) before saving the file. When you call the **test** or any other file from the command prompt of MATLAB, the command prompt must be in the same directory where the file is in or its path must be defined to MATLAB.

♦ ♦ The input and output arguments of a function file

MATLAB is a collection of thousands of M-files. Some files are executed without any return and some return results which are called function files (appendix E). You have seen the use of function sin(x) before which has one input argument x. The statement test(x,y) means that the test is a function file which has two input arguments – x and y. Again the test(x,y,z) means the test is a function file which needs three input arguments – x, y, and z. Similar style also follows for the return but under the third brace. The [a,b]=test(x,y) means there are two output arguments from the test which are a and b and the [a,b,c]=test(x,y) means three returns from the test which are a, b, and c.

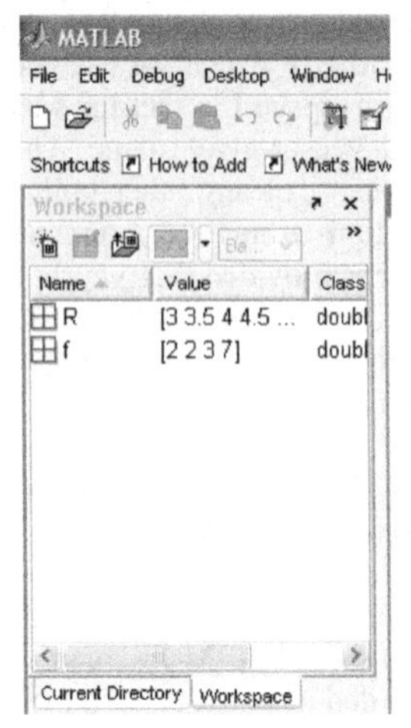

Figure 1.2(d) Workspace browser displays variable information

♦ ♦ How one can plot a graph

MATLAB is very convenient for plotting different sorts of graphs. The graphs are plotted either from the mathematical expression or from the data. Let us plot the function $y = -2\sin 2x$. The MATLAB function ezplot plots y versus x type graph taking the expression as its input argument. The MATLAB code (appendix B) for the function $-2\sin 2x$ is -2*sin(2*x). The functional code is input argumented by using the single inverted comma hence we conduct the following at the command prompt:

```
>>ezplot('-2*sin(2*x)') ↵
```

Figure 1.2(c) presents the outcome from above execution. The window in which the graph is plotted is called the MATLAB figure window. Any graphics is plotted in the figure window, which has its own menu (such as File, Edit, etc) as shown in the figure 1.2(c).

1.1.3 Some queries about MATLAB environment

Users need to know the answers to some questions when they start working in MATLAB. Some MATLAB environment related queries are presented as follows:

How to change the numeric format?

When you perform any computation at the command prompt, the output is returned up to the four decimal display due to the short numeric format which is the default one. There are other numeric formats also. To reach the numeric format dialog box, the clicking operation sequence is MATLAB command window ⇒ File ⇒ Preferences ⇒ Command Window ⇒ Numeric Format under Text Display ⇒ popup menu for available options.

How to change the font or background color settings?

One might be interested to change the background color or font color while working in the command window. The clicking sequence is MATLAB command window ⇒ File ⇒ Preferences ⇒ Fonts or Colors.

How to delete some/all variables from the workspace?

In order to delete all variables present in the workspace, the clicking sequence is MATLAB command window ⇒ Edit ⇒ Clear Workspace (figure 1.1(c)). If you want to delete a particular workspace variable, select the concern variable by using the mouse pointer in the workspace browser (assuming that it is open like the figure 1.2(d)) and then rightclick ⇒ delete.

How to clear workspace but not the variables?

Once you conduct some sessions at the command prompt, monitor screen keeps all interactive sessions. You can clear the screen contents without removing the variables present in the workspace by the command clc or performing the clicking operation MATLAB command window ⇒ Edit ⇒ Clear Command Window (figure 1.1(c)).

How to know the current path?

In the upper portion of the figure 1.1(a), the current directory bar is located which indicates in which path the command prompt is or execute cd (abbreviation for the current directory) at the command prompt.

How to see different variables in the workspace?

There are two ways of viewing this – either use the command who or look at the workspace browser (like figure 1.2(d)) which exhibits information about workspace variables for example R is the name of the variable which holds some values and their data class is double precision. One can view, change, or edit the contents of a variable by doubleclicking the concern variable situated in the workspace browser as conducted in Microsoft Excel.

How to enter a long command line?

MATLAB statements can be too long to fit in one line. Giving a break in the middle of a statement is accomplished by the ellipsis (three dots are called ellipsis). We show that considering the entering of the vector x=[1:3:10] as follows:

```
>>x=[1:3: . . . ↵
          10] ↵
x =
          1    4    7    10
```

Typing takes place in two lines and there is one space gap before the ellipsis.

Editing at the command prompt

This is advantageous specially for them who work frequently in the command window without opening an M-file. Keyboard has different arrow keys marked by ← ↑ → ↓. One may type a misspelled command at the command prompt causing error message to appear. Instead of retyping the entire line, press uparrow (for previous line) or downarrow (for next line) to edit the MATLAB statement. Or you can reexecute any past statement this way. For example we generate a row vector 1 to 10 with increment 2 and assign the vector to x. The necessary command is x=1:2:10. Mistakenly you typed x+1:2:10. The response is as follows:

```
>>x+1:2:10 ↵
??? Undefined function or variable 'x'.
```

You discovered the mistake and want to correct that. Press ↑ key to see,

```
>>x+1:2:10
```

Edit the command going to the + sign by using the left arrow key or mouse pointer. At the prompt, if you type x and press ↑ again and again, you see used commands that start with x.

Saving and loading data

User can save workspace variables or data in a binary file having the extension .mat. Suppose you have the matrix A= $\begin{bmatrix} 3 & 4 & 8 \\ 0 & 2 & 1 \end{bmatrix}$ and wish to save A in a file by the name data.mat. Let us carry out the following:

```
>>A=[3 4 8;0 2 1]; ↵ ← Assigning the A to A
```

Now move on to the workspace browser (figure 1.2(d)) and you see the variable A including its information located in the subwindow. Bring the mouse pointer on the A, rightclick the mouse, and click the Save As. The Save dialog window appears and type only data (not the data.mat) in the slot of File name. If it is necessary, you can save all workspace variables by using the same action but clicking File ⇒ Save Workspace As (figure 1.1(b)). One retrieves the data file by clicking the menu File ⇒ Import Data (figure 1.1(b)). Another option is use the command load data at the command prompt.

How to delete a file from the command prompt?

Let us delete just mentioned data.mat executing the command delete data.mat at the command prompt.

How to see the data held in a variable?

Figure 1.2(d) presents some variable information in which you find R. Doubleclick the R or your variable in the workspace browser and you find the matrix contents of R in a data sheet.

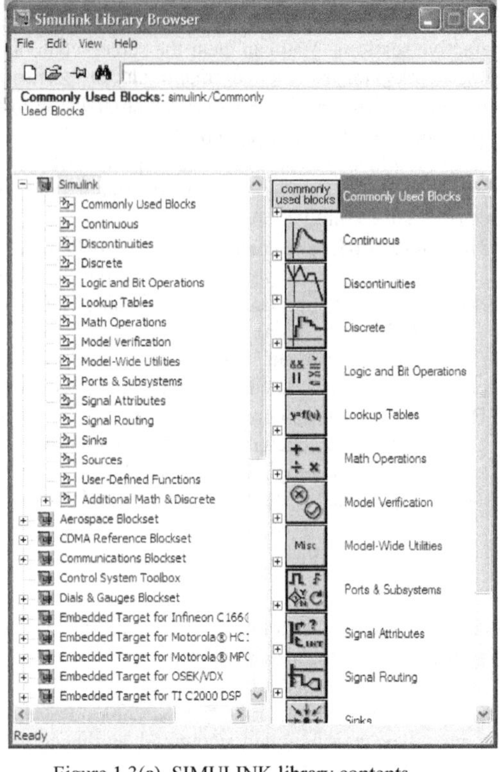

Figure 1.3(a) SIMULINK library contents

1.2 What is SIMULINK?

SIMULINK is an additional part of MATLAB which provides an easeful way to model, simulate, and analyze many dynamic systems which are characterized by some inputs and outputs. Without opening MATLAB we can not turn SIMULINK operational.

Elaboration of SIMULINK is simulation and link. A particular input-output relationship can be assigned to some block. One can interpret that SIMULINK is a vast collection of this kind of blocks. Although the blocks stand for simple mathematical relationship but being concatenated they build a much complicated system. Initially SIMULINK was intended particularly to handle the linear time invariant continuous systems. With the progress of time the discrete time systems as well as the hybrid ones surfaced in SIMULINK to adapt it to more pragmatic modeling of the real world's dynamic systems. In a simplistic way if MATLAB is a world of matrices, then SIMULINK is a world of blocks. In most scientific and engineering systems three types of constituent elements are seen – source, system, and sink. For example in electrical engineering, applied DC voltage to a circuit, R-C filter, and voltmeter correspond to the source-system-sink terminology. SIMULINK is the best tool for technical analysis if we can characterize a scientific problem in terms of source, system, and sink.

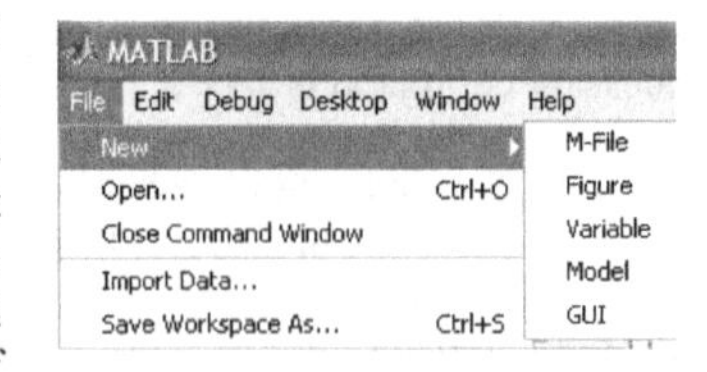

Figure 1.3(b) Different new file options in MATLAB

1.2.1 How can I get into SIMULINK?

Since SIMULINK is an extension of MATLAB, first we have to get into MATLAB. Both the MATLAB command prompt and its menu bar provide means of getting into SIMULINK. Figure 1.1(a) shows the indication of command prompt. Either you type simulink at the command prompt and press enter or click the SIMULINK library browser icon shown in figure 1.1(a). SIMULINK is an aggregation of functional blocks arranged in a tree structure which you see like the figure 1.3(a) by just mentioned either action.

In the figure 1.3(a) you find different families of block set. For example Aerospace Blockset keeps source/system/sink analysis blocks for the aerospace engineering. Again the Communications Blockset keeps source/system/sink analysis blocks for the communications engineering and so on. If you scroll down the library by using the mouse, you find one family by the name SignalProcessingBlockset which is solely dedicated for the robust signal processing problems – this family is mainly used for professional problems.

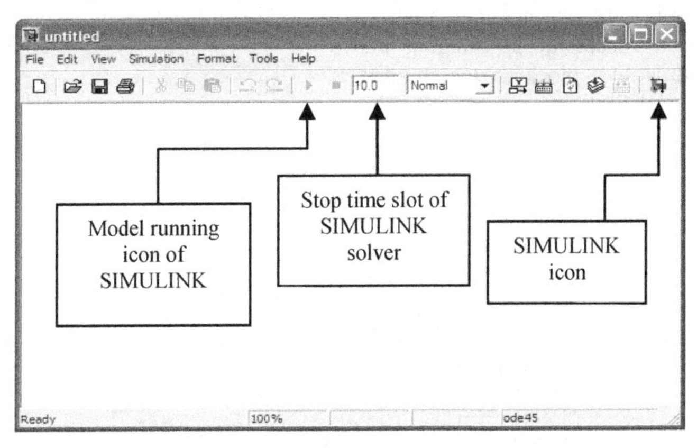

Figure 1.3(c) A newly opened SIMULINK model file

1.2.2 Where can I build a SIMULINK model?

Like every software there should be a file where we can build our SIMULINK model. The model we intend to build is entirely problem dependent. Referring to the MATLAB Command Window of figure 1.1(a) if you click the File menu down New of MATLAB, you see the pulldown menu as shown in the figure 1.3(b). Click the Model in the pulldown menu and you see the newly untitled SIMULINK model file opened like the figure 1.3(c). This is the platform where we build any SIMULINK model.

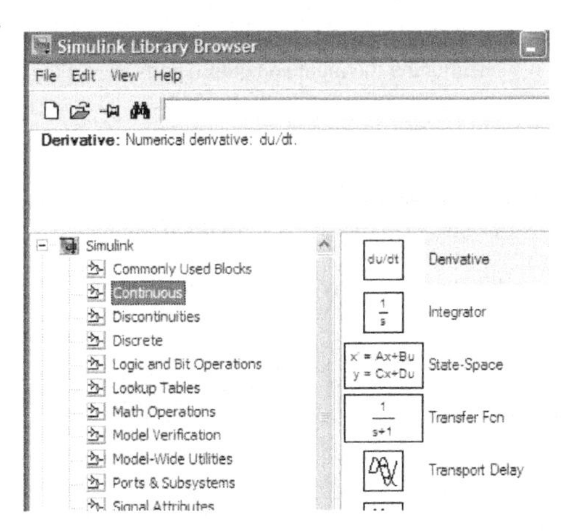

Figure 1.3(d) Different blocks availability under the subfamily Continuous

The SIMULINK library browser icon is also found in the menu bar of the untitled model file (seen in figure 1.3(c)). The reader should have the knowledge about any particular block's function and its input-output descriptions before bringing it from SIMULINK library to the untitled model file. A model is defined as the interconnected blocks found in SIMULINK library to work out a scientific/engineering problem.

Let us bring a block from the SIMULINK library in the untitled model file. To perform such action, click the SIMULINK library browser icon located in the menu bar of the untitled model file thus the SIMULINK library of figure 1.3(a) appears and the cursor is residing in SIMULINK. The right half part of the window in figure 1.3(a) displays the subfamily blocksets for example Commonly Used Blocks, Continuous, and so on. Click the Continuous down the SIMULINK on the left part of the window in figure 1.3(a). We see various blocks available under the subclass Continuous like the figure 1.3(d) for example Derivative, Integrator, State-Space, etc. Referring to the figure 1.3(d), we see a block called **Derivative** which we intend to bring in our untitled model file. Bring the mouse pointer on the **Derivative** block, move the mouse pointer keeping your finger pressed in the left button of the mouse to any convenient area of the untitled model, and release the left button of the mouse. Now you see the **Derivative** block in the untitled model file as shown in figure 1.3(e). Another way of bringing the block is click the **Derivative**, rightclick the mouse, see the **Add to untitled** in the popup, click the **Add to untitled**, and find the block in your model file. Therefore we say the link for the **Derivative** block is **SIMULINK** → **Continuous** → **Derivative**. We maintain this style of locating a block in the appendix A. However we have been successful in bringing the **Derivative** block in the untitled model file. Now we can save the model by any convenient name in the working directory. By the way MATLAB source code file has the extension **.m** but the SIMULINK model file does **.mdl**. Keep in mind that **one should know the link of a block to bring it from the SIMULINK library to a model file.**

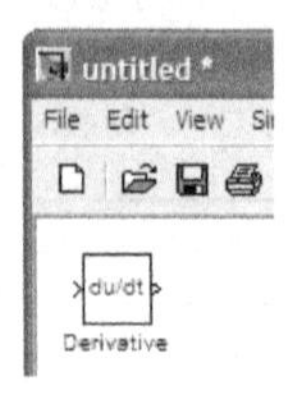

Figure 1.3(e) **Derivative** block in the untitled model file

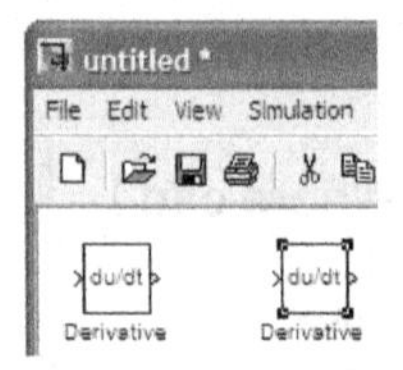

without selection   with selection

Figure 1.3(f) **Derivative** block with and without selection

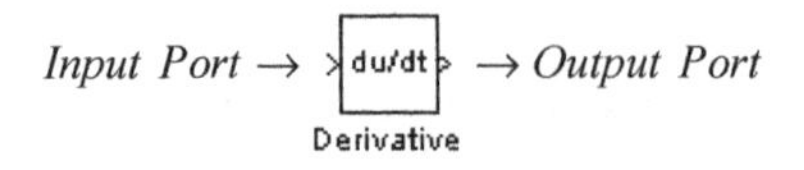

Figure 1.3(g) **Derivative** block mentioning the input and output ports

|u|
∠u
Complex to
Magnitude-Angle

Figure 1.3(h) Blocks with multiple input and output ports

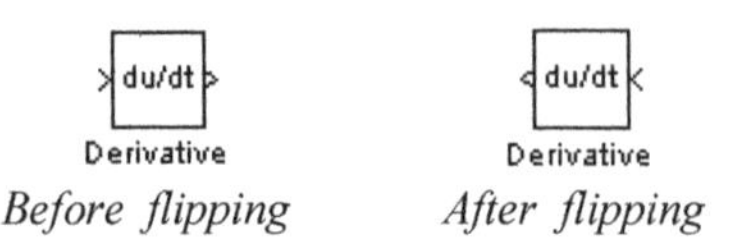

Figure 1.4(a) Flipping the **Derivative** block

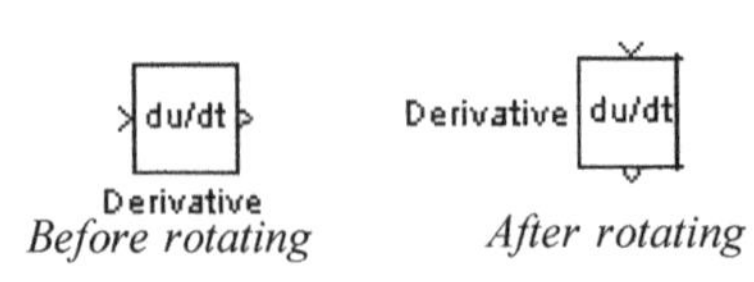

Figure 1.4(c) Rotating the **Derivative** block by 90^0 clockwise

Figure 1.4(b) Pulldown menu of the **Derivative** following the rightclick on the block

1.2.3 Block manipulation in SIMULINK

During a SIMULINK model building process, we need some manipulations of the blocks to construct a seemly, well-placed, and well-devised model, most frequently encountered ones of which are the following.

How to select a block?

Let us say you brought the **Derivative** block in an untitled model file following the link **Simulink** → **Continuous** → **Derivative**. Bring the mouse pointer on the **Derivative** block and click the left button of the mouse to see the selection as shown in figure 1.3(f).

🗗 How to detect the input and output ports of a block?

Referring to the Derivative block of the figure 1.3(e), we see that the left and right sides of the block contain the symbol >. One can identify the input and output ports of the block as presented in figure 1.3(g). The number of the input and output ports is not always 1, the blocks Real-Imag to Complex and Complex to Magnitude-Angle of the figure 1.3(h) have two input ports and two output ports respectively.

🗗 How to delete a block?

Let us say you brought the Derivative block in an untitled model file but you want to delete that. There are several options for this. First select the block, then press the Delete button from the keyboard, click the Cut icon in the model menu bar, or click the Cut followed by the rightclick of the mouse on the block.

🗗 How to flip a block?

Let us say we have the ongoing Derivative block in our model file. We wish to flip the block like in the figure 1.4(a). First we bring the mouse pointer on the block, click the right button of the mouse, and then click the Flip Block via Format. The action is shown in the figure 1.4(b).

🗗 How to rotate a block?

Suppose we wish to rotate just mentioned Derivative block. Bring the mouse pointer on the block, click the right button of the mouse, and click the Rotate Block via Format (figure 1.4(b) shows the action too). You see the change as shown in figure 1.4(c). Figure 1.4(c) seen rotation indicates the operation that the block is rotated clockwise 90^0 at a time. If you intend to rotate the block 270^0 clockwise (indicated in figure 1.4(d)), you need the operation three times.

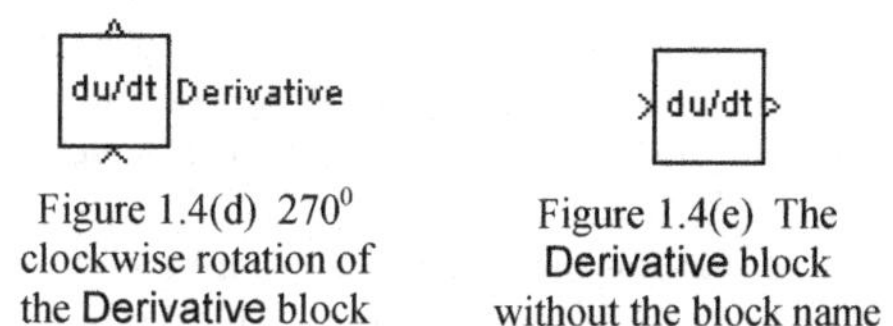

Figure 1.4(d) 270^0 clockwise rotation of the Derivative block

Figure 1.4(e) The Derivative block without the block name

🗗 How to remove the block name?

The name of the Derivative block can be removed as presented in figure 1.4(e) by clicking the Hide Name of the figure 1.4(b). A clumsy model might give better look on removing the block names if the reader is well acquainted with the blocks' operation.

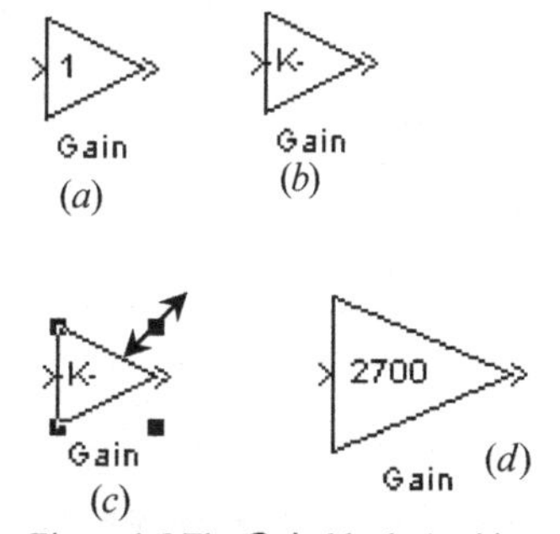

Figure 1.5 The Gain block a) with default gain, b) with gain 2700, c) with selection, and d) after enlargement

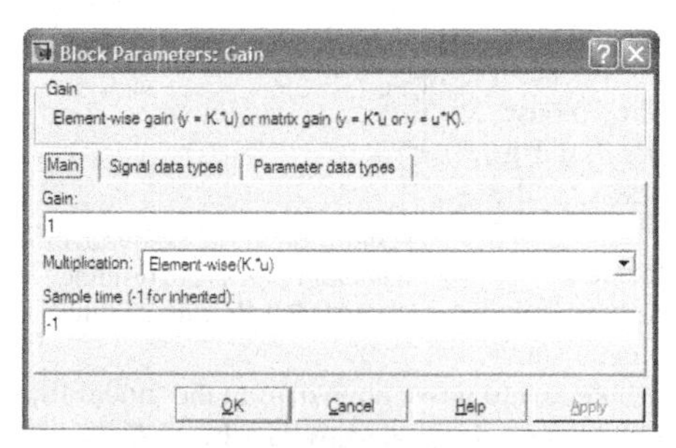

Figure 1.5(e) Block parameter window of the Gain block

🗗 How to enlarge or contract a block?

Let us bring a Gain block as shown in figure 1.5(a) following the link Simulink → Math Operations → Gain in an untitled SIMULINK model file, Gain has the default gain 1. If we have five or six digits gain, the default size will not allow to display that. Doubleclick the block to see its parameter window like in figure 1.5(e), let us enter the Gain slot value of the figure 1.5(e) from default 1 to 2700 (as a four digit example) by using the keyboard, and click OK. The block displays the inside gain as shown in figure 1.5(b). Select the block, bring your mouse pointer on the upper right square target to see the figure 1.5(c), move the mouse pointer to the right keeping the left button of the mouse pressed, and release the left button of the mouse. You should see the enlarged Gain block of the figure 1.5(d). In a similar way for the oversize block, we can reduce the block size by moving towards the inside of the block after selection.

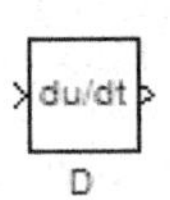

Figure 1.5(f) Derivative block by the name D

How to rename a block?

During SIMULINK model building process, it may be necessary to use the same kind of block twice or more. Then it requires renaming the block for identification. Let us say we brought a Derivative block in an untitled model file as shown in the figure 1.3(e). We wish to write just D as the block name instead of the Derivative (like the figure 1.5(f)). Bring the mouse pointer on the word Derivative, click the left button of the mouse (word is selected), and delete the other letters except D by using the Delete button of the keyboard, bring the mouse pointer outside the block, and leftclick it. After completely deleting the name Derivative, you can even enter any word of your choice from the keyboard.

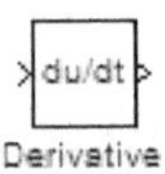

Derivative block in a model

Figure 1.5(g) Derivative block with the annotation

How to include the annotation to a block?

Let us say we have the Derivative block in the untitled model file as shown in figure 1.3(e) and wish to write the line Derivative block in a model down the block in the model file. Bring the mouse pointer at the desired position in the model file, doubleclick the mouse to see the blinking cursor, type the Derivative block in a model from the keyboard, bring mouse pointer out of the block, and click the left button of the mouse. Figure 1.5(g) shows the action we performed. Once typed, we can even drag the whole text to move anywhere in the model file.

1.6(a) Without drop shadow

1.6(b) With drop shadow

Figures 1.6(a)-(b) Derivative block without and with the drop shadow

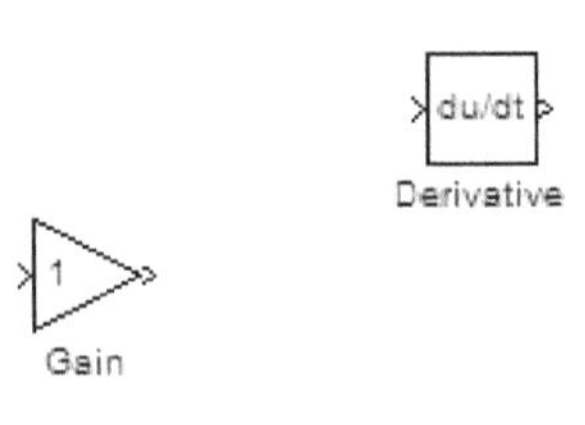

Figure 1.6(c) The Derivative and Gain blocks are residing in a SIMULINK model file

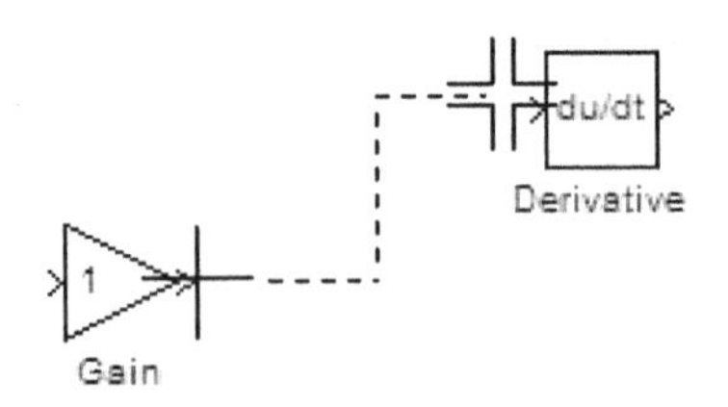

Figure 1.6(d) Connection phase of the two blocks

How to add drop shadow to a block?

Some reader might be interested to see the drop shadow form of the SIMULINK block rather than the plain shape. Let us say we have the Derivative block in an untitled model file (figure 1.3(e)). Rightclick on the block to see the pulldown menu of the figure 1.4(b) and click the Show Drop Shadow via Format. The necessary change is depicted in the figures 1.6(a)-(b). To remove the shadow, again rightclick on the block and click the Hide Drop Shadow via Format.

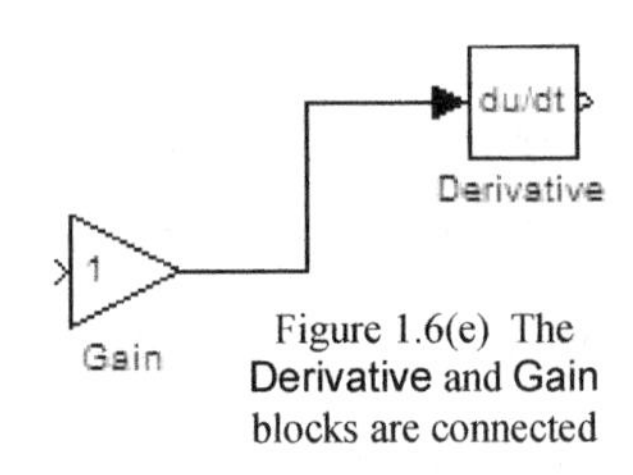

Figure 1.6(e) The Derivative and Gain blocks are connected

How to change SIMULINK model file background color?

Rightclick anywhere in the model file and see the Screen Color in the prompt menu. From the popup of the Screen Color, you can choose any background color for the model file.

How to copy a block within a SIMULINK model?

Select the block, click the Copy icon in the model menu bar (this action is called copy in the clipboard), and paste it as many times as you want.

How to connect two blocks?

This manipulation is very important in the sense that we frequently need to connect blocks in a model while working in SIMULINK in the subsequent chapters. The reader is familiar with the Gain and Derivative blocks from previous discussions. Let us say that the two blocks are residing in a SIMULINK model file as shown in the figure 1.6(c). We intend to connect the two blocks. The connection of the blocks must be correct syntactically. The output port of any block can only be connected to the input port of any other block but not to the output port of others. The same syntax is also true for the input port. However bring the mouse pointer on the output port of the Gain block, see the single cross target as shown in the figure 1.6(d), press the left button of the mouse, move the single cross target anywhere in the model keeping your finger pressed, bring the single cross target close to the input port of the Derivative block, see the double cross target as shown in the figure 1.6(d), and release the left button of the mouse. You should find the two blocks connected as shown in the figure 1.6(e).

What is a parameter or block parameter window?

In the following chapters we are going to mention the term parameter window many times. After bringing any block in a SIMULINK model file and then doubleclicking it, always do we see a prompt or dialog window in which we find one or more slots for value or parameter taking depending on the purpose of the block. That dialog window is termed as the parameter window for example the window for the Gain block in figure 1.5(e).

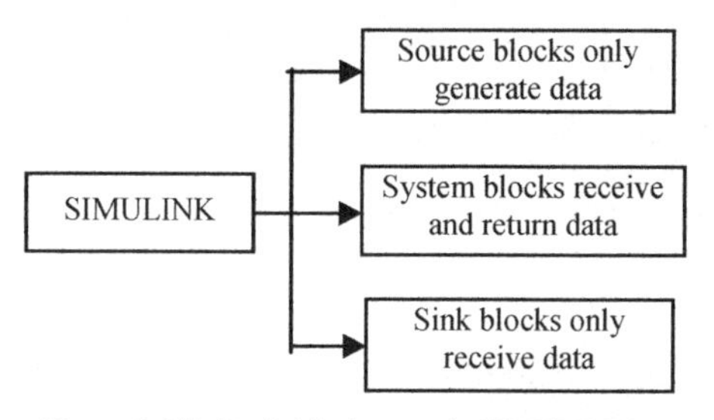

Figure 1.6(f) Basic block types in SIMULINK

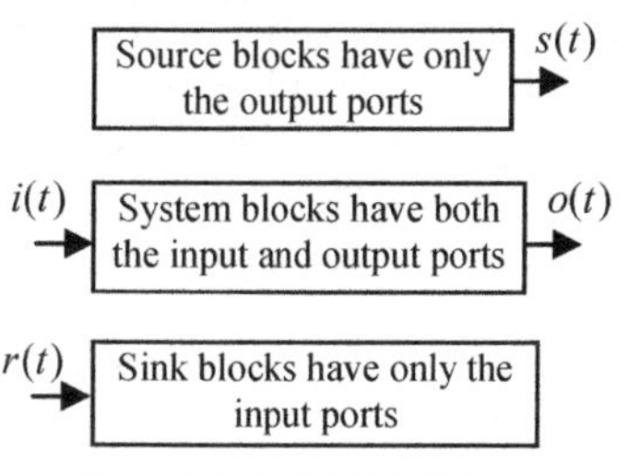

Figure 1.6(g) SIMULINK data flow in various blocks

1.2.4 Basic block categories of SIMULINK

By now we know that SIMULINK is a vast collection of blocks. These blocks follow three basic characteristics. Some blocks only generate data which are called source blocks, some blocks receive data, perform mathematical operation depending on the problem, and then return data which are called system blocks, and the third types only receive data which are called sink blocks. Figure 1.6(f) presents the basic block types found in SIMULINK. Figure 1.6(g) shows the data flow as a function of t where $s(t)$ generated, $i(t)$ input and $o(t)$ output, and $r(t)$ received correspond to source, system, and sink blocks respectively.

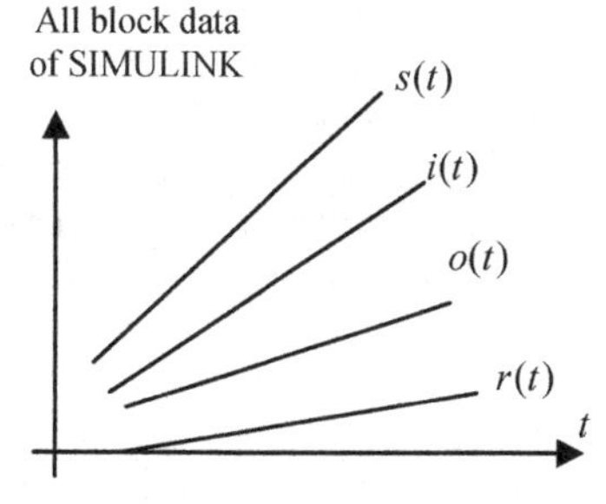

Figure 1.6(h) State or t dependency of SIMULINK data

It is extremely important to mention that all block functional data whether source, system, or sink present in SIMULINK model shares the common t variation. This is called the state or t dependency of the SIMULINK block data. Figure 1.6(h) illustrates this sort of dependency assuming all linear data.

1.2.5 Display and Scope blocks of SIMULINK

A practical model contains dozens of blocks which are interconnected by the functional lines in SIMULINK. When a model is being run, the blocks Display and Scope (appendix A for link and outlook) show how the functions are changing. The functional data flow or computation may or may not be seen during the run time because it happens so rapidly – in a fraction of second. It also depends on the problem whether it is time consuming.

The Display block is convenient only for showing a single scalar or few matrix data output at the end of the simulation. The block is designed to show the instantaneous value flowing through the functional line which it is connected to. Once SIMULINK has finished a simulation, the block shows the last value. The default size of the block is for a single scalar. For matrix data output one needs to enlarge the block to view all in it.

If the turnout of a SIMULINK block is in the form of a long row or column matrix which may hold hundreds of data elements, it is not feasible to see the results through the Display block. The graphical plot is a better way to observe the output. The Scope in all sense mimics an oscilloscope that essentially displays the signal variation with time. The Scope has two axes – horizontal and vertical. The horizontal and vertical axes simulate the independent and dependent variables respectively. The horizontal axis does not have to be time even though it is originated in that name, any physical quantity such as displacement, frequency, speed or other can be assigned to the horizontal axis.

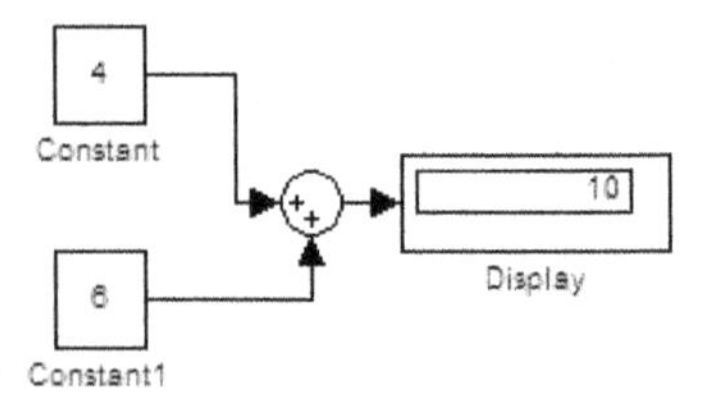

Figure 1.7(a) Adding two constants and displaying the result

1.2.6 How to get started in SIMULINK?

In previous sections the reader has gone through bringing and connecting blocks in a SIMULINK model file. Now we present simple modeling lessons for beginners in SIMULINK aiming to illustrate simulation style in this platform. Whatever operations such as manipulations, computations, assignments, or comparisons are carried out in conventional software can be conducted in SIMULINK through various blocks and functional lines. Since most algorithms are hidden in functional lines and blocks of SIMULINK, initially one

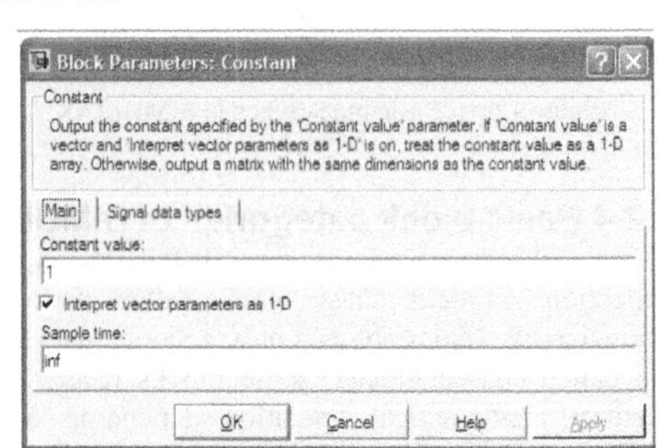

Figure 1.7(b) Block parameter window for the Constant

might feel it complicated. Most of the model building in SIMULINK happens through the mouse operation rather than writing the source codes.

Let us go through the following three tutorials as a quick start in SIMULINK.

◆◆ Tutorial one

Two numbers are to be added – 4 and 6. The output should be 10. This is the problem statement.

The first question is where we should keep the numbers. In SIMULINK every programming aspect happens through blocks. You find a block called Constant through the link SIMULINK → Sources → Constant (figure 1.3(a)). Open a new SIMULINK model file (subsection 1.2.2) and bring the Constant block in the untitled model file as we did before. The default value in the block is 1. Doubleclick the block to see its parameter window like the figure 1.7(b) and enter 4 in the slot of Constant value from the keyboard after deleting the default 1 but leaving the other parameters unchanged in the window. It means the first number 4 is going to be generated by the Constant block. Similarly bring another Constant block from the same link and enter 6 as the Constant value after doubleclicking it. You see the latter block by the name Constant1. If you bring one more Constant block in the model file, SIMULINK names that as Constant2. This style of naming is followed for all other blocks.

However we need a Sum block to add the two numbers that can be reached via SIMULINK → Math Operations → Sum. Bring the Sum block in the model file. To see the computation, we need a Display block whose link is SIMULINK → Sinks → Display and also bring the block in the model file. Place the four blocks relatively and connect them (subsection 1.2.3 for connection) according to the figure 1.7(a). Referring to the figure 1.3(c), you find an inactivated start simulation icon whose symbol is ►. The symbol remains inactivated when no blocks are present in the model. Click the start simulation icon

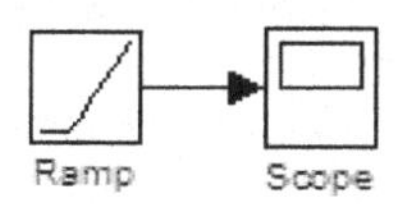

Figure 1.7(c) Ramp block connected with the Scope block

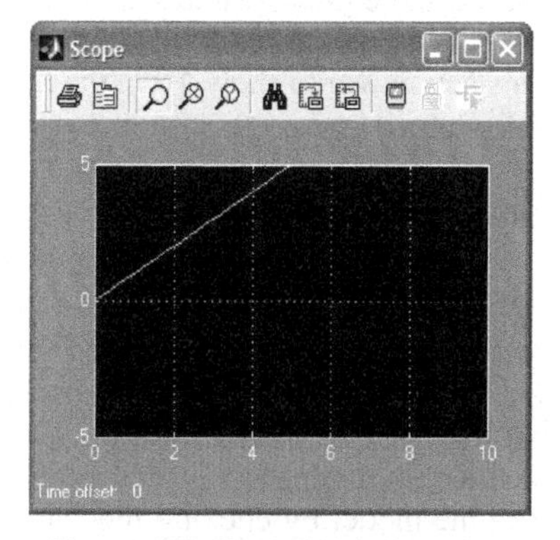

Figure 1.7(d) Scope block shows the ramp function

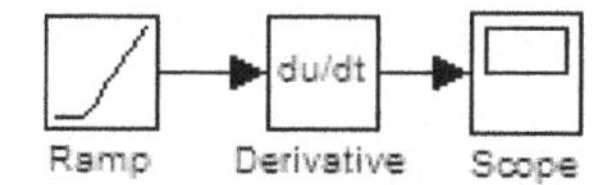

Figure 1.7(e) Derivative block differentiates the output generated by the Ramp and Scope shows the Derivative output

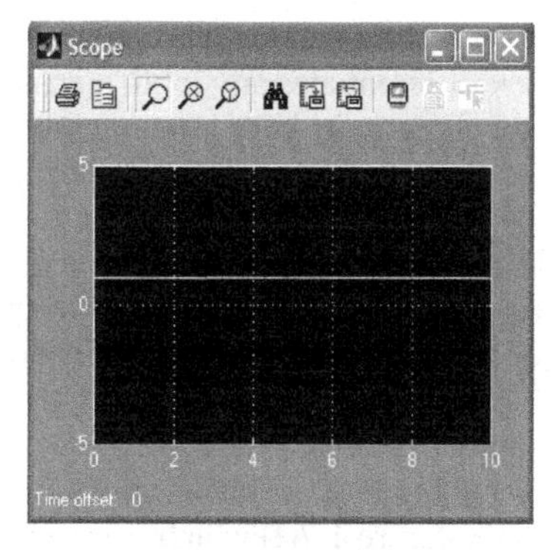

Figure 1.7(f) Scope output for the model of the figure 1.7(e)

▶ in the icon bar of the model and SIMULINK responds showing the summation 10 in the Display block like the figure 1.7(a). You can also run the model file from the menu bar of SIMULINK by clicking first the Simulation and then the Start in the pulldown menu.

MATLAB command window provides another alternative for running a SIMULINK model from its command prompt. Let us say we have the SIMULINK model saved by the name test.mdl. To run it from the command prompt, we carry out the following (sim is a built-in command for running a SIMLINK model file and the file name must be under quote):

```
>>sim('test') ↵
```

This action would show 10 in the Display block too.

⬥⬥ Tutorial two

We intend to generate the function $y(t)=t$ and view the generation in earlier mentioned Scope – this is the problem statement.

The function is a straight line passing through the origin and has a unity slope which is also known as the ramp function. There is a block by the name Ramp (appendix A for link and outlook) in SIMULINK which generates t data by default. That means it is a source block as depicted in the figure 1.6(f) or 1.6(g). Bring one Ramp and one Scope blocks in a new SIMULINK model file following the link mentioned in appendix A and connect the two blocks as shown in the figure 1.7(c). Run the model by clicking the start simulation icon ▶ in the icon bar of the model and doubleclick the Scope to view the figure 1.7(d) presented curve which displays our wanted straight line plot. The horizontal and vertical axes of the Scope correspond to the t and $y(t)$ data respectively so the Ramp generated the function $y(t)$ and the Scope just displayed that. You can inspect that the line of the figure 1.7(d) is passing through the points (0,0) and (5,5) confirming the generation with the correct slope.

⬥⬥ Tutorial three

If we differentiate the ramp function $y(t)=t$ with respect to t, we should get $\frac{dy(t)}{dt}=1$. We intend to simulate this mathematical operation.

The reader is familiar with the Derivative block from subsection 1.2.3. The block differentiates any input signal to its input port and returns the derivative signal to its output port both in continuous sense. Insert the Derivative block between the Ramp and Scope blocks of the figure 1.7(c) so that we have the model in figure 1.7(e). Select the Ramp or Scope block of the model in figure 1.7(c), use the left or right arrow key from the keyboard so that the space is enough between the two blocks to accommodate the Derivative block. When you bring the Derivative block, drop the block keeping its input and output ports in line with the connection line of the Ramp and Scope. SIMULINK is so smart that it connects the Derivative block automatically like the figure 1.7(e). On forming the model, click the start simulation icon ▶ in the icon bar of the model and doubleclick the Scope to view the output like the figure 1.7(f) which is essentially a straight line parallel to the horizontal axis and located at 1 in the vertical axis of the Scope. In other words the vertical and horizontal axes of the Scope refer to the $\frac{dy(t)}{dt}$ and t data respectively.

That is what we expected from SIMULINK.

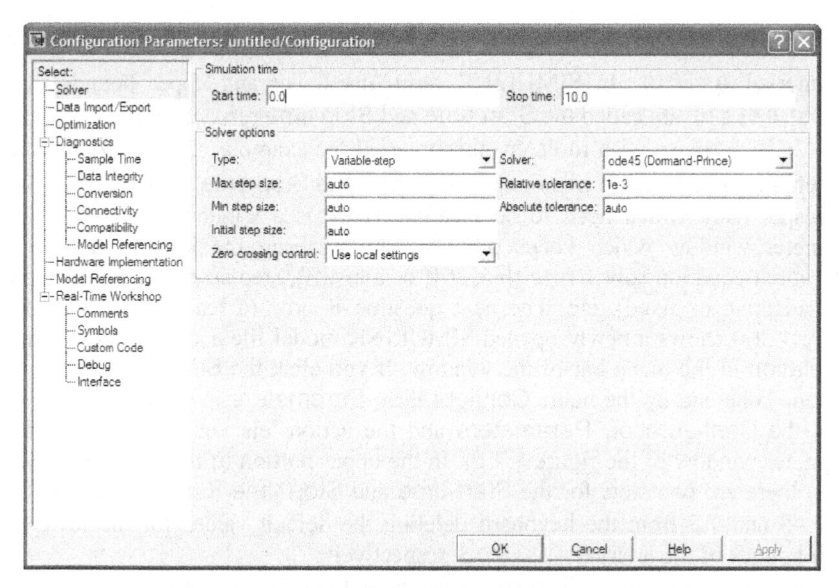

Figure 1.7(g) Simulation parameter window of SIMULINK

** The models in figures 1.7(a), 1.7(c), or 1.7(e) have connecting lines between various blocks. The data flowing through these lines is the functional data not the t data. It is the style of SIMULINK that the t data in all these blocks is common. For example in the figure 1.7(e), $y(t)$ data flows from the Ramp to Derivative and $\frac{dy(t)}{dt}$ data flows from the Derivative to Scope blocks.

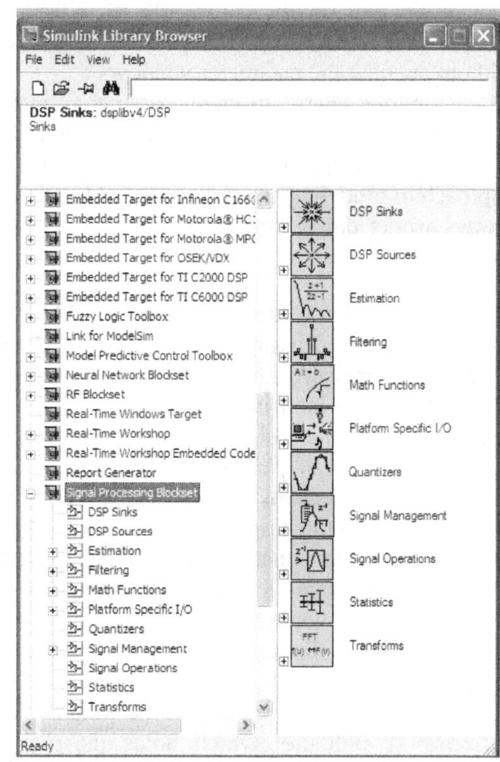

Figure 1.8(a) Signal Processing Blockset library of SIMULINK for signal and system analysis

❖❖ Interval entering or t information in SIMULINK

In the tutorial 2 the Scope (figure 1.7(d)) shows $y(t)=t$ versus t graph and in the tutorial 3 the Scope (figure 1.7(f)) shows $\frac{dy(t)}{dt}=1$ versus t graph. We mentioned that the horizontal axis of both the Scopes represents t variation. The t variation seen in both Scopes is between 0 and 10 because that is the default setting. Mathematically we can say that the $y(t)$ or $\frac{dy(t)}{dt}$ is graphed over the interval $0 \le t \le 10$. In SIMULINK term, the lower and upper bounds of the interval $0 \le t \le 10$ are called the Start time and Stop time respectively.

What if we wish to enter other interval for example $-3 \le t \le 7.5$ instead of the default $0 \le t \le 10$? Just mentioned Start time and Stop time then become –3 and 7.5 respectively which need to be entered. There is a window called simulation parameter window which keeps provision for entering the interval description, differential equation solver type (like stiff or nonstiff), step size of the computation (like adaptive or fixed), etc. The next question is how to reach to that window? Figure 1.3(c) shows a newly opened SIMULINK model file and you find the menu Simulation in the menu bar of the window. If you click the Simulation menu, you find one submenu by the name Configuration Parameters in the pulldown menu. Click the Configuration Parameters and the action lets you see the simulation parameter window of the figure 1.7(g). In the upper portion of the window in figure 1.7(g) there are two slots for the Start time and Stop time respectively. There we enter –3 and 7.5 from the keyboard deleting the default values for the lower and upper bounds of the interval $-3 \le t \le 7.5$ respectively.

If the lower bound is 0 and upper bound is other than 10, you do not even need to open the parameter window. In the untitled model file of the figure 1.3(c), there is the slot for the Stop time as indicated by the arrow. For example if the interval is $0 \le t \le 15$, we just enter 15 in the Stop time slot of the figure 1.3(c) without opening the parameter window.

Note: There are many parameters in the simulation parameter window of the figure 1.7(g) whose discussions are beyond the scope of the text. We suggest that you do not change any other parameter unless you know about it. SIMULINK approach of modeling any dynamic problem is completely numerical. Also computer always works on discrete data instead of continuous one even though we do the simulation in continuous sense.

1.2.7 Signal processing library in MATLAB and SIMULINK

Signal processing is a vast topic and the library contents of MATLAB and SIMULINK are so. Signal processing library is included both in MATLAB and SIMULINK. The built-in functions of MATLAB and block-sets of SIMULINK are devised to handle robust signal processing problems for professional use. Professional way of explaining theoretical matters may not be appreciative to a beginner. We select the functions and blocks whose executions or operations are closely related to the introductory electrical signal and system text terms.

The library that contains signal related functions has the name signal in MATLAB. But that library alone is not enough to solve all signal processing problems. Other libraries for example control, sigtools, sptoolgui, etc are also equally applicable for solving signal associated problems in MATLAB. Again in the figure 1.3(a) you find the SIMULINK library browser window. If you scroll down using the mouse in the left half of the window, you find a library by the name Signal Processing Blockset which holds most signal related professional blocks in SIMULINK. Doubleclick the Signal Processing Blockset, wait for a while, and find the subfamilies of the Signal Processing Blockset as shown in figure 1.8(a).

But for the elementary level we will be using figure 1.3(d) shown Continuous and other necessary subfamilies for the signal and system problems.

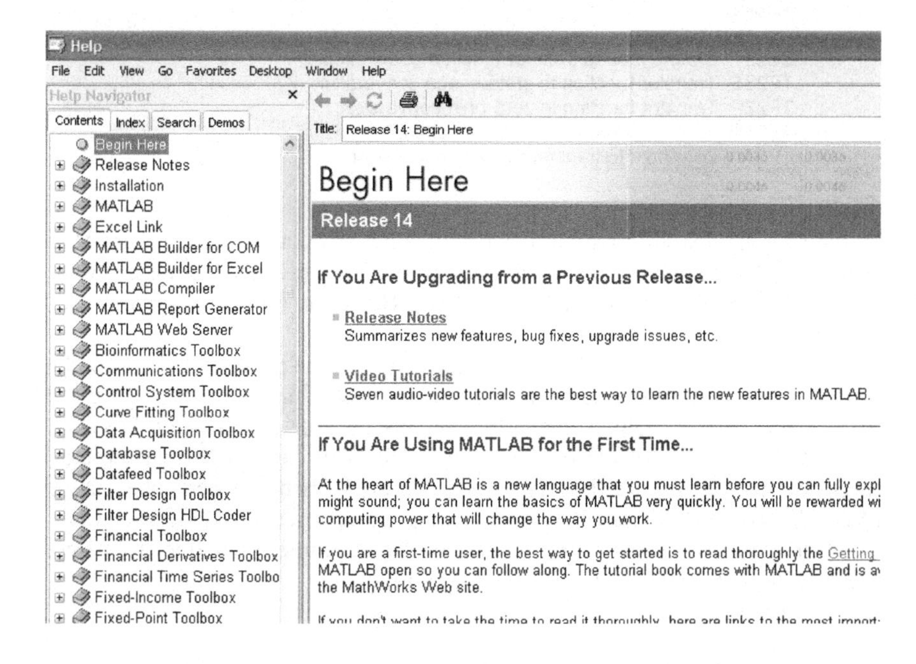

Figure 1.8(b) General Help window for MATLAB and SIMULINK

1.3 How to get help?

Help facilities in MATLAB are plentiful. One can access to the information about a MATLAB function or a SIMULINK block in a variety of ways. Command help finds the help of a particular function file. You are familiar with the function sin(x) from earlier discussion and can have the command prompt help regarding the sin(x) as follows:

```
>>help sin ↵                    ← Function name without the argument
SIN   Sine.
    SIN(X) is the sine of the elements of X.

    See also asin, sind.

    Overloaded functions or methods (ones with the same name in other directories)
    help sym/sin.m

    Reference page in Help browser
    doc sin
```

One disadvantage of this method is that the user has to know the exact file name of a function. For a novice, this facility may not be appreciative.

Casually we know the partial name of a function or try to check whether the function exists by that name. Suppose we intend to see whether any function by the name tf exists. We execute the following by the intermediacy of the command lookfor (no space gap between look and for) to see all possible functions bearing the file name tf or having the file name tf partly:

```
>>lookfor tf ↵

SS2TF  State-space to transfer function conversion.
TF2SS  Transfer function to state-space conversion.
TF2ZP  Transfer function to zero-pole conversion.
TFCHK  Check for proper transfer function.
ZP2TF  Zero-pole to transfer function conversion.

        ⋮
```

The return is having all possible matches of functions containing the word tf. Now the command help can be conducted to go through a particular one for example the first one is SS2TF and we execute help SS2TF to see its description at the command prompt.

In order to have window form help, click the Help icon (i.e. ?) of the figure 1.1(a) and MATLAB responds with the opening Help window of the figure 1.8(b). As the figure shows, help is available content-based or index-based. If you have some search word for MATLAB or SIMULINK, you can search that through the Search of the figure 1.8(b). This help form is better when one navigates MATLAB/SIMULINK's capability not looking for a particular function/block.

If you just execute help at the command prompt, you find the names of dozens of libraries embedded in MATLAB as follows:

```
>>help ↵

HELP topics

matlab\general      - General purpose commands.
matlab\ops          - Operators and special characters.
matlab\lang         - Programming language constructs.

        ⋮
```

MATLAB exhibits a long list of libraries in which the first one is general. If you execute help general at the command prompt, you see all functions under the family general.

Hidden algorithm or mathematical expression is often necessary whose assistance we can have through the search option from MathWorks Website provided that our PC is connected to the Internet (figure 1.1(g)).

However we close the introductory discussion on MATLAB and SIMULINK with this.

Chapter 2
SIGNAL SAMPLE GENERATION

Introduction

In this chapter we aim at introducing elementary signal sample generation in MATLAB command window, which is needed in different courses of electrical engineering in senior level years. Keeping in mind that a computer never generates a continuous signal, we produce sufficient samples of a signal so that the signal appears to be continuous. Our topic outline is the following:

- Simplistic definition of a continuous signal and its intrinsic generational approach in MATLAB
- Signal handling techniques both for the expression and graphical representation based
- Way to generate a signal by user-defined function file and signal shape verifying techniques

2.1 What is a signal?

A signal is conceptually a functional variation of one quantity with respect to other. The dependent quantity in electrical engineering context can be voltage, current, frequency, or power and the independent quantity is mostly time. Realistic signals are always continuous but a computer never generates a continuous signal due to finite memory reason. What we exercise is we take sufficient samples of a signal so that the sample envelope follows actual signal variation.

2.2 Signal generation in MATLAB

MATLAB does not have any knowledge of your signal. All it can do is respond to some written command. A signal can be voltage, current, power, displacement, or other. As a user, you are the one who decides the type of signal. MATLAB only provides convenience of generation by programming statements and pre-written function files. We must not forget that we always generate discrete values of a continuous signal not the continuous signal itself. For example a voltage versus time (i.e. $v(t)$ versus t) signal is to be generated so it is all about $v(t)$ and t data generation. In signal and system course different cases arise regarding signal generation whose discussion follow next.

2.3 Signals from mathematical expressions

This type of signal is described in terms of mathematical functions within some give interval. Involvement of mathematical expression requires that we should be able to write the code of a function in MATLAB whose reference is in appendix B. Suppose we intend to generate $v(t)$ versus t first we must select some step size for the t, generate the t vector (subsection 1.1.2) based on chosen step size and given interval, and then write the scalar code of the functional expression for $v(t)$ according to appendix B.

Example 1

Generate the ramp voltage signal $f(t)=t$ over the interval $0 \le t \le 3$ sec.

Since we have to select the t step size let it be 0.1sec. Based on that the signal generation is as follows:

```
>>t=0:0.1:3; ↵     ← t holds the t variation as a row matrix
>>f=t; ↵           ← Workspace variable f holds the signal as a row matrix
```

The t and f are user-chosen variable names and both are identical size row matrix. If we had the shifted and scaled ramp signal like $f(t)=\frac{4}{3}t-6$, the functional code command would be f=4/3*t-6.

Example 2

Generate the voltage signal $f(t)=2\,V$ over the interval $0 \le t \le 3$ sec.

We can not generate this $f(t)$ signal value by example 1 mentioned technique. The main problem here if the t changes, we can not relate this change to f by writing functional code. If we write f=2, only one value will be assigned to f whereas the t vector of example 1 contains many t point values. Which element value of the row matrix t does this f correspond to? Obviously there is no answer.

Let us choose the t step size as 0.2sec from 0 to 3sec then the t assumes the values 0, 0.2, 0.4...3sec. At each of these values, the $f(t)$ must be 2. We generate the t as a row vector by writing the code t=0:0.2:3;. Let MATLAB find the number of elements in the vector t by the command length(t). Then we generate the number of ones equal to the number of elements in t and multiply each element by 2 by using the command 2*ones(1,length(t)) (appendix C.1 for ones). However the formal procedure is as follows:

```
>>t=0:0.2:3; ↵                  ← t holds the t variation as a row matrix
>>f=2*ones(1,length(t)); ↵      ← Workspace f holds the signal as a row matrix
```

Example 3

Let us generate the parabolic voltage signal $f(t)=1-\frac{2}{5}t+t^2\ V$ over the interval $0 \le t \le 3$ sec with a step size 0.1sec as follows:

>>t=0:0.1:3; ↵ ← t holds the t variation as a row matrix
>>f=1-2/5*t+t.^2; ↵ ← Workspace variable f holds the signal as a row matrix

It is compulsory that we write the scalar code of a signal (appendix B) related to t during the code writing not the vector code on account of that the 1-2/5*t+t^.2 is the scalar code of function $1-\frac{2}{5}t+t^2$.

Example 4

Generate the voltage signal $v(t)=0.3\sin 2\pi ft$ V over the interval 0 to 20 $m$sec where the frequency f =200 Hz.

Let us choose the time step as 0.1 $m$sec and the last time point 20 $m$sec is written as 20×10^{-3}sec whose MATLAB code is 20e-3 (1e-3 is equivalent to 10^{-3} not the natural number). The function $0.3\sin 2\pi ft$ is coded as 0.3*sin(2*pi*200*t) and the implementation is as follows:

>>t=0:0.1e-3:20e-3; ↵ ← t holds the t variation as a row matrix
>>v=0.3*sin(2*pi*200*t); ↵ ← Workspace v holds the signal as a row matrix, the v is user-chosen variable name

Table 2.A Signal data generation from mathematical expression

Signal to be generated	Command we need
$f(t)=5e^{-t}$ for $0\le t\le 0.1$sec with a step 1 $m$sec	>>t=0:1e-3:0.1; ↵ >>f=5*exp(-t); ↵
$f(t)=10(1-e^{-3t})$ for $0\le t\le 0.1$sec with a step 1 $m$sec	>>t=0:1e-3:0.1; ↵ >>f=10*(1-exp(-3*t)); ↵
$f(t)=\|t\|$ for $-2\le t\le 2$sec with a step 0.1sec	>>t=-2:0.1:2; ↵ >>f=abs(t); ↵
$f(t)=\frac{3-\|t\|}{6}$ for $-2\le t\le 2$sec with a step 0.1sec	>>t=-2:0.1:2; ↵ >>f=(3-abs(t))/6; ↵
$f(t)=e^{-t^2}$ for $-2\le t\le 2$sec with a step 0.1sec (called Gaussian)	>>t=-2:0.1:2; ↵ >>f=exp(-t.^2); ↵
$f(t)=\frac{1}{\sqrt{2\pi}}e^{-\left(\frac{t-4}{3}\right)^2}$ for $-2\le t\le 2$sec with a step 0.1sec (Gaussian with different mean and scale)	>>t=-2:0.1:2; ↵ >>f=exp(-((t-4)/3).^2)/sqrt(2*pi); ↵
$f(t)=\frac{\sin \pi t}{\pi t}$ for $-2\le t\le 2$sec with a step 0.1sec (called sinc function)	>>t=-2:0.1:2; ↵ >>f=sinc(t); ↵ (sinc is MATLAB built-in with the same definition)
$f(t)=\frac{\sin t}{t}$ for $-2\le t\le 2$sec with a step 0.1sec (also called sinc function, different text defines differently)	>>t=-2:0.1:2; ↵ >>f=sinc(t/pi); ↵ (Using MATLAB built-in sinc)
$f(t)=3e^{-t}\sin 5t$ for $0\le t\le 5$sec with a step 0.1sec (called damped sine signal)	>>t=0:0.1:5; ↵ >>f=3*exp(-t).*sin(5*t); ↵
$f(t)=\begin{cases}1 & for\ t>0\\ -1 & for\ t<0\end{cases}$ for $-2\le t\le 2$sec with a step 0.1sec (called signum function)	>>t=-2:0.1:2; ↵ >>f=sign(t); ↵ (sign is MATLAB built-in with the left side functional definition)

If the sinusoidal signal had some phase for example $v(t)=0.3\sin(2\pi ft-60^0)$, the command would be v=0.3*sin(2*pi*200*t-pi/3); because the sine input argument must be in radian.

Example 5

In electronic and power system circuits sometimes we need to generate sinusoidal with harmonics. Let us say a voltage signal is composed of fundamental $50\,Hz$ and its 2nd and 4th harmonics. Their amplitudes are $2V$, $0.7V$, and $0.2V$ respectively. We intend to generate the signal over the interval $0 \le t \le 0.1\sec$.

The specification indicates that there are three frequencies related with the signal which are $50\,Hz$, $100\,Hz$, and $200\,Hz$ respectively. On that account the expression of the complete signal is given by $v(t)=2\sin 2\pi 50t+0.7\sin 2\pi 100t+0.2\sin 2\pi 200t$ whose code is written as 2*sin(2*pi*50*t)+0.7*sin(2*pi*100*t)+0.2*sin(2*pi*200*t). However following is the implementation:

```
>>t=0:0.001:0.1; ↵          ← t holds the t variation as a row matrix (chosen step size is 0.001sec)
>>v=2*sin(2*pi*50*t)+0.7*sin(2*pi*100*t)+0.2*sin(2*pi*200*t); ↵
```

In above implementation the workspace v holds the signal values as a row matrix where the v is a user-chosen name.

Miscellaneous examples

In the last five examples we introduced the idea behind the functional expression based signal generation. We presented a collection of concise signal codes in the table 2.A maintaining ongoing symbology and function.

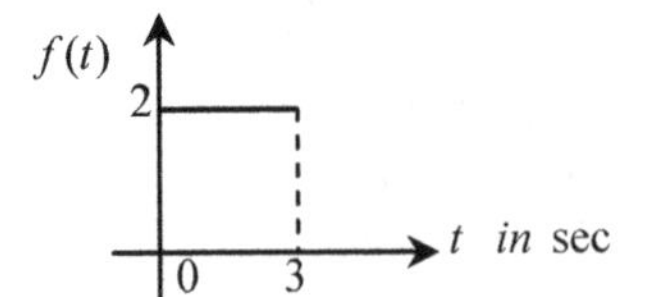

Figure 2.1(a) A finite duration rectangular voltage pulse

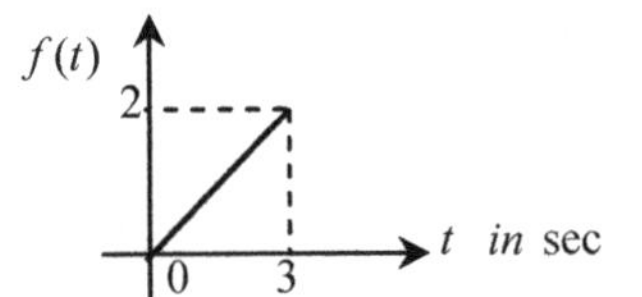

Figure 2.1(b) A finite duration ramp voltage pulse

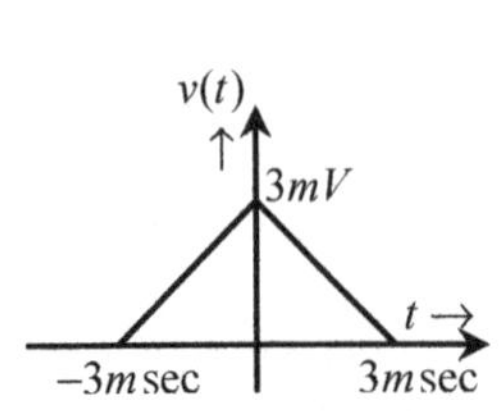

Figure 2.1(c) A finite triangular pulse

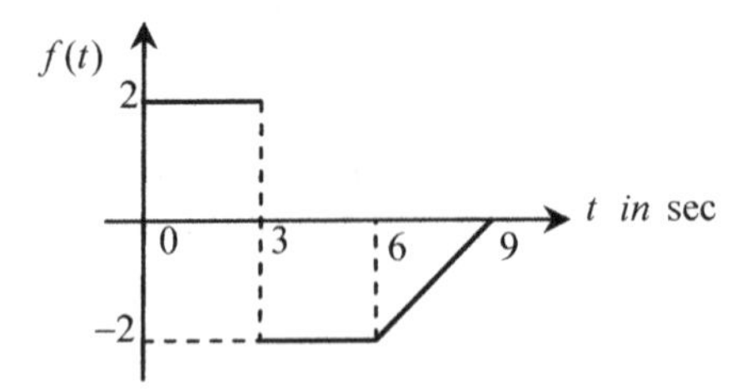

Figure 2.1(d) A piecewise continuous signal

2.4 Signals from graphical representations

Given the graphical representation, we generate a signal from its functional properties. Even though the graph is available, we need to write the expression of the graph for the computation or code writing reason. Once the expression has been available, then section 2.3 cited technique is applied. Let us proceed with the following examples.

Example 1

Generate the finite duration rectangular voltage pulse of the figure 2.1(a). We solved the problem as example 2 in last section.

Example 2

Generate the finite duration ramp voltage pulse of the figure 2.1(b). The ramp function passes through two points whose coordinates are (0,0) and (3,2) and the equation of finite ramp pulse is $f(t)=\frac{2}{3}t$. Choosing a step size 0.1sec within the interval $0 \le t \le 3$ sec, we generate the signal (example 1 of last section) as follows:

```
>>t=0:0.1:3; ↵    ← t holds the t variation as a row matrix
>>f=2/3*t; ↵      ← Workspace variable f holds the signal as a row matrix
```

Example 3

Generate the symmetric triangular voltage $v(t)$ of the figure 2.1(c) taking a step size of $0.1\ m\sec$.

The signal is composed of two ramp or straight line functions. From the figure the left and right straight lines pass through the points $\begin{Bmatrix}(-3m\sec,0)\\(0,3mV)\end{Bmatrix}$ and $\begin{Bmatrix}(3m\sec,0)\\(0,3mV)\end{Bmatrix}$ respectively. Applying the technique of appendix C.6, the equation of the two lines are found as follows:

```
>>maple('with(geometry)'); ↵        ← Activating the geometry package
```

The left line (will be called L1) finding from the two points are as follows:

```
>>maple('line(L1,[point(A,-3/1000,0),point(B,0,3/1000)],[t,v])'); ↵
>>E1=maple('Equation(L1)'); ↵       ← Assigning equation of L1 to E1
>>solve(E1,'v') ↵                   ← Expressing v(t) in terms of t

ans =

3/1000+t
```

← Meaning the left line equation is $v(t)=\frac{3}{1000}+t$ with standard unit

The right line (will be called L2) finding from the two points are as follows:

```
>>maple('line(L2,[point(A,3/1000,0),point(B,0,3/1000)],[t,v])'); ↵
>>E2=maple('Equation(L2)'); ↵       ← Assigning equation of L2 to E2
>>solve(E2,'v') ↵                   ← Expressing v(t) in terms of t

ans =

3/1000-t
```

← Meaning the right line equation is $v(t)=\frac{3}{1000}-t$ with standard unit

Obviously the line intervals are over $-3m\sec \le t < 0$ and $0 \le t \le 3m\sec$ respectively. Since there are two equations for two segments of the signal $v(t)$, writing one equation for the whole signal can not be performed. Sample generation takes place one straight line at a time. Let us generate the whole time vector in standard unit with the given step size as follows:

```
>>t=[-3:0.1:3]*1e-3; ↵        ← t holds the whole t variation as a row matrix (on
                                 the step 0.1 msec )
>>v=ones(1,length(t)); ↵      ← v is a row matrix whose all elements are 1 (section
              2.3), the number of ones is exactly equal to the number of elements in t
>>r1=find(-3e-3<=t&t<0); ↵    ← r1 holds only the position indexes of t at which
                  -3msec ≤ t < 0 is satisfied (appendix C.3), r1 is user-chosen name
```

```
>>v(r1)=t(r1)+3/1000; ↵
>>r2=find(0<=t&t<=3e-3); ↵
```
← r2 holds only the position indexes of t at which $0 \le t \le 3m\sec$ is satisfied, r2 is user-chosen name

```
>>v(r2)=3/1000-t(r2); ↵
```

The logical condition $-3m\sec \le t < 0$ is split in two parts, $-3m\sec \le t$ and $t < 0$ for programming reason (appendix C.2). The two logical expression is connected by the AND operator. The command (v(r1)=t(r1)+3/1000) in the fourth line of the last execution implements $v(t) = \frac{3}{1000} + t$. We applied t(r1) not t. If we had t+3/1000, 3/1000 would be added to all elements in t which is not how the function is defined. Again the assignment took place only to the indexes of the first interval by writing v(r1). Similar explanation goes for the other interval. However the whole signal $v(t)$ is stored in the workspace variable v. The junction point of two intervals should be properly defined. For example the intervals $-3m\sec \le t < 0$ and $0 \le t \le 3m\sec$ have the junction point $t = 0$ which is considered in the second interval not in the first one. Operator writing also needs proper attention. If you implement $0 \le t$ instead of $0 < t$, you end up with erroneous signal or wave shape.

Example 4

Generate the piecewise signal of the figure 2.1(d) over the interval $0 \le t \le 9$ sec with a step size 0.1sec. There are three segments in the signal whose descriptions are $f(t) = \begin{cases} 2 & for \;\; 0 \le t < 3 \\ -2 & for \;\; 3 \le t < 6 \\ \frac{2}{3}t - 6 & for \;\; 6 \le t < 9 \end{cases}$. Drawing the idea and functions of the example 3, we implement the following:

```
>>t=0:0.1:9; ↵
```
← t holds the whole t variation $0 \le t \le 9$ sec as a row matrix with the step 0.1sec

```
>>f=ones(1,length(t)); ↵
```
← f is a row matrix whose all elements are 1, the number of ones is exactly equal to the number of elements in t

```
>>r1=find(0<=t&t<3); ↵
```
← r1 holds only the position indexes of t at which $0 \le t < 3$ sec is satisfied

```
>>f(r1)=2; ↵
```
← assigns 2 only to the position indexes stored in r1 corresponding to $0 \le t < 3$ sec

```
>>r2=find(3<=t&t<6); ↵
```
← r2 holds only the position indexes of t at which $3 \le t < 6$ sec is satisfied

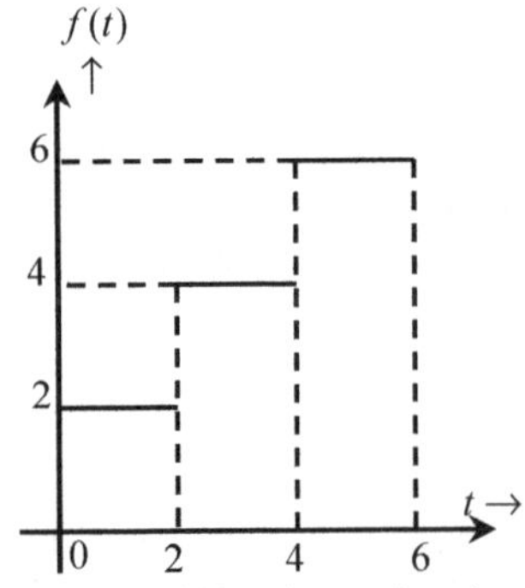

Figure 2.1(e) A staircase function

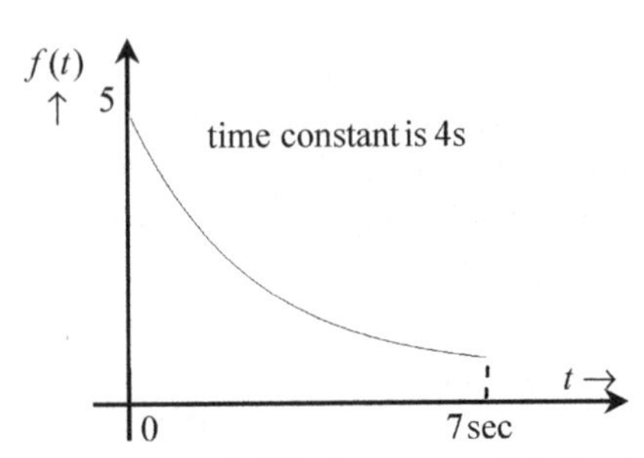

Figure 2.1(f) An exponential signal

>>f(r2)=-2; ↵ ← assigns –2 only to the position indexes stored in r2 corresponding to $3 \le t < 6$ secs

>>r3=find(6<=t&t<=9); ↵ ← r3 holds only the position indexes of t at which $6 \le t \le 9$ secs is satisfied

>>f(r3)=2/3*t(r3)-6; ↵ ← assigns $\frac{2}{3}t-6$ values only to the position indexes stored in r3 corresponding to $6 \le t \le 9$ secs

The r1, r2, and r3 are user-chosen variable names. The workspace variable f holds the whole signal as a row matrix.

Example 5

The staircase signal of the figure 2.1(e) needs to be generated. Reading off the signal from the graph and writing the signal as a function, we have $f(t)=\begin{Bmatrix} 2 & for & 0 \le t < 2 \\ 4 & for & 2 \le t < 4 \\ 6 & for & 4 \le t < 6 \end{Bmatrix}$. Referring to example 2 of section 2.3, the signal needs multiple constant generation on different intervals. In last two examples we illustrated how the multiple intervals can be handled. The complete code of the signal is as follows:

>>t=0:0.1:6; ↵ ← t holds the whole t variation $0 \le t \le 6$ as a row matrix choosing a step 0.1sec

>>f=ones(1,length(t)); ↵ ← f is a row matrix whose all elements are 1, the number of ones is exactly equal to the number of elements in t

>>r1=find(0<=t&t<2); ↵ ← r1 holds only the position indexes of t at which $0 \le t < 2$ is satisfied

>>f(r1)=2; ↵ ← assigns 2 only to the position indexes stored in r1 corresponding to $0 \le t < 2$

>>r2=find(2<=t&t<4); ↵ ← r2 holds only the position indexes of t at which $2 \le t < 4$ is satisfied

>>f(r2)=4; ↵ ← assigns 4 only to the position indexes stored in r2 corresponding to $2 \le t < 4$

>>r3=find(4<=t&t<=6); ↵ ← r3 holds only the position indexes of t at which $4 \le t < 6$ is satisfied

>>f(r3)=6; ↵ ← assigns 6 only to the position indexes stored in r3 corresponding to $4 \le t < 6$

When you graph the signal by using the function plot of section 2.9, you find the transition lines slightly deviated from perfect vertical line. Reducing the step size from 0.1 to 0.05 or less solves the problem. In the last interval, mathematically we say $4 \le t < 6$ but for the closing point we write $t \le 6$. If we do not do so, the last value of f is taken as 1 (because we assigned before). Anyhow the workspace variable f holds the whole signal as a row matrix.

Example 6

The exponential signal of the figure 2.1(f) is to be generated. If the initial value (value at t=0) and the time constant of an exponential signal are A and t_c respectively, the signal is given by the expression $f(t)=A\,e^{-\frac{t}{t_c}}$. With the given specification, the signal expression is going to be $f(t)=5\,e^{-\frac{t}{4}}$. We generate the signal considering a step size 0.1sec as follows:

>>t=0:0.1:7; ↵ ← t holds the whole t variation $0 \le t \le 7$ sec as a row matrix on the step 0.1sec

```
>>f=5*exp(-t/4); ↵
```
← Workspace variable f holds the signal as a row matrix (code of $5e^{-\frac{t}{4}}$ is 5*exp(-t/4))

2.5 Function file or M-file based signal

The reader is referred to subsection 1.1.2 and appendix E for the M-file and function file details respectively. Examples illustrated for signal generation so far have been implemented at the command prompt. If a signal is piecewise continuous like the figure 2.1(d), code writing for each segment of the signal might be lengthy specially for a signal with multiple pieces. An M-file function is better in this regard for which the following four examples are provided.

Example 1

Write an M-file that generates the samples of the piecewise continuous signal of figure 2.1(d).

Its piecewise definition is $f(t)=\begin{cases} 2 & for \quad 0 \le t < 3 \\ -2 & for \quad 3 \le t < 6 \\ \frac{2}{3}t-6 & for \quad 6 \le t < 9 \end{cases}$. There are two approaches for the M-file writing. The first approach assumes the input argument of the function file to be a single scalar (single t value) and the other approach assumes the input argument to be the whole t vector. We address both of which in the following:

Approach 1: Input is a single time value

Figure 2.2(a) shows the complete code of the generation for the signal in figure 2.1(d). In this approach we check the single t value for every piece of the signal by using the if-else-end statement (appendix C.7) to which interval it belongs. The interval appears as the logical expression of the if-else-end statement. We split the interval $0 \le t < 3$ as $0 \le t$ and $t < 3$ and connect them by the AND operator (appendix C.2). For example $0 \le t < 3$ is written as 0<=t&t<3. We assign the functional code to some user-chosen variable (we chose y) corresponding to the interval followed by elseif or else. The return data from the function file is accumulated outside the function file and the accumulation happens in another M-file or in the command window. The first step is to write the piecewise definition as shown in the figure 2.2(a) in a new M-file and save the file by the name f in your working path of MATLAB. We would like to verify that $f(t)$=–2 at t=4 for which we execute the following at the command prompt:

```
function y=f(t)
  if 0<=t&t<3
     y=2;
  elseif 3<=t&t<6
     y=-2;
  elseif 6<=t&t<9
     y=2/3*t-6;
  else
     y=0;
  end
```

Figure 2.2(a) Function file describing the signal of figure 2.1(d) assuming the t is a single scalar

```
>>f(4) ↵                ← calling the function for t=4

ans =

        -2              ← single input and single output
```

The next is to generate the complete signal over the whole interval. Again one needs to decide the step size say 0.1sec on that we exercise the following:

```
>>s=[ ]; for t=0:0.1:9 s=[s f(t)]; end ↵
```

Data accumulation and for-loop references are seen in appendices C.8 and C.9 respectively. For every t value we get the $f(t)$ value by the command f(t) and gather the return sequentially in a row matrix form by the command s=[s f(t)]. The workspace variable s holds the signal samples of the figure 2.1(d) over the whole interval on the chosen step size where the s is a user-chosen variable name.

Approach 2: Input is the whole time vector

The reader must go through the approach 1 of this example. Now we assume that the input to the function file is the whole t vector on chosen step size not a single scalar. The necessary code for this approach is presented in the figure 2.2(b). Write the codes in a new M-file editor and save the file by the name f (you can choose your own file name, for simplicity we used again the same name f to be consistent with $f(t)$) in your working path. Input argument of f(t) in figure 2.2(b) (that is t) is a vector this time. The single scalar t in the figure 2.2(a) is now replaced by the t(k) in the logical expression in the figure 2.2(b) considering the k-th element in the time vector t. The for-loop of the figure 2.2(b) has the last counter as length(t) that makes the continuation until the total number of elements in t is taken care of. For every k, the content of y of the figure 2.2(b) is a scalar and we accumulate this y to v side by side by the command v=[v y] where v is a user-chosen name. The t vector is a row matrix, so is the return from f(t).

```
function v=f(t)
v=[ ];
for k=1:length(t)
    if 0<=t(k)&t(k)<3
        y=2;
    elseif 3<=t(k)&t(k)<6
        y=-2;
    elseif 6<=t(k)&t(k)<9
        y=2/3*t(k)-6;
    else
        y=0;
    end
    v=[v y];
end
```

Figure 2.2(b) Function file describing the signal of figure 2.1(d) assuming the t is the whole time vector

From the figure 2.1(d), the signal values at t =1, 4, and 9 are 2, –2, and 0 respectively. Let us verify that as follows:

```
>>t=[1 4 9]; ↵    ← Assigning the required t values as a row matrix
>>f(t) ↵          ← Calling the function for vector t

ans =
       2  -2   0     ← Returns are the functional values of f(t)
                       as a row matrix respectively
```

You can even call the f(t) from another M-file as long as they are in the same working path. Suppose we intend to generate the signal from 0 to 9sec with a step 0.1sec and do so by the following:

```
>>t=0:0.1:9; ↵    ← t holds the whole t variation 0≤t≤9 as a row
                    matrix with the step 0.1sec
>>s=f(t); ↵       ← The workspace variable s holds the whole signal as
                    a row matrix
```

The s and t are any user-chosen variable names in the last implementation.

Example 2

We assume that the reader has gone through the example 1 thoroughly. Drawing the idea and notation of the example 1, we generate the signal of the figure 2.1(e) by using both approaches as follows:

Approach 1: Input is a single time value

The functional definition of the signal is seen in example 5 of section 2.4. Figure 2.2(c) presents the code of the signal sample generation. Type the codes in a new M-file editor and save the file by the name f. We call the function for the generation of the signal over the interval $0 \le t \le 6$ sec with a step size 0.1sec as follows:

>>s=[ ]; for t=0:0.1:6 s=[s f(t)]; end ↵ ← The variable s holds the whole signal as a row matrix

Approach 2: Input is the whole time vector

Figure 2.2(d) holds the necessary codes for the signal generation. Type the codes in a new M-file and save the file by the name f. Generation of signal samples over interval $0 \le t \le 6$ sec with step size 0.1sec is as follows:

>>t=0:0.1:6; ↵ ← t holds the whole t variation $0 \le t \le 6$ sec as a row matrix with the step 0.1sec

>>s=f(t); ↵ ← The workspace variable s holds the whole signal as a row matrix

```
function y=f(t)
   if 0<=t&t<2
        y=2;
   elseif 2<=t&t<4
        y=4;
   elseif 4<=t&t<=6
        y=6;
   else
        y=0;
   end
```

Figure 2.2(c) Function file describing the signal of figure 2.1(e) assuming the t is a single scalar

```
function v=f(t)
v=[ ];
for k=1:length(t)
    if 0<=t(k)&t(k)<2
        y=2;
    elseif 2<=t(k)&t(k)<4
        y=4;
    elseif 4<=t(k)&t(k)<=6
        y=6;
    else
        y=0;
    end
    v=[v y];
end
```

Figure 2.2(d) Function file describing the signal of the figure 2.1(e) assuming the t is the whole time vector

Example 3

The command elseif is not required for simple function like $u(t)$ or its derivative signals. If the number of interval checkings is 3 or more, then we need elseif. For two interval checking only else is used. For example the shifted unit step function $u(t-c)$ has the amplitude 1 and shifted at c. Figures 2.2(e) and 2.2(f) show the codes on the approaches 1 and 2 respectively. This time we have two input arguments t and c where c is the user-defined step value and a single number. For the $u(t)$, the c or c is 0.

```
function y=f(t,c)
   if t>=c
        y=1;
   else
        y=0;
   end
```

Figure 2.2(e) Function file for a shifted unit step function $u(t-c)$ assuming the t is a single scalar

Approach 1: Input is a single time value

Type the codes of the figure 2.2(e) in a new M-file editor and save the file by the name f. For example $u(t-1.4)=1$ when $t=2$. How do we verify that? Just call f(2,1.4) at the command prompt. Let us say we need $u(t-1.4)$ with a step size 0.1sec over the interval $0 \le t \le 6$ sec and do so as follows:

```
>>s=[ ]; for t=0:0.1:6 s=[s f(t,1.4)]; end ↵
```

Approach 2: Input is the whole time vector

Similar to the other two examples but the functional calling will be s=f(t,1.4) instead of s=f(t).

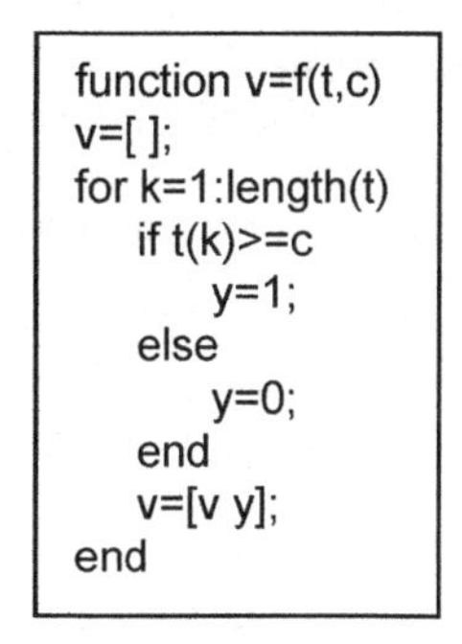

```
function v=f(t,c)
v=[ ];
for k=1:length(t)
    if t(k)>=c
        y=1;
    else
        y=0;
    end
    v=[v y];
end
```

Figure 2.2(f) Function file for a shifted unit step function $u(t-c)$ assuming the t is the whole time vector

Example 4

Scaled and shifted functions can also be derived from the function file just devised. For example $12.5u(t-1.4)$ needs the command 12.5*f(t,1.4) for both approaches as regard to the example 3. Again $12.5u(3t-1.4)$ needs the command 12.5*f(3*t,1.4) for the same example.

*** Note**

We saved the file by the name f just for simplicity in each of the figures 2.2(a)-(f). One problem is there if you save the file by the same name again and again, you lose the last file. That is why it is better to change the file name from f to other. For example if figure 2.2(f) shown file is renamed and saved as f1, then you should write function v= f1(t,c) in the first line of the figure where f1 is user-chosen name. The calling also occurs likewise i.e. 12.5*f1(t,1.4) for the example 4.

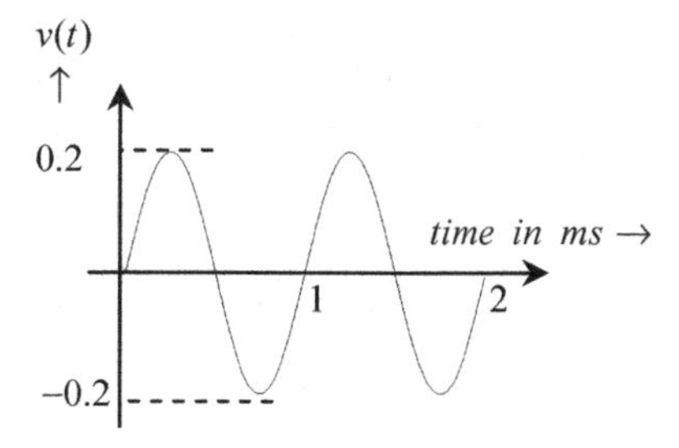

Figure 2.3(a) Plot of a two cycle sine wave

2.6 Periodic signals

A periodic signal repeats its wave shape every after the period of the signal. If the signal $v(t)$ is periodic over time period T, the relationship $v(t)=v(t+T)$ exists at every point t and the signal has the frequency $f=\frac{1}{T}$.

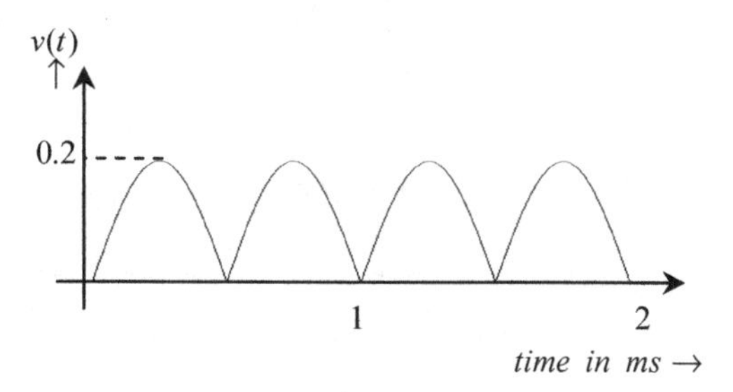

Figure 2.3(b) A full rectified sine wave

To generate a periodic wave, the definition of the signal in one period is enough. Like the signals mentioned in previous sections, the definition or description of a periodic signal can be expression, graphical representation, or M-file function based. We address few periodic signal generations in the following.

Example 1

Generate the wave of the figure 2.3(a). Sometimes our ability to pick up the information from a given graph is important. Referring to the figure, the time period and amplitude of the wave are 1 $m$sec and 0.2 V respectively. The frequency of the sine wave is then $f=\frac{1}{T}=1000\ Hz$ hence equation of the wave is $v(t) = 0.2\sin 2\pi 1000t$. To generate the wave by choosing a step 0.01 $m$sec, we exercise (section 2.3) the following commands:

```
>>t=0:0.01e-3:2e-3; ↵        ← t holds t variation 0≤t≤2 msec as a row
                                matrix with a step 0.01 msec
>>v=0.2*sin(2*pi*1000*t); ↵  ← code of 0.2 sin 2π1000t is 0.2*sin(2*pi*1000*t)
```

Workspace v holds the signal sample as a row matrix where the v is a user-chosen variable.

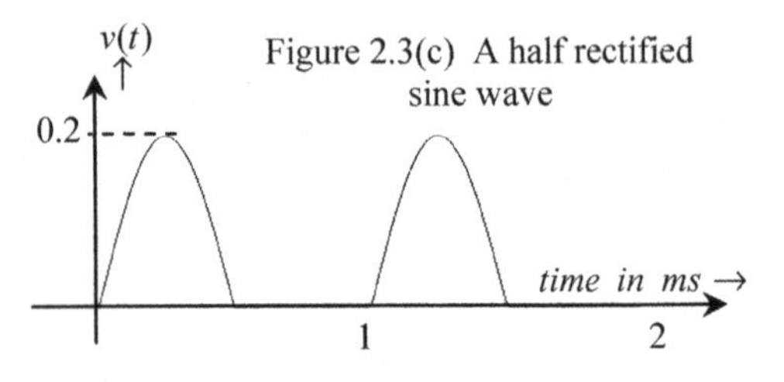

Figure 2.3(c) A half rectified sine wave

Example 2

Generate the full rectified sine wave of the figure 2.3(b). It is exactly the wave of the figure 2.3(a) but the negative halves are turned to positive. The command abs turns the negative value to a positive one. In example 1, we need to write v=abs(0.2*sin(2*pi* 1000*t)); to generate the wave assuming the same step size.

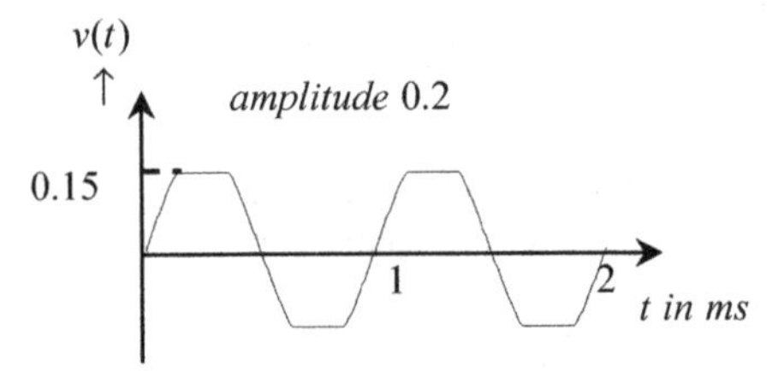

Figure 2.3(d) The sine wave of figure 2.3(a) is clipped at ±0.15

Example 3

Generate the half rectified sine wave of the figure 2.3(c). This wave is also derived from the wave of the figure 2.3(a) in which the negative part of the wave is turned to zero. In order to generate this signal, first we generate the signal of the figure 2.3(a) as in example 1, find the indexes of the t vector at which the signal is less than zero by using the find (appendix C.3) function, and set the signal values corresponding to these indexes to zero as follows:

```
>>t=0:0.01e-3:2e-3; ↵        ← t holds t variation 0≤t≤2 msec as a row
                                matrix with a step 0.01 msec
>>v=0.2*sin(2*pi*1000*t); ↵  ← v holds the signal of the figure 2.3(a) where v is a
                                user-chosen name
>>r=find(v<0); ↵             ← r holds the t indexes when the signal has negative value
                                where r is user-chosen variable name
>>v(r)=0; ↵                  ← setting the v elements of negative indexes to 0, v holds the
                                signal as a row matrix
```

Example 4

We intend to generate the clipped sine wave of the figure 2.3(d). This wave is also derived (example 1) from the wave of the figure 2.3(a):

```
>>t=0:0.01e-3:2e-3; ↵        ← t holds t variation 0≤t≤2 msec as a row
                                matrix with a step 0.01 msec
>>v=0.2*sin(2*pi*1000*t); ↵  ← v holds the signal samples of the figure 2.3(a)
```

Once the wave is generated, we check the value of signal at every point. If $v(t)>0.15$, we set $v(t)=0.15$ and if $v(t)<-0.15$, we set $v(t)=-0.15$ that is how the clipping is implemented:

```
>>r1=find(v>0.15); ↵  ← r1 holds the t indexes at which the signal value>0.15
>>v(r1)=0.15; ↵       ← 0.15 is assigned only to r1 index values of signal vector v
>>r2=find(v<-0.15); ↵ ← r2 holds the t indexes at which the signal value<–0.15
>>v(r2)=-0.15; ↵      ← –0.15 is assigned only to the r2 index values of the
                                                    signal vector v
```

The clipped signal samples of the figure 2.3(d) is available in the workspace variable v in the form of a row matrix.

Example 5

In one period T, a square wave $v(t)$ is defined as $v(t)=\begin{Bmatrix} A & for & 0\le t\le D \\ -A & for & D<t\le T \end{Bmatrix}$ where D is the duty cycle of the wave as a percentage of T. Figure 2.3(e) depicts the plot of one cycle square wave in which the duty cycle varies from 0 to 100%. MATLAB format for the generation of the wave is square($2\pi ft$, D) where $f=\frac{1}{T}$ is the frequency of the square wave, t is the desired time interval vector in the form of a row matrix over

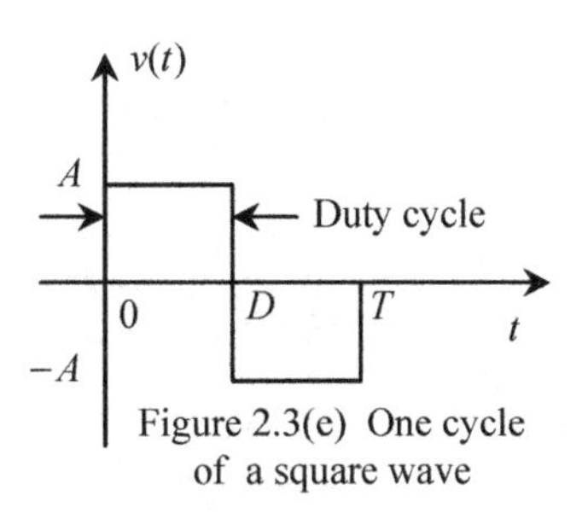

Figure 2.3(e) One cycle of a square wave

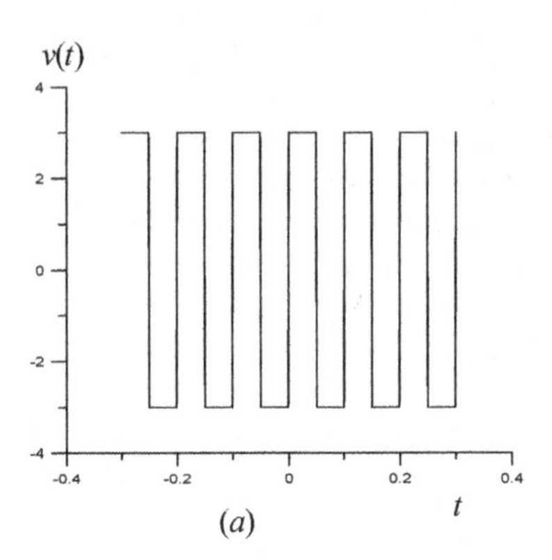

(a)

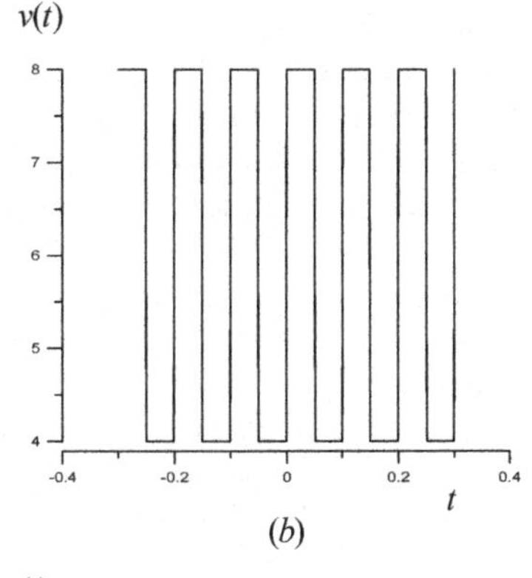

(b)

Figures 2.4(a)-(c) Different kinds of square waves: (a) amplitude $\pm 3\,V$, frequency $10\,Hz$, duty cycle 50%, (b) amplitude swing $4\,V$ to $8\,V$, frequency $10\,Hz$, duty cycle 50%, and (c) amplitude swing $4\,V$ to $8\,V$, frequency $10\,Hz$, duty cycle 80%

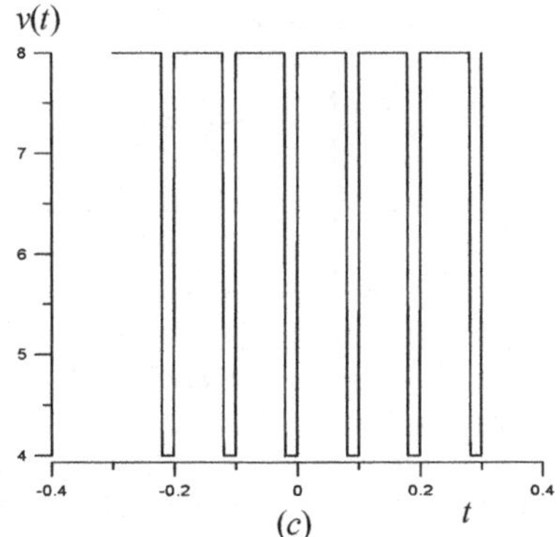

(c)

which we want to see the square wave, and the square is a built-in function.

Let us generate a 10 Hz square wave with amplitude variation ± 1 volt, with duty cycle 50%, and over the interval $-0.3 \le t \le 0.3$ sec. Again the selection of the step size is mandatory as well as user-supplied and let it be 0.01sec. The wave generation is as follows:

```
>>t=-0.3:0.01:0.3; ↵             ← t holds t variation −0.3 ≤ t ≤ 0.3 sec as a row
                                   matrix with a step 0.01sec
>>v=square(2*pi*10*t,50); ↵      ← The workspace variable v holds the square wave
                                   as a row matrix
```

Figures 2.4(a)-(c) show the square waves of different characteristics. Each of the waves has the time period (duration of one cycle) T =0.1sec and frequency $f = \frac{1}{T}$ =10 Hz and exists over the interval $-0.3 \le t \le 0.3$ sec. Choosing a step size 0.01sec, the commands for the generation of the square waves are presented as follows:

```
>>t=-0.3:0.01:0.3; ↵            ← t holds t variation −0.3 ≤ t ≤ 0.3 sec as a row matrix
                                  with a step 0.01sec
>>v=3*square(2*pi*10*t,50); ↵       ← Command for the figure 2.4(a)
>>v=6+2*square(2*pi*10*t,50); ↵     ← Command for the figure 2.4(b)
>>v=6+2*square(2*pi*10*t,80); ↵     ← Command for the figure 2.4(c)
```

In each case the workspace variable v holds the signal as a row matrix. The default swing of the square is ±1 so just multiplying by 3 achieves the required swing of the wave in figure 2.4(a).

In figure 2.4(b) a linear mapping $y = mx + c$ is necessary to make the function square sweep from 4 to 8 where x and y correspond to the former and latter signal values respectively. The related parameters m and c are to be found from the specification of the given signal. When the value of square is −1 (means $x = -1$), the signal value of the figure 2.4(b) should be 4 (means y=4) so $4 = -m + c$. Again if the value of square is 1 (means $x = 1$), the signal value of the figure 2.4(b) should be 8 (means y=8) on that $8 = m + c$. Solving the two equations, one obtains $m = 2$ and $c = 6$. Treating the square as x, the equation of the wave of the figure 2.4(b) should be $y = 2x + 6$ or v=6+2*square(2*pi*10*t,50) – that is how we wrote the command.

This kind of linear transformation is often necessary for the same type of wave shape for instance square to square, triangular to triangular, sine to sine, Gaussian to Gaussian, etc. In the periodic wave generation, the notion of the cycle and period must be transparent. With the time period 0.1sec and duration 0.6sec for each wave, there must be 0.6/0.1=6 cycles in the generated wave.

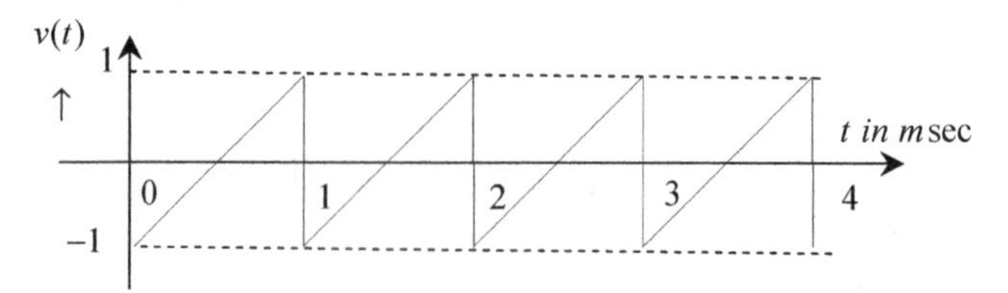

Figure 2.4(d) A sawtooth wave

Example 6

A sawtooth wave is generated by using the built-in function sawtooth. The general format for the wave generation is sawtooth($2\pi ft$) where T is the time period of the sawtooth wave, $f=\frac{1}{T}$ is the frequency of the wave, and t is the desired time interval as a vector over which we want to see the wave. The default swing of the wave generated by **sawtooth** is from -1 to 1. We intend to generate the sawtooth wave of the figure 2.4(d). Looking into the figure, the time period T of the wave is $1\,m\sec$ therefore frequency $f=1000\,Hz$. The wave exists over the interval

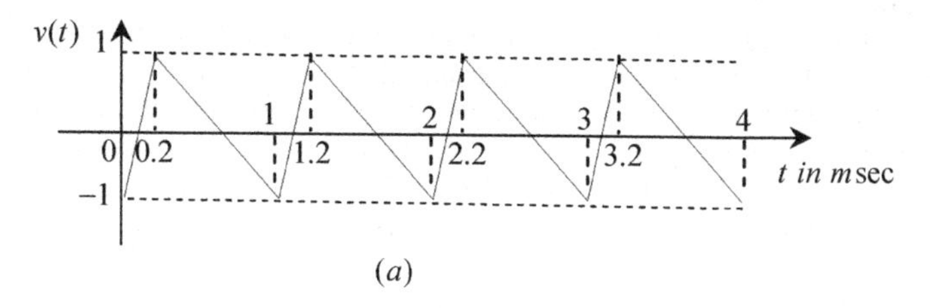

(a)

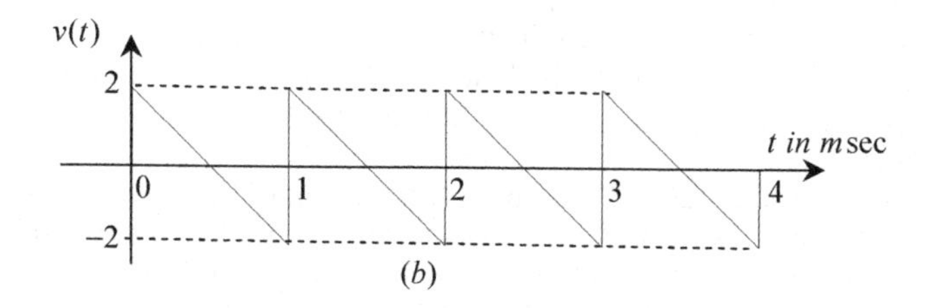

(b)

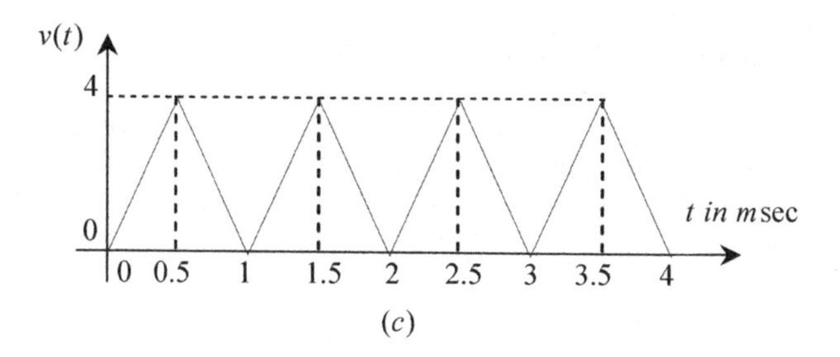

(c)

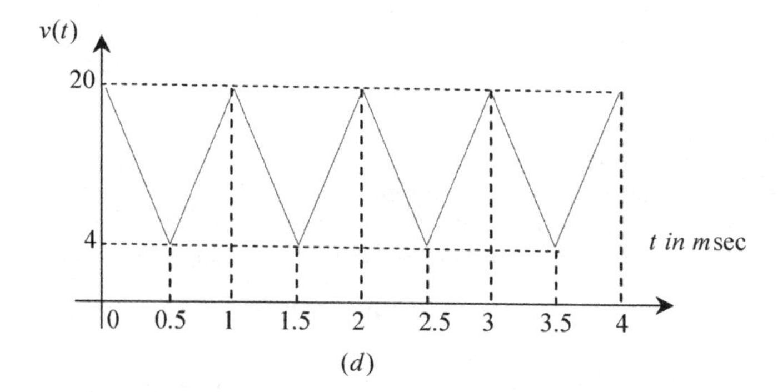

(d)

Figures 2.5(a)-(d) Periodic triangular waves of different characteristics

$0 \le t \le 4$ $m\sec$. The step size selection comes from the user and let us say it is 0.001 $m\sec$. The implementation is as follows:

```
>>t=0:0.001e-3:4e-3; ↵
```

Above t holds the t variation $0 \le t \le 4$ $m\sec$ as a row matrix with a step 0.001 $m\sec$.

```
>>v=sawtooth(2*pi*1000*t); ↵
```

Above workspace variable v holds the signal as a row matrix where v is user-chosen.

Example 7

The wave of the figure 2.5(a) is a variant of the wave of the figure 2.4(d). In the figure 2.4(d), the maximum 1 of the triangular wave is occurring at the 100% of the period point on the t axis (for example at 1 $m\sec$, 2 $m\sec$, etc). On the contrary the maximum 1 of the figure 2.5(a) is occurring at 20% period of the wave for example at 0.2 $m\sec$, 1.2 $m\sec$, etc. The function sawtooth is still effective to generate the signal of the figure 2.5(a) as follows (assuming the step size of example 6):

```
>>t=0:0.001e-3:4e-3; ↵
>>v=sawtooth(2*pi*1000*t,0.2); ↵
```

Now there are two input arguments in the function sawtooth (another syntax of sawtooth), the first and second of which are the $2\pi ft$ description and the peak point occurrence point in terms of 0-1 of the period respectively.

Example 8

The wave of the figure 2.5(b) is another variant of the wave in figure 2.4(d). Contrary to example 7, the peak value 2 is occurring at the 0% point on the period of the wave but the amplitude swing is ± 2 instead of ± 1. Multiplying the function sawtooth by 2 achieves the change. However using the step size of example 6, the wave generation is as follows:

```
>>t=0:0.001e-3:4e-3; ↵
>>v=2*sawtooth(2*pi*1000*t,0); ↵
```

Example 9

The wave in the figure 2.5(c) is also a variant of the wave in figure 2.4(d). The maximum point 4 in the wave is taking place at the 50% of the period point unlike example 7. The amplitude swing is from 0 to 4 instead of -1 to 1. Referring to the example 5, a linear mapping is required to decide the functional relationship from the sawtooth to the one in figure 2.5(c). We have the necessary equations $\begin{Bmatrix} 4 = m + c \\ 0 = -m + c \end{Bmatrix}$ and whose solution is $\begin{Bmatrix} m = 2 \\ c = 2 \end{Bmatrix}$ (treating the sawtooth as example 5 mentioned x). Therefore we have the required equation $y = 2x + 2$. Anyhow the formal implementation (using the step size of example 6) is as follows:

```
>>t=0:0.001e-3:4e-3; ↵
>>v=2*sawtooth(2*pi*1000*t,0.5)+2; ↵
```

Example 10

Figure 2.5(d) presented wave is also a variant of the wave in figure 2.4(d). It should be pointed out that the linear mapping is associated only with the amplitude swing not with the time information. As the graph says the swing of the wave in figure 2.5(d) is from 4 to 20 (in lieu of default -1 to 1) therefore requiring to solve the equation $\begin{Bmatrix} 20 = m + c \\ 4 = -m + c \end{Bmatrix}$ according to the example 5. On that the solution is $\begin{Bmatrix} m = 8 \\ c = 12 \end{Bmatrix}$ and the equation we need is $y = 8x + 12$. The time information should be read out from the given graph. The first cycle of the graph exists from 0 to 1 $m\sec$. If we want to fit the given wave by using the sawtooth of ongoing examples, the first cycle of the wave had better be stated from $0.5\,m\text{sec}$ to $1.5\,m\text{sec}$ making the sense that the maximum is occurring at the 50% period of the wave (means the second input argument of sawtooth should be 0.5 as regard to example 7). To be consistent with the given graph, we say that the wave so mentioned is shifted by $0.5\,m\text{sec}$ on the time axis and the shifting is accounted for considering sawtooth($2\pi f(t-0.5m\text{sec})$) instead of sawtooth($2\pi ft$). Anyhow the complete code of the generation by using the step of example 6 is as follows:

```
>>t=0:0.001e-3:4e-3; ↵
```
← t holds t variation $0 \le t \le 4\ m\text{sec}$ as a row matrix with a step $0.001\,m\text{sec}$

```
>>v=8*sawtooth(2*pi*1000*(t-0.5e-3),0.5)+12; ↵
```
← Workspace variable v holds the signal as a row matrix

2.7 Electric voltage, current, and power signals

Electric voltage, current, or power signal is primarily a sinusoidal signal. Generation of a voltage or current signal is the generation of a sinusoidal signal like the example 1 of section 2.6. The instantaneous electrical power signal is the product of the instantaneous voltage and current signals. A machine does not have any knowledge about a signal. It is the user who has to decide the signal type.

Let us assume that the instantaneous voltage and current signals are given by the expressions $v(t) = 2\sin(2\pi ft + 10^0)$ and $i(t) = -3\cos(2\pi ft - 30^0)$ respectively (the i is used for the imaginary unit of complex number as well as the instantaneous current). The instantaneous electrical power signal is given by $p(t) = 2\sin(2\pi ft + 10^0)$ $\times -3\cos(2\pi ft - 30^0)$.

To obtain the power signal, we generate the voltage signal samples as a row matrix, generate the current signal samples as a row matrix, and take their scalar product (appendix B) to obtain the power signal. Let us assume that the power supply frequency f =60 Hz and we intend to have the signal over $0 \le t \le 0.05\text{sec}$ with a step 0.0001sec. The step by step command is as follows:

```
>>t=0:0.0001:0.05; ↵
```
← t holds the t variation $0 \le t \le 0.05\text{sec}$ as a row matrix

```
>>v=2*sin(2*pi*60*t+10*pi/180); ↵
```
← v holds the voltage signal $v(t)$ as a row matrix over $0 \le t \le 0.05\text{sec}$

```
>>c=-3*cos(2*pi*60*t-30*pi/180); ↵
```
← c holds the current signal $i(t)$ as a row matrix over $0 \le t \le 0.05\text{sec}$

```
>>p=v.*i; ↵
```
← p holds the power signal $p(t)$ as a row matrix over $0 \le t \le 0.05\text{sec}$

The t, v, c, and p are user-chosen variable names. The reader should not use i as the variable name for current. When a complex number is processed, usage of i may create some problems because i is defined as the unit imaginary number in MATLAB.

2.8 Unit step and Dirac delta signals

We wish to pay special attention to these two signals in this section. In signal and system course these two signals play very important role. In any system particularly in electrical system if the input to the system is one unit, what the output should be? This is known as step response which requires the use of the function $u(t)$. Again if the input to the system is all possible frequencies, what the output should be? This is known as impulse response. This requires the usage of Dirac delta or $\delta(t)$ function. Let us see the following examples.

Example 1

The $u(t)$ is to be generated over $0 \le t \le 4$ sec.

Apply example 2 cited technique of section 2.3 or apply modeling approach of section 3.2.

Example 2

The $12.5u(t-1.5)$ is to be generated over $0 \le t \le 4$ sec.

Apply function file approach of section 2.5 specially examples 3 and 4 or apply modeling approach of section 3.2.

Example 3

The $\delta(t-2)$ is to be generated over $0 \le t \le 4$ sec.

Truly speaking we can not generate this function because at t =2 the value of $\delta(t-2)$ is infinite. In order to be practical for numerical simulation let us say 10^{20} represents the infinity. We need to choose a step size over the interval $0 \le t \le 4$ sec (sections 2.2-3) say 0.1sec. Let us generate the t vector as follows:

```
>>t=0:0.1:4; ↵
```

The integer position index in t corresponding to t =2 is found by the find function of appendix C.3 as follows:

```
>>I=find(t==2); ↵        ← I holds the integer index where I is user-chosen
```

We generate a row matrix v whose all elements are zeroes (appendix C.1) but the number of elements in v is equal to the number of elements in the t as follows:

```
>>v=zeros(1,length(t)); ↵
```

In the v only the I index element should be 10^{20} which we assign as follows:

```
>>v(I)=1e20; ↵
```

Therefore the last v holds the numerical $\delta(t-2)$ samples as a row matrix over $0 \le t \le 4$ sec with a step 0.1sec.

Example 4

The $23\delta(t-2)$ is to be generated by another approach over $0 \le t \le 4$ sec.

We know that $\delta(t)$ has 0 width, unity area, and infinite amplitude. When we have $23\delta(t-2)$ the impulse strength 23 is a matter of concern. In this regard we simulate the unit area by choosing a short duration pulse. Let us say the pulse exists for 0.01sec after t =2. Unity area says that the amplitude of the finite pulse should be 1/0.01=100. What we need is the constant value 2300 is to be generated over $2 \le t \le 2.01$ sec like the example 2 of section 2.3 but sampling step size must be much smaller than 0.01.

Or, apply the modeling approach of section 3.2.

2.9 How to graph a signal?

We addressed how to generate various types of signals in previous sections. Now we concentrate on how these generated signals can be viewed in MATLAB figure window just to make sure that our signal generation is correct. Before we plot any signal, we have to identify the signal generation technique. Graphing techniques are different for expression based and sample value based signals. Let us see the following examples.

Expression based signal

For expression based single signal, the built-in function ezplot of appendix F is applied with the syntax ezplot(signal code under quote, interval information as a two element row matrix). We do not have to compute the signal value from the given functional expression but MATLAB automatically computes the functional expression and graphs the function in continuous sense despite internally discrete data generation. Let us see the following examples for the use of the function.

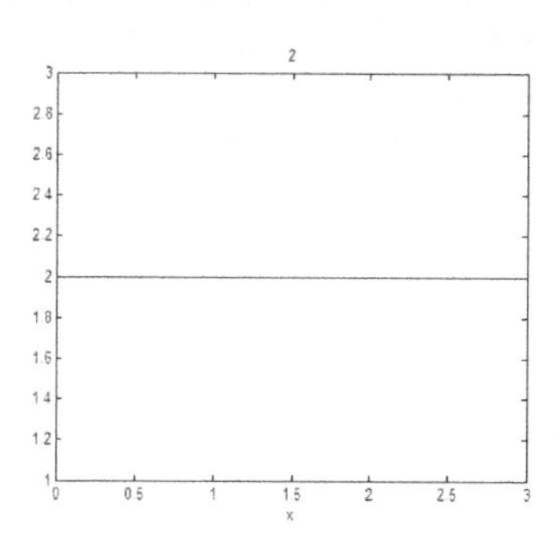

Figure 2.6(a) Plot of the signal $f(t)=2V$ in MATLAB

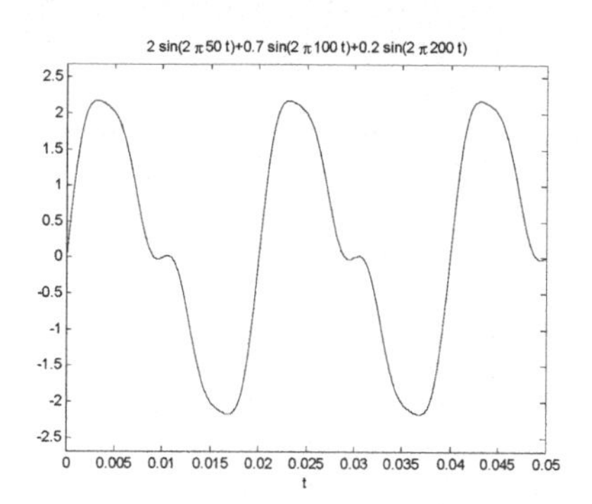

Figure 2.6(b) Plot of the multi-frequency signal

Example 1:

Let us graph the signal of the example 2 of section 2.3. The function $f(t)=2V$ has the code '2' and the interval $0 \le t \le 3$ sec is argumented by taking the beginning and ending bounds of the interval so we have [0 3]. With this information, the signal is graphed as follows:

```
>>ezplot('2',[0 3]) ↵
```

On execution you see the graph of the figure 2.6(a). The horizontal and vertical axes of the figure refer to t and $f(t)$ respectively. Because of the auto range setting of MATLAB, the graph may not be better perceived. The default independent variable of the plotter ezplot is x that is why the x axis of the figure 2.6(a) is labeled as x. The figure shows the x axis variation from 0 to 3 which is basically the t variation.

Example 2:

Graph the multiple frequency signal $v(t) = 2\sin 2\pi 50t + 0.7\sin 2\pi 100t + 0.2\sin 2\pi 200t$ over $0 \le t \le 0.05$ sec. The signal code can be assigned to any user-chosen variable (let us say v) if it is long which we perform in the following:

```
>>v='2*sin(2*pi*50*t)+0.7*sin(2*pi*100*t)+0.2*sin(2*pi*200*t)'; ↵
```

Based on the given interval, calling of the ezplot is as follows:

```
>>ezplot(v,[0 0.05]) ↵
```

Figure 2.6(b) is the result from the last line execution.

Signal from its sample values

If we have signal sample values in some row or column matrix, appendix F cited plot is applied to graph the signal with the syntax plot(x or t data as a row matrix, y or signal sample values as a row matrix). Let us see the following examples for the use of the function.

Example 1:

Graph the staircase signal of the figure 2.1(e). Example 5 of section 2.4 elaborately explained the signal sample data generation. Re-execute all commands of the example. We know that the workspace t and f hold the sample values of the t and $f(t)$ respectively. All we need is execute plot(t,f) at the command prompt. MATLAB response is identical with the figure 2.1(e) which we did not show for space reason.

Example 2:
We intend to graph the half rectified wave of the figure 2.3(c). We addressed the generation of the periodic wave as example 3 in section 2.6 wherefrom the workspace t and v hold the sample values of the t and $v(t)$ respectively. We just have to exercise the command plot(t,v) to see the wave shape of the figure 2.3(c), the graph is not shown for space reason.

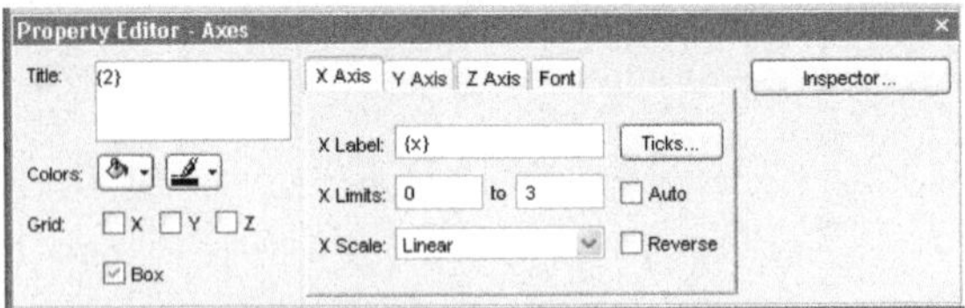

Figure 2.6(c) Property Editor Window of the MATLAB figure window

Graphical property manipulation on a drawn signal

After drawing a signal in the figure window of MATLAB, graphical property manipulation sometimes gives a better look to the drawn graphics. For example figure 2.6(a) is the plot of $f(t)=2V$ which we did at the beginning by using the ezplot. The graph will be better perceived if we see 0 value of the function beyond the interval $0 \le t \le 3$. To do so, click the Edit menu in the figure window, click the Axes Properties in the pulldown menu of Edit, and find the figure 2.6(c) appended with the figure 2.6(a). In figure 2.6(c) we find X Limits as 0 and 3 for $0 \le t \le 3$. Let us make it –1 and 4 by using the keyboard and then close the Property Editor Window. Figure 2.6(d) is the response – certainly the look is better than in figure 2.6(a).

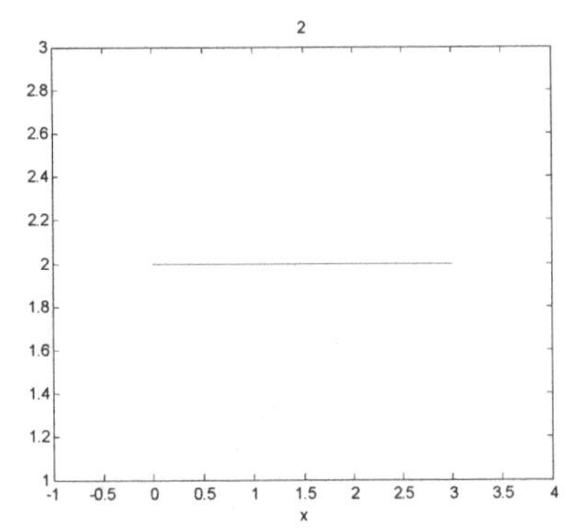

Figure 2.6(d) Figure 2.6(a) is redrawn on different horizontal axis setting

We changed the horizontal axis setting, for some signal vertical axis setting might be necessary. In that case click the Y axis of the figure 2.6(c) and enter necessary limit values like we did for the horizontal axis.

If you want to thicken a drawn curve, click the Edit plot icon of the figure 1.2(c), select the curve by the mouse, rightclick on the mouse, find different options in the popup (one of which is the Line Width), and click some value to see the change. Once you finish the curve editing, click the Edit plot icon again.

Similarly you can make the drawn curve color different and line style dotted than the default displayed one.

However we bring an end to the chapter with this discussion.

Chapter 3
SIGNAL MODELING

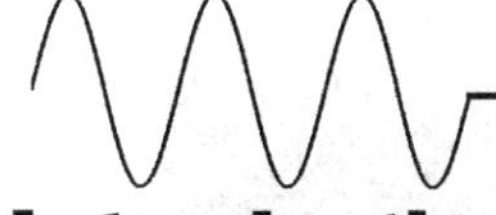

Introduction

Code writing of the chapter 2 cited examples is replaced by employing pre-designed SIMULINK blocks to a large extent in this chapter. A particular engineering system by and large follows certain model to this context modeling approach is extremely essential on whose account this chapter is contemplated. The SIMULINK blocks are so effective and well-designed that our computer virtually turns to an analog machine although behind the simulation the data processing is discrete. Our implementations outline the following:

- Signal modeling basics in SIMULINK
- Modeling an elementary signal including its derivatives
- Techniques to modeling the periodic, nonperiodic expression based, or other common signals

3.1 Signal modeling in SIMULINK

In chapter 2 mostly we wrote the code of a signal and generated the signal samples in the command window of MATLAB. As an additional part of MATLAB, SIMULINK keeps provision for signal generation by employing modeling approach for which following sections are dedicated. The reader needs SIMULINK basics which are addressed in chapter 1. From a given signal characteristic we pick up signal information or its parameters and enter those parameters to pre-designed SIMULINK blocks to generate the signal in continuous sense despite discrete generation. In MATLAB we view a generated signal by the command ezplot or plot (section 2.9) contrarily the block Scope (subsection 1.2.5) of SIMULINK exhibits the signal so generated in a model.

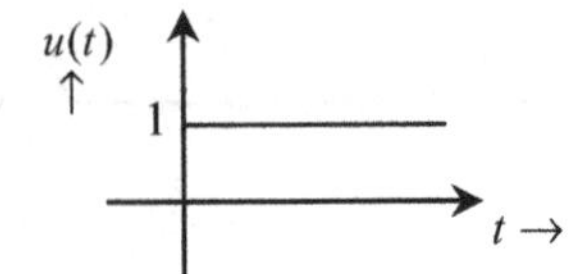

Figure 3.1(a) The unit step function

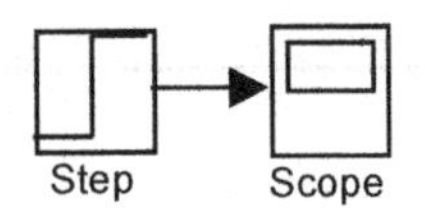

Figure 3.1(b) Step block connected with Scope

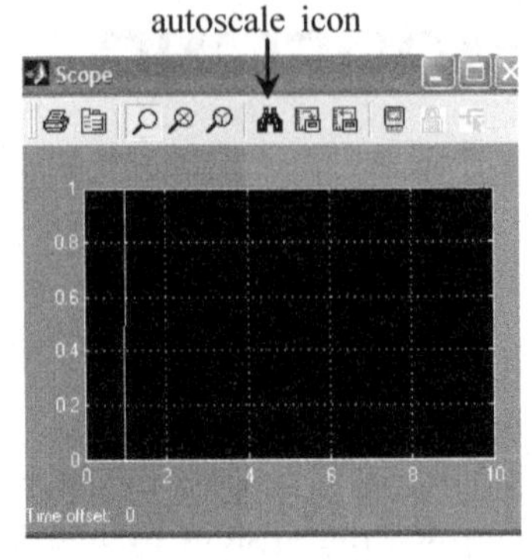

Figure 3.1(c) Scope output for the default Step block

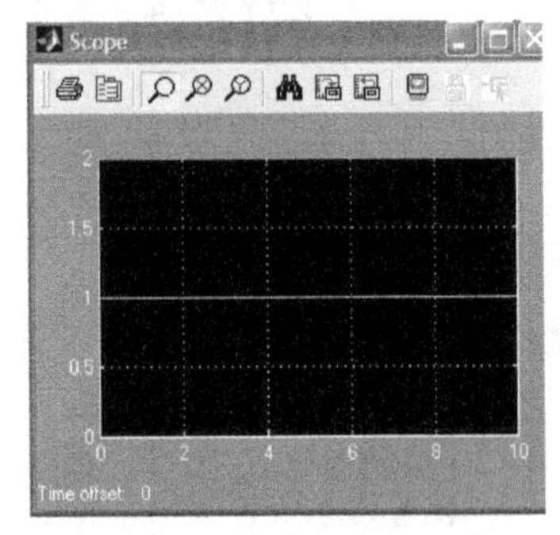

Figure 3.1(d) Scope output for the $u(t)$

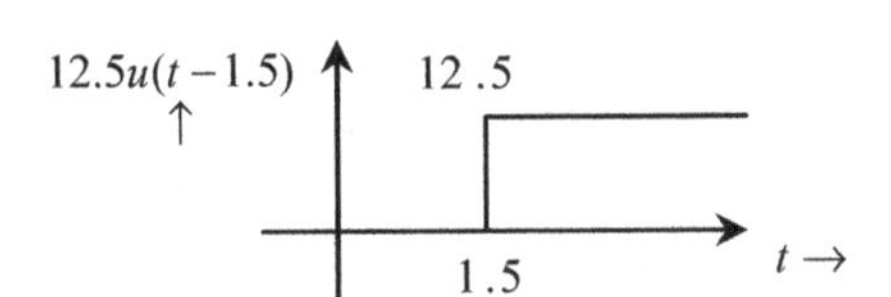

Figure 3.1(e) The unit step function shifted at 1.5 and of final value 12.5

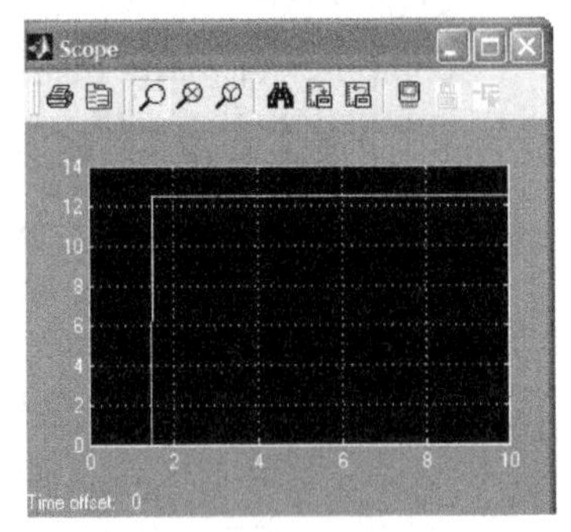

Figure 3.1(f) Scope output for the $12.5u(t-1.5)$

3.2 Modeling the unit step and its derivative signals

The unit step or Heaviside function $u(t)=\begin{Bmatrix} 1 & for\ t \ge 0 \\ 0 & for\ t < 0 \end{Bmatrix}$ which has the plot in figure 3.1(a) is simulated by the Step (appendix A for outlook and link) block. Open a new SIMULINK model file (subsection 1.2.2), bring one Step and one Scope blocks in the model file, connect (subsection 1.2.3) the blocks as shown in figure 3.1(b), run the model by clicking ▶ icon in the menu bar, and doubleclick the Scope. The autoscale icon as indicated in figure 3.1(c) has some default settings both in horizontal and vertical axes of the Scope. Clicking the autoscale icon adapts the default axes setting to current one. On clicking the autoscale icon we find the Scope output as seen in figure 3.1(c). The displayed output has the step value at 1 or functionally the default return is $u(t-1)$ but the $u(t)$ or figure 3.1(a) has the step value 0. Doubleclick the **Step** block in the model to see its parameter window like the figure 3.1(g), change the **Step time** from default 1 to 0 in the parameter window, run the model, and doubleclick the **Scope** to see the correct unit step function with the autoscale setting like the figure 3.1(d) i.e. figure 3.1(d) shown output refers to $u(t)$ of figure 3.1(a).

As another example the signal $12.5u(t-1.5)$ has the plot in figure 3.1(e) which has the final value 12.5 and is shifted at 1.5 to the right on the time axis. We need to enter the setting in the parameter window of figure 3.1(g) as $\begin{Bmatrix} \text{Step value to 1.5} \\ \text{Final value to 12.5} \end{Bmatrix}$ and the **Scope** output with the autoscale setting is shown in figure 3.1(f).

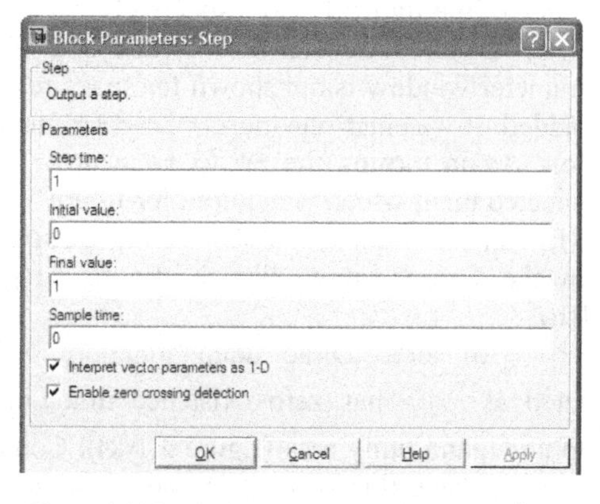

Figure 3.1(g) Block parameter window of the **Step**

A finite duration constant value signal on and off is used in electrical system analysis whose generation needs using two **Step**

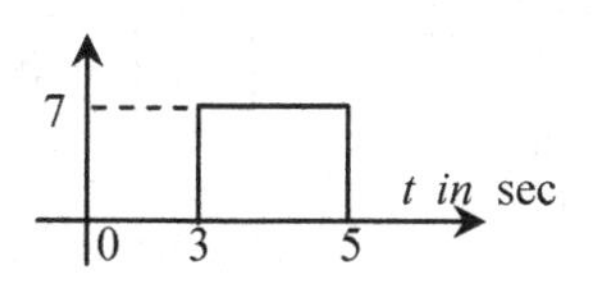

Figure 3.1(h) A finite duration pulse

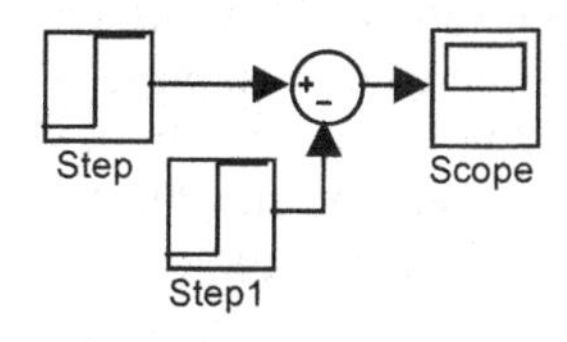

Figure 3.1(i) Two **Step** blocks model the signal in figure 3.1(h)

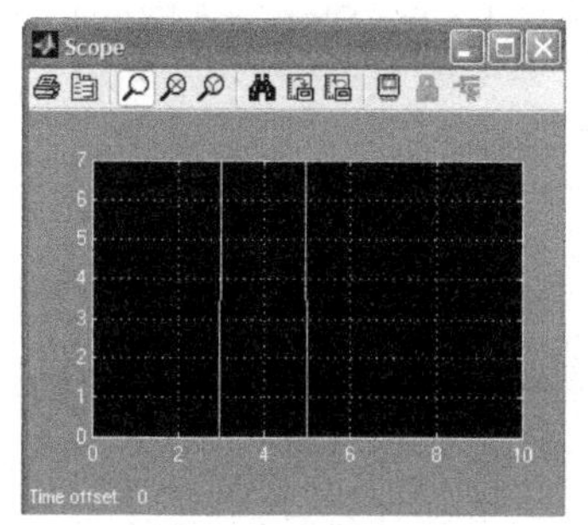

Figure 3.1(j) **Scope** output for the $7u(t-3)-7u(t-5)$

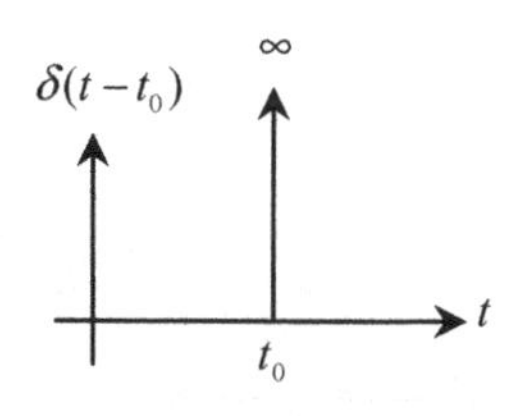

Figure 3.1(k) Ideal Dirac delta function

blocks. Figure 3.1(h) presents a finite duration constant signal. Mathematically we read off the signal as $7u(t-3)-7u(t-5)$. The components $7u(t-3)$ and $7u(t-5)$ are simulated by two different **Step** blocks. The subtraction of $7u(t-5)$ from $7u(t-3)$

takes place by using the Sum block (appendix A for outlook and link). Figure 3.1(i) shows the model for generating the finite pulse in figure 3.1(h). Bring two Step (appear as Step and Step1), one Sum, and one Scope blocks in a new SIMULINK model file. Doubleclick each of the Step blocks at a time and enter the settings $\left\{\begin{array}{l}\text{Step value to 3}\\ \text{Final value to 7}\end{array}\right\}$ and $\left\{\begin{array}{l}\text{Step value to 5}\\ \text{Final value to 7}\end{array}\right\}$ for the Step and Step1 in the parameter window of the figure 3.1(g) respectively.

Doubleclick the Sum block and find its List of signs as default ++ (parameter window is not shown for space reason) which indicates two inputs are to be added. If we enter one more + i.e. +++ meaning three inputs are to be added, and so on. Again turning the ++ to +- means - connected input is subtracted from + connected input which is required for figure 3.1(i). Connect the blocks like the figure 3.1(i), run the model, doubleclick the Scope block, click the autoscale icon, and view the designed pulse like the figure 3.1(j) which is consistent with the figure 3.1(h).

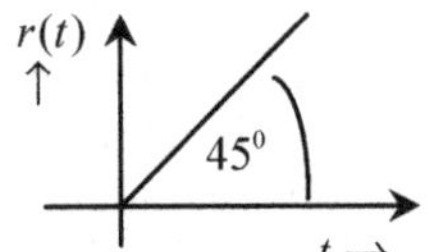

Figure 3.2(a) The ramp function

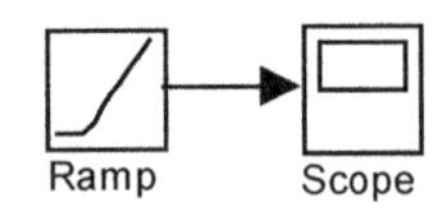

Figure 3.2(b) Ramp block connected with Scope

An ideal Dirac delta function $\delta(t-t_0)$ located at $t=t_0$ has zero existence time, infinite amplitude, and unity area (figure 3.1(k)). Computer never simulates this type of hypothetical function. Numerically we assume some large value with short existence which serves the purpose of the Dirac delta function generation. That means this is a special case of the finite duration pulse of the figure 3.1(h). As an example let us choose the duration of the pulse to be 0.01sec and the function starts at 3sec i.e. t_0=3sec. By function we write the numerical $\delta(t-t_0)$ as $100u(t-3)-100u(t-3.01)$ maintaining the unit area of the pulse therefore modeling of the $\delta(t-3)$ is very similar to that of the pulse in figure 3.1(h) (i.e. $\left\{\begin{array}{l}\text{Step value to 3}\\ \text{Final value to 100}\end{array}\right\}$ and $\left\{\begin{array}{l}\text{Step value to 3.01}\\ \text{Final value to 100}\end{array}\right\}$ for the Step and Step1 of figure 3.1(i) respectively).

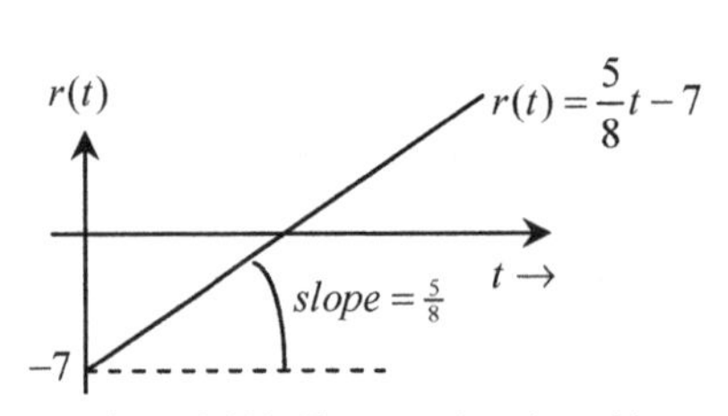

Figure 3.2(c) The ramp function with different slope and initial output

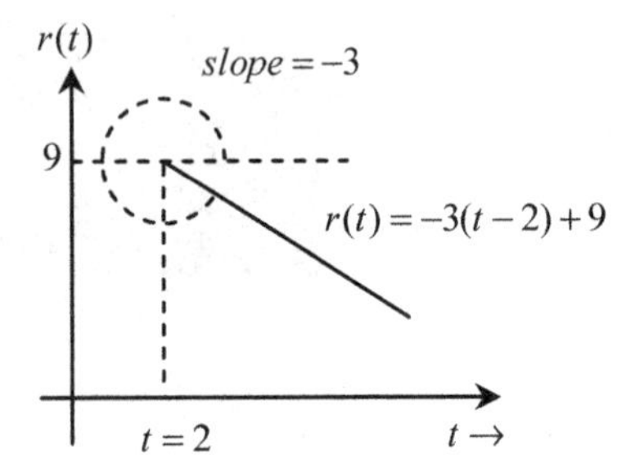

Figure 3.2(d) The ramp function $r(t) = -3(t-2)+9$ with different starting time

3.3 Modeling the ramp and its derivative signals

The ramp function graphed in figure 3.2(a) is defined as $r(t)=t$ which is a straight line with unity ($\tan 45^0=1$) slope and starts from $t=0\,\text{sec}$. The block **Ramp** (appendix A for outlook and link) generates the $r(t)$ by default. Let us bring one **Ramp** and one **Scope** blocks in a new SIMULINK model file (subsection 1.2.2), connect (subsection 1.2.3) them according to the figure 3.2(b), run the model by clicking ▶ icon in the menu bar, and doubleclick the **Scope** to see the output as shown in figure 3.2(f) with the autoscale setting (figure 3.1(c)). Ramp derived functions are simulated with a slight change in the parameter window data of the **Ramp**.

Let us doubleclick the **Ramp** block to see its parameter window like the figure 3.2(e). You find there the parameters as $\left\{\begin{array}{l}\text{Slope}\\ \text{Start time}\\ \text{Initial output}\end{array}\right\}$. The function $r(t)=t$ has the slope 1, the start time 0 (meaning beginning of the interval), and the initial output 0 (meaning value of the function at the beginning bound).

Let us see another ramp function whose equation is $r(t)=\frac{5}{8}t-7$ and its

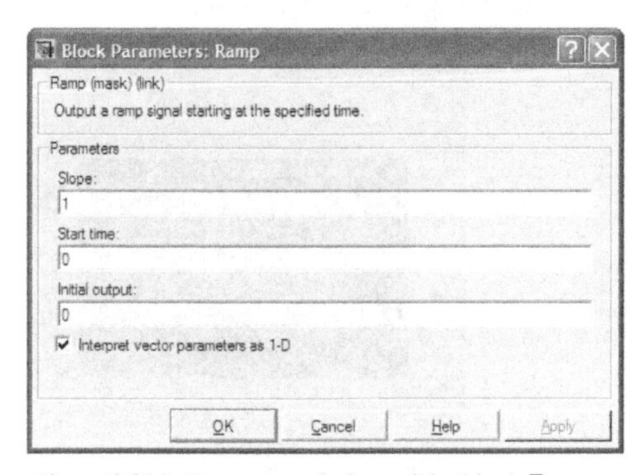

Figure 3.2(e) Parameter window of the block **Ramp**

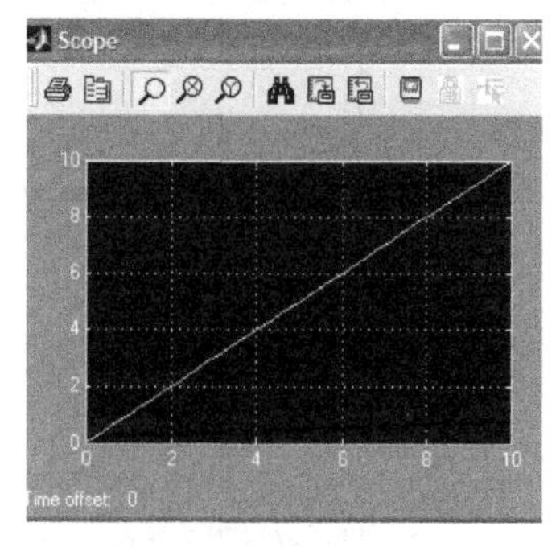

Figure 3.2(f) **Scope** output for the default **Ramp** block i.e. $r(t)=t$

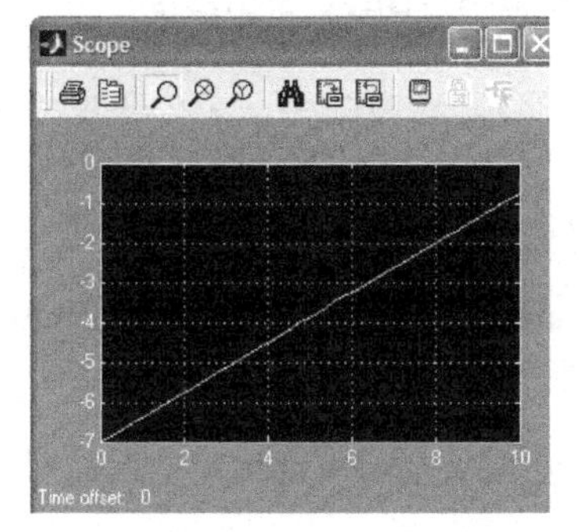

Figure 3.2(g) **Scope** output for the function $r(t)=\frac{5}{8}t-7$

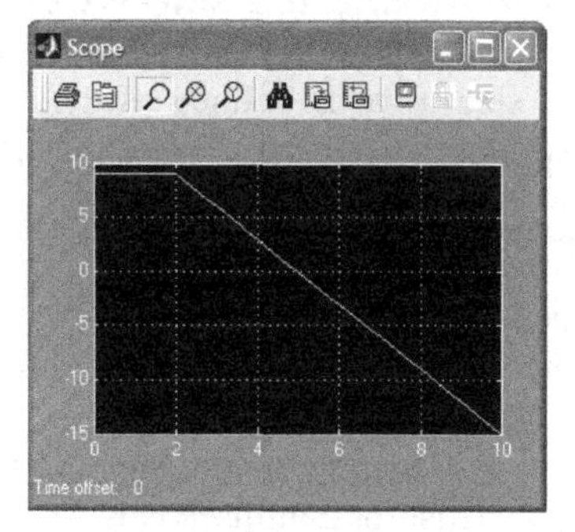

Figure 3.2(h) **Scope** output for the function $r(t)=-3(t-2)+9$

graph is the figure 3.2(c). Doubleclick the Ramp block in model in figure 3.2(b), enter the setting of Ramp as $\begin{Bmatrix} \text{Slope} = 5/8 \\ \text{Start time} = 0 \\ \text{Initial output} = -7 \end{Bmatrix}$ in the parameter window, run the model, and see the output as shown in the figure 3.2(g) with the autoscale setting.

The function $r(t) = -3(t-2)+9$ is also a ramp which has the parameters $\begin{Bmatrix} \text{Slope} = -3 \\ \text{Start time} = 2 \\ \text{Initial output} = 9 \end{Bmatrix}$ and which is detailed in figure 3.2(d). The corresponding Scope output with the autoscale setting is depicted in figure 3.2(h). Referring to the figure, the block assumes the functional value for $t = 2$ over the interval $0 \le t \le 2$.

If the reader persists in being the function strictly 0 over $0 \le t \le 2$, the functional description had better be $r(t) = [-3(t-2)+9]u(t-2)$ and you need the model of the figure 3.2(i) whose Scope output with the autoscale setting is the figure 3.2(j). The $u(t-2)$ in $r(t)$ is modeled by the Step block of section 3.2 with the parameter setting $\begin{Bmatrix} \text{Step value to 2} \\ \text{Final value to 1} \end{Bmatrix}$. The functional multiplication of $-3(t-2)+9$ and $u(t-2)$ takes place using the Product block (appendix A for link).

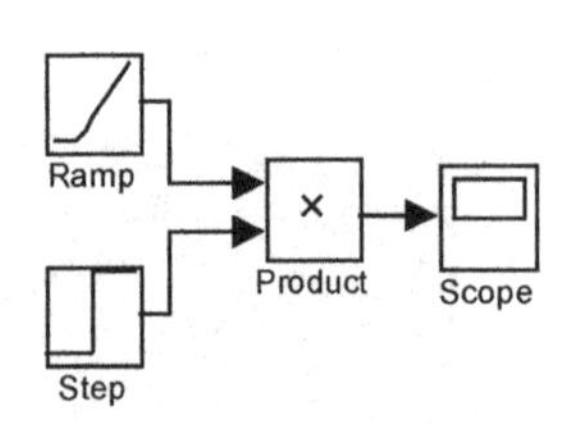

Figure 3.2(i) The model for the signal $r(t) = [-3(t-2)+9]u(t-2)$

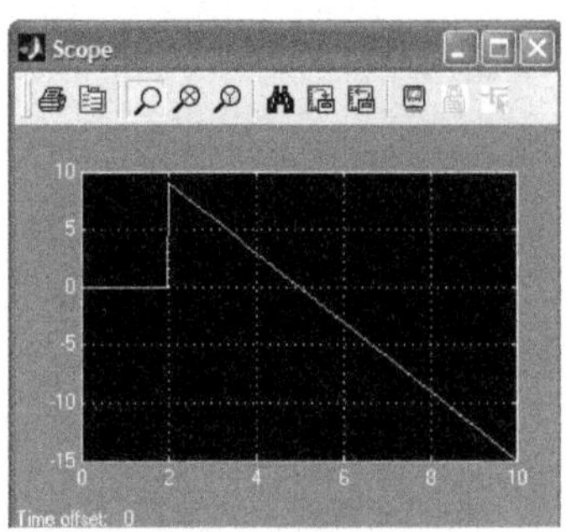

Figure 3.2(j) Scope output for the model of the figure 3.2(i)

3.4 Modeling the sine wave and its derivative signals

Perhaps the sine wave is the most addressed one in electrical engineering. The sine wave is defined as $y(t) = A\sin(2\pi ft + \theta)$ where A, f, θ, and t are the amplitude, frequency, phase angle, and the time interval over which the wave is to be simulated respectively. Also the time period of the wave is given by $T = \frac{1}{f}$. The block Sine Wave (appendix A for outlook and link) generates different sinusoids for which a number of examples are presented in the following.

Example 1

Generate a sine wave of frequency $1KHz$ and amplitude ±0.2. The wave should exist for 2 milliseconds.

The amplitude of the wave can represent any physical quantity such as displacement, voltage, or current. The wave has the time period $T = \frac{1}{10^3}$ secs or

1 $m$sec hence in the given time interval we expect the wave to have two cycles. The wave and the model for implementation are shown in figures 2.3(a) and 3.3(a) respectively. Bring one Sine Wave and one Scope blocks in a new SIMULINK model file (subsection 1.2.2) and connect (subsection 1.2.3) them according to the figure 3.3(a). Doubleclick the block Sine Wave to see its parameter window like the

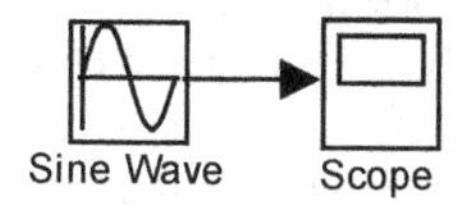

Figure 3.3(a) The model for the sine wave simulation for example 1

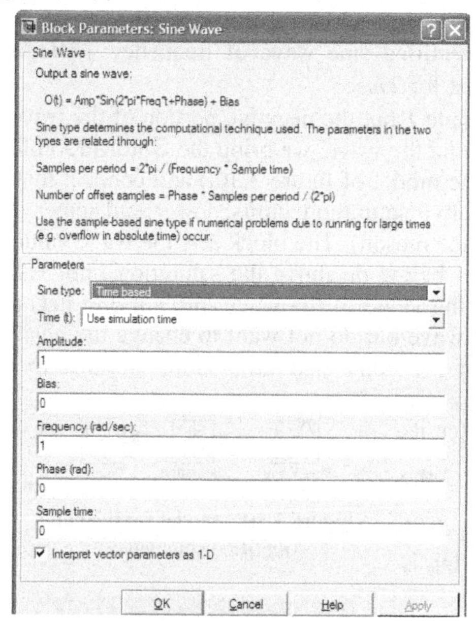

Figure 3.3(b) Block parameter window of the Sine Wave

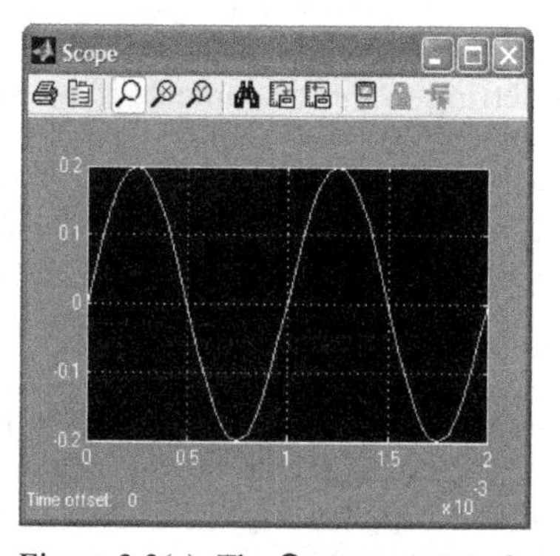

Figure 3.3(c) The Scope output of the model for the example 1

figure 3.3(b) and enter the settings as $\begin{Bmatrix} \text{Amplitude: 0.2} \\ \text{Frequency (rad/sec): 2 * pi * 1e3} \end{Bmatrix}$ keeping the others as default ($\omega = 2\pi f$ is used for the frequency, appendix B for coding). For the time interval entering (subsection 1.2.6), click the menu Simulation down the Configuration Parameters, be prompted with the figure 1.7(g), and enter the Stop time as 2e-3 (for 2 $m$sec duration starting from 0). Run the model by clicking ► icon in the menu bar and the Scope output should look like the figure 3.3(c) with the autoscale setting (figure 3.1(c)).

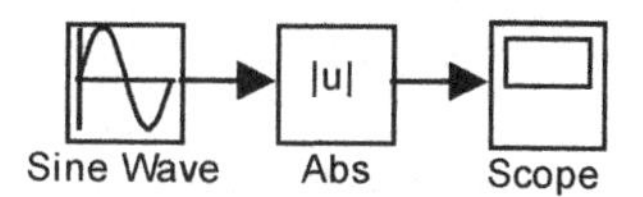

Figure 3.3(d) Model for the full rectified sine wave

Example 2

Let us generate a full rectified sine wave of frequency $1KHz$, amplitude ±0.2, and duration 2 $m$sec.

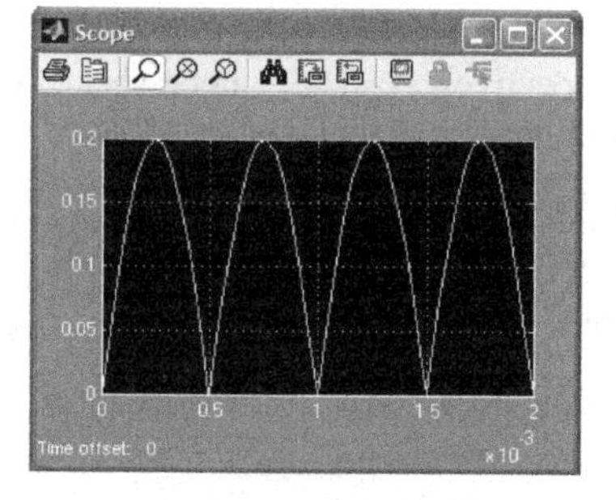

Figure 3.3(e) Scope return for the full rectified sine wave

This is basically the example 1 with

the exception that the negative peaks become positive as shown in figure 2.3(b). What extra simulation we do is we bring one Abs block (the block finds the absolute value of its input signal, appendix A for outlook and link) and insert that between the Sine Wave and the Scope like in figure 3.3(d). After running the model, the reader should see the Scope output as shown in figure 3.3(e) with the autoscale setting.

Example 3

We wish to generate a half rectified sine wave of frequency $1KHz$ and amplitude ±0.2 and the wave should exist for $2\,m\sec$.

This wave is the wave of example 1 but the negative portion of the wave is set to zero like the figure 2.3(c). To model the wave, we bring the Saturation block (appendix A for outlook and link) in the model of figure 3.3(a) and connect that as shown in figure 3.3(f). The block has two saturation limits: lower and upper (the parameter window is not shown for space reason). The block sets the wave value to the saturation limit if the wave reaches below or above the saturation limit value. Referring to the example 1, if we set the value of the wave to 0 for any negative value, we have the half wave rectified wave but do not want to change the positive portion of the wave which happens if the upper saturation limit is equal to the positive amplitude of the wave. On doubleclicking the Saturation, we enter its settings as $\left\{\begin{matrix}\text{Upper limit}: 0.2 \\ \text{Lower limit}: 0\end{matrix}\right\}$ in the parameter window and run the model. The Scope response with the autoscale setting is shown in figure 3.3(g).

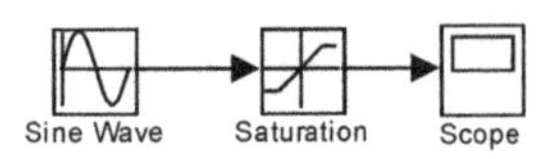

Figure 3.3(f) Model for the half rectified sine wave

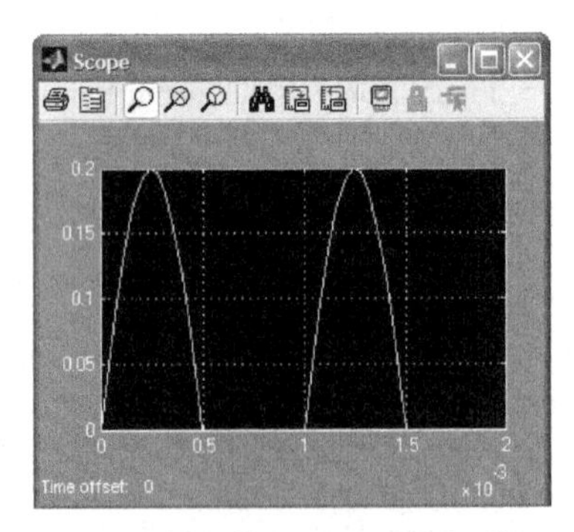

Figure 3.3(g) Scope output for the half rectified sine wave

Example 4

Let us generate the two sinusoidal frequency wave $y(t) = 0.2\sin 2\pi ft + 0.1\sin 4\pi ft$ over $0 \le t \le 1ms$ where $f = 2KHz$.

The given $y(t)$ has two sine wave components: the first one with amplitude 0.2 and frequency $2KHz$ and the second one with amplitude 0.1 and frequency $4KHz$. Figures 3.3(h) and 3.4(a) show the model and its Scope output with the autoscale setting respectively. The necessary parameter settings for the Sine Wave and Sine Wave1 are $\left\{\begin{matrix}\text{Amplitude}: 0.2 \\ \text{Frequency (rad/sec)}: 2 * \text{pi} * 2\text{e}3\end{matrix}\right\}$ and $\left\{\begin{matrix}\text{Amplitude}: 0.1 \\ \text{Frequency (rad/sec)}: 2 * \text{pi} * 4\text{e}3\end{matrix}\right\}$ respectively. Also the Stop time should be 1e-3 (subsection 1.2.6).

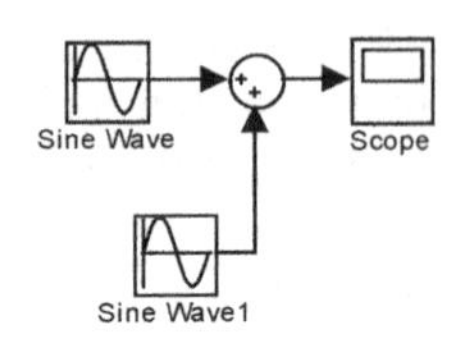

Figure 3.3(h) Model for the two frequency wave

Example 5

In this example we generate the three frequency wave $y(t) = 0.2\sin 2\pi ft + 0.1\sin 4\pi ft + 0.08\sin 10\pi ft$ over $0 \le t \le 1ms$ where $f = 2KHz$.

Obviously one needs three Sine Wave and a three input Sum (doubleclick the Sum block and change its List of signs to +++ to turn its input port number to three from the default two) blocks as depicted in the model 3.4(d). The settings for the Sine Wave, Sine Wave1, and Sine Wave2 blocks are $\begin{Bmatrix} \text{Amplitude: 0.2} \\ \text{Frequency (rad/sec): 2 * pi * 2e3} \end{Bmatrix}$, $\begin{Bmatrix} \text{Amplitude: 0.1} \\ \text{Frequency (rad/sec): 2 * pi * 4e3} \end{Bmatrix}$, and $\begin{Bmatrix} \text{Amplitude: 0.08} \\ \text{Frequency (rad/sec): 2 * pi * 10e3} \end{Bmatrix}$ respectively. The Stop time of the solver is 1e-3. Figure 3.4(b) is the Scope output for the wave with the autoscale setting.

Example 6

Adding some phase in each wave which is $y(t) = 0.2\sin(2\pi ft - 60^0) + 0.1\sin(4\pi ft + 10^0)$ modifies the wave of the example 4. In the parameter windows of the Sine Waves just append the phases as –pi/3 and 10*pi/180 respectively. The Scope returns the output with the autoscale setting as depicted in figure 3.4(c).

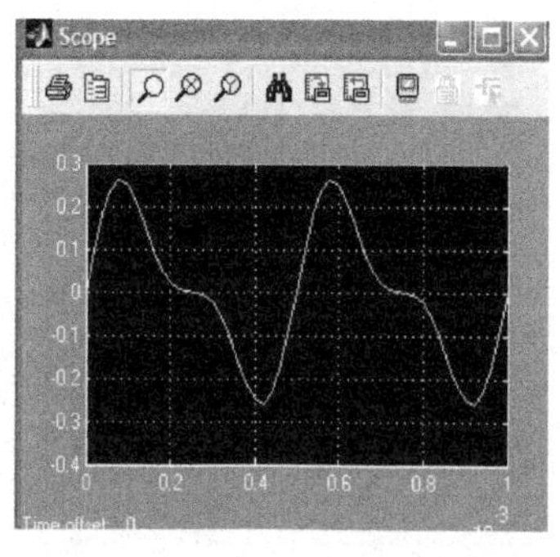

Figure 3.4(a) Two frequency wave from the Scope

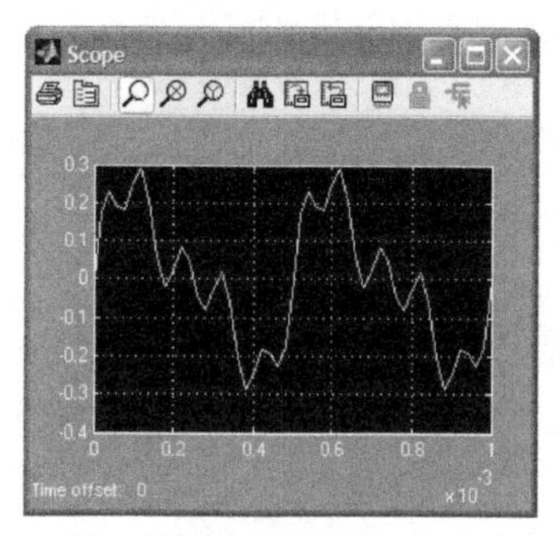

Figure 3.4(b) Scope output for the three frequency wave

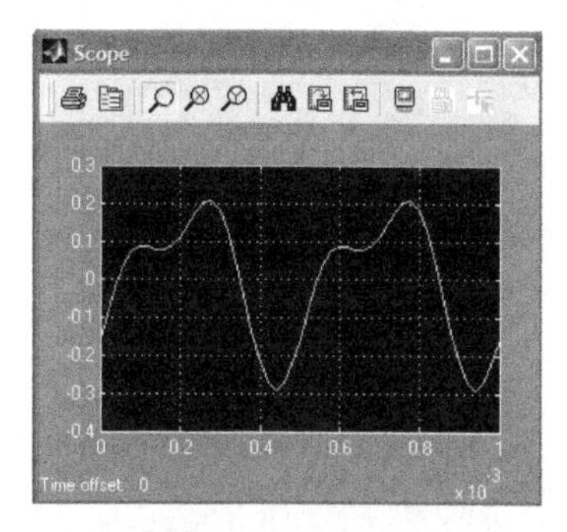

Figure 3.4(c) Scope response for two phase angle sine waves

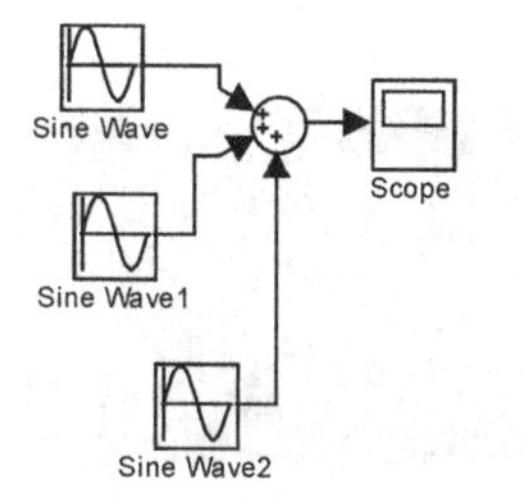

Figure 3.4(d) Model for the three frequency wave

Example 7

An example of the damped sine wave is $y(t) = 0.2e^{-1000t}\sin(2\pi ft - 60^0)$ where $f = 2KHz$. Let us generate the wave over $0 \le t \le 5m\sec$.

One separates the function $y(t)$ as e^{-1000t} and $0.2\sin(2\pi ft - 60^0)$, the latter part is resembling to the sine wave of the example 1. The exponent part e^{-1000t} can be modeled by using the blocks Ramp and Fcn (appendix A for outlook and link). The Ramp simulates the t and the Fcn performs the user-defined mathematical operation on its input port signal assuming that the input variable is in terms of u.

Let us bring one Ramp, one Fcn, one Sine Wave, one Product, and one Scope blocks in a new SIMULINK model file. Connect the blocks as shown in figure 3.4(e). Doubleclick the Fcn, enter the code of e^{-1000t} as exp(-1000*u) considering the independent variable u in its parameter window (not shown for space reason), and enlarge the block to display its contents. Doubleclick the Sine Wave, enter its settings as $\begin{cases} \text{Amplitude: } 0.2 \\ \text{Frequency (rad/sec): } 2*\text{pi}*2\text{e}3 \\ \text{Phase (rad): } -\text{pi}/3 \end{cases}$, change the solver stop

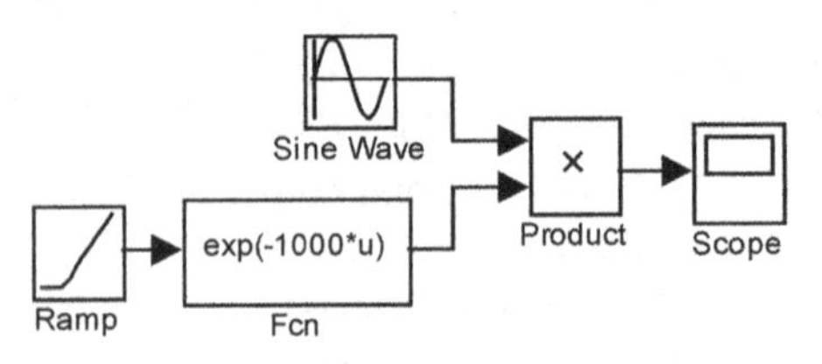

Figure 3.4(e) Model for the damped sine wave generation

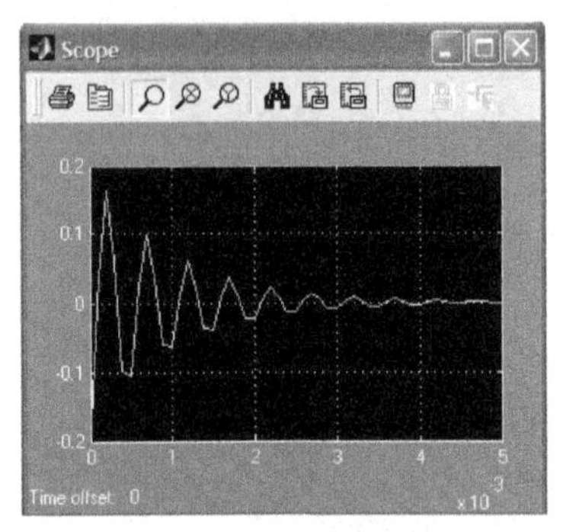

Figure 3.4(f) Damped sine wave returned by SIMULINK with the adaptive setting

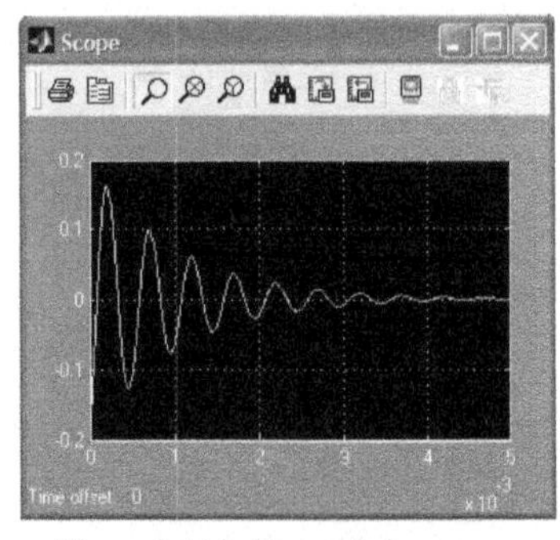

Figure 3.4(g) Damped sine wave when the solver is set for the fixed step

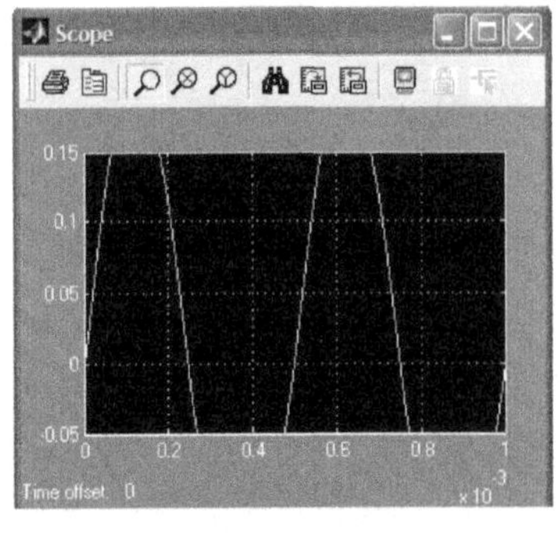

Figure 3.4(h) The clipped sine wave for the example 8

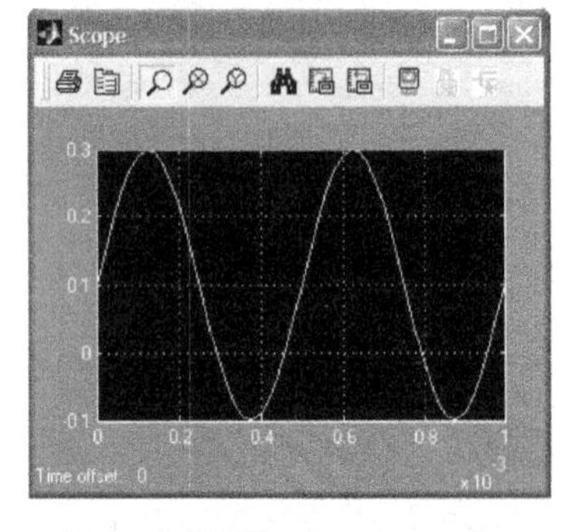

Figure 3.4(i) The sine wave of the example 1 with a bias 0.1

time to 5e-3 for $0 \le t \le 5m\sec$ (figure 1.7(g)), run the model, and doubleclick the Scope to see the output as shown in figure 3.4(f) with the autoscale setting.

Looking into the figure 3.4(f), the successive maxima of the wave are not so smooth. The reason for this is SIMULINK solver adaptively selects the time step of the wave which is nonuniform in general. There is provision for changing the t step. Since the wave existence is in the millisecond range, let us choose the fixed step size as $0.01\,m\sec$. Change the Solver option Type from the Variable step to the Fixed step in the figure 1.7(g), enter the Fixed step size as 0.01e-3, and run the model. The output is depicted in the figure 3.4(g) with the autoscale setting.

Example 8

A clipped sine wave is generated by employing the Saturation block like the model of the figure 3.3(f). Clipping means that the wave is constant below and over some user-defined value of the wave. Considering the sine wave of the example 1, let us clip the wave to 0.15 towards the positive peak and to –0.05 towards the negative peak. Doubleclick the block Saturation and enter the settings as $\begin{Bmatrix} \text{Upper limit}: 0.15 \\ \text{Lower limit}: -0.05 \end{Bmatrix}$. The Scope output with the autoscale setting is shown in figure 3.4(h).

Example 9

When a sine wave is raised up or pushed down from the horizontal axis by some user-defined value, the operation is called adding a bias to the sine wave.

Let us consider the wave of the example 1. We wish to raise the sine wave by 0.1 that is the equation of the wave becomes $y(t) = 0.2\sin 2\pi ft + 0.1$ so the swing from the +0.2 to –0.2 should be from the +0.3 to –0.1 due to the bias inclusion. However let us doubleclick the Sine Wave in the model in figure 3.3(a) and enter the bias 0.1 in the parameter window in figure 3.3(b). The Scope output with the autoscale adjustment is shown in the figure 3.4(i).

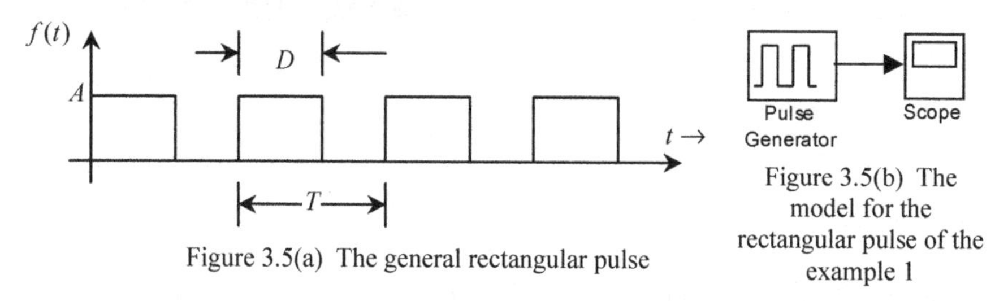

Figure 3.5(a) The general rectangular pulse

Figure 3.5(b) The model for the rectangular pulse of the example 1

3.5 Modeling the rectangular wave and its derivative signals

In a rectangular wave the wave value is constant over some interval. Even though the wave value is constant, different kinds of the waves are seen in electrical engineering courses. We illustrate some examples on the wave generation in the following.

Example 1

Figure 3.5(a) presents the general rectangular pulse $f(t)$ where A is the amplitude of the wave, D is the duty cycle (or the time when it exists) of the wave measured as the percentage of the period, and t is the desired time interval over which we intend to see the square wave.

Let us generate a rectangular pulse of amplitude 0.3, frequency $5KHz$, duty cycle 60%, and duration $0.6\,m\sec$. The shape of the wave is similar to that in the figure 3.5(a). The model of the figure 3.5(b) is constructed in this regard in which the block Pulse Generator (appendix A for outlook and link) generates the rectangular wave. The time period of the wave should be $T=\frac{1}{f}=\frac{1}{5KHz}=0.2\,m\sec$. Upon modeled like figure 3.5(b), we doubleclick the Pulse Generator and enter its settings as $\left\{\begin{array}{l}\text{Amplitude: 0.3}\\ \text{Period(secs): 0.2e}-3\\ \text{Pulse Width (\% of period): 60}\end{array}\right\}$ in the parameter window keeping the others as default (the parameter window is not shown for space reason). For the duration $0.6\,m\sec$ entering, change the solver stop time to 0.6e-3 (figure 1.7(g)). Figure 3.5(c) shows the simulation output on autoscale setting (figure 3.1(c)).

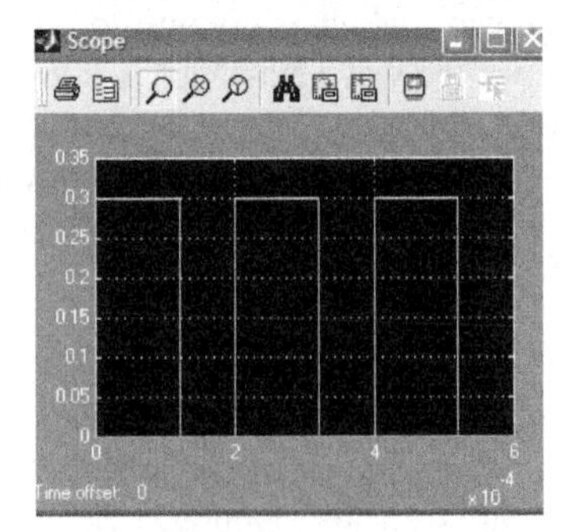

Figure 3.5(c) The Scope output for the model of the example 1 with the autoscale setting

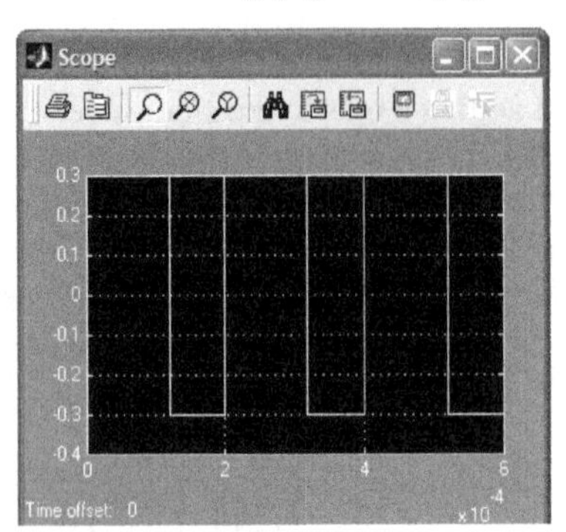

Figure 3.5(d) Scope output for the model of the example 2 with the autoscale setting

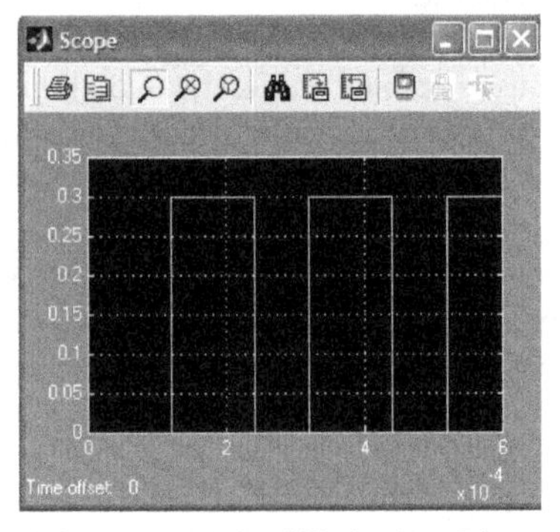

Figure 3.5(e) The shifted pulse of the example 4 with the autoscale setting

Example 2

In the last example, the pulse amplitude is 0.3 and the minimum is 0. What if we would like to have a wave of the same frequency and duration but the swing is from –0.3 to 0.3. The difference between the maximum and minimum is 0.6. We doubleclick the block Pulse Generator of the figure 3.5(b), change its amplitude to 0.6, and add a constant value of –0.3 to obtain the expected swing. The model of the figure 3.5(f) depicts the implementation. In addition to the blocks of the model in figure 3.5(b), you need one Constant and one Sum blocks (appendix A). We enter –0.3 on doubleclicking

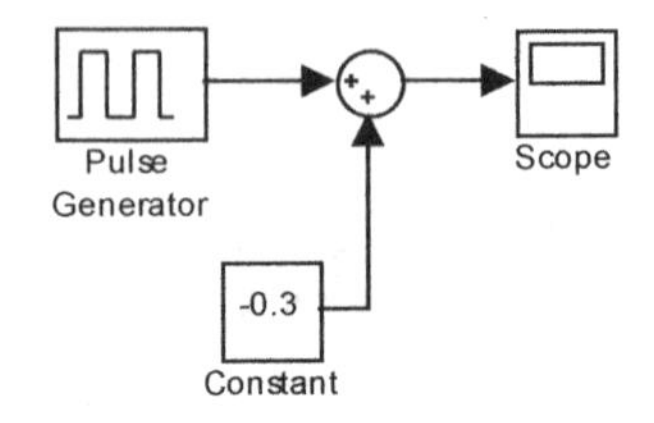

Figure 3.5(f) The model for equal positive and negative swings

the Constant block however the model's Scope output is presented in the figure 3.5(d).

Example 3

Generate a square wave of the same frequency and duration as the wave of the example 1 holds. The wave should be having the duty cycle 50% and equal positive and negative swings. We need the techniques of last two examples to simulate this wave.

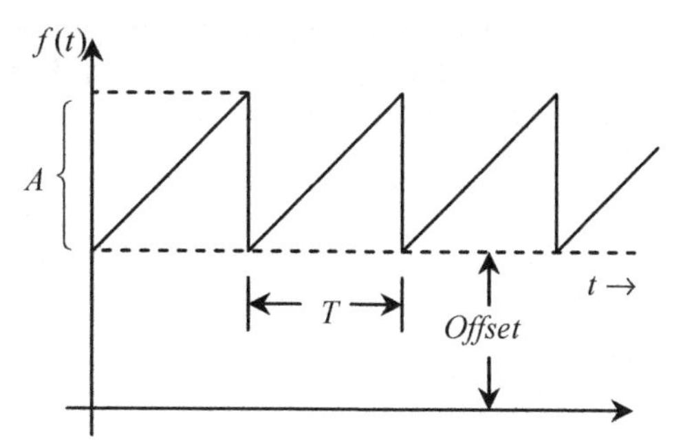

Figure 3.6(a) A triangular wave of period T seconds and amplitude swing from 0 to A

Example 4

In the examples 1 and 2, the wave started at time t =0. Let us say we intend to shift the wave of the example 1 to the right by one duty cycle. In absolute unit, the time shift becomes $0.12\,m\text{sec}$. Doubleclick the Pulse Generator of the model in figure 3.5(b) and enter the Phase delay (secs) as 0.12e-3 in the parameter window of the block. Figure 3.5(e) is the outcome of the simulation on autoscale setting.

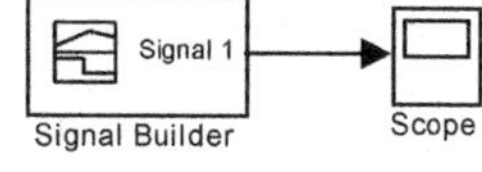

Figure 3.6(b) Signal Builder connected with the Scope block

3.6 Modeling the triangular wave and its derivative signals

Like the rectangular and sine waves, a triangular wave has a lot of varieties too. A triangular wave is basically a ramp function over one period with different slope, swing, and offset. Figure 3.6(a) shows a typical triangular wave whose equation is given by $f(t)=\left\{\frac{At}{T}\ for\ 0\le t\le T\right\}$ excluding the offset. The offset just shifts the wave up or down.

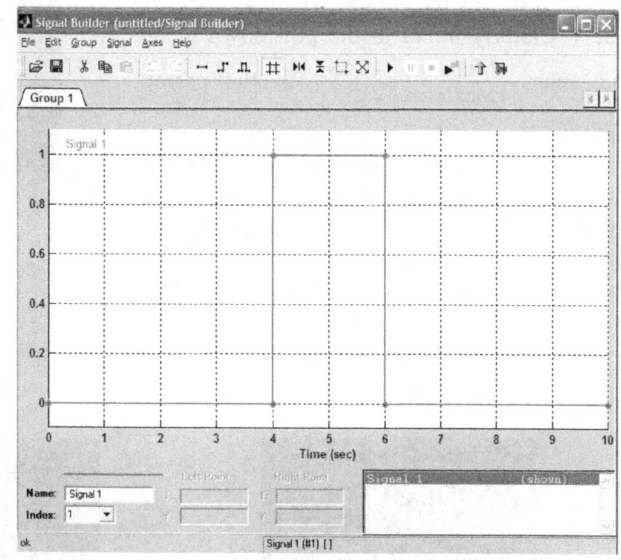

Figure 3.6(c) Design window of the Signal Builder block

Example 1

Let us generate a triangular wave which has the frequency $500\ Hz$, the amplitude swing from –0.05 to 0.05, and the duration 0.01secs.

We wish to introduce the Signal Builder block of SIMULINK here which keeps the provision for window interface signal design. The time period of the wave is $T = \frac{1}{f}$ =0.002secs indicating five cycles in the given duration.

Let us bring one Signal Builder (appendix A for outlook and link) and one Scope blocks in a new SIMULINK model file (subsection 1.2.2) and connect (subsection 1.2.3) them like the figure 3.6(b). SIMULINK responds with the design window of the figure 3.6(c) on doubleclicking the Signal Builder in which a finite duration rectangular pulse exists as the default one. Since the wave frequency or time period and the duration must be consistent in the wave modeling, we first enter the wave duration in the Signal Builder design window. To do so, click the Axes from the menu of the design window, and you find the option Change time range in the pull down menu. Let us click that and enter $\begin{Bmatrix} \text{Min time}: 0 \\ \text{Max time}: 0.01 \end{Bmatrix}$ in the prompt window for entering the wave duration 0.01 secs (the design window is updated on account of the change). The required amplitude swing is from –0.05 to 0.05 hence we click again the Axes in the menu bar of the design window, click the Set Y Display limits in the pull down menu, and enter $\begin{Bmatrix} \text{Minimum}: -0.05 \\ \text{Maximum}: 0.05 \end{Bmatrix}$ in the prompt window.

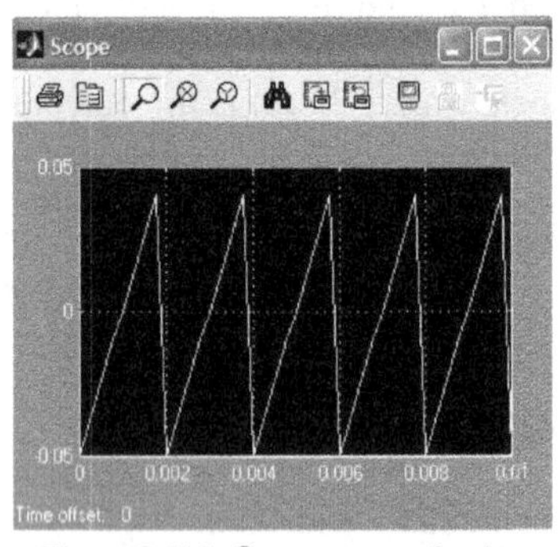

Figure 3.6(d) Scope output for the triangular wave of the example 1

In the design window menu bar, you also find the menu for Signal, click the Signal menu, and find Replace with option in the pull down menu and Triangle in the second stage pull down menu. Click the Triangle and enter

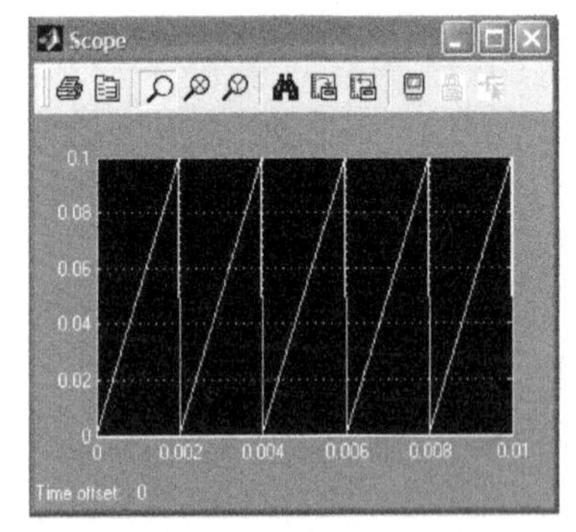

Figure 3.6(e) Scope output for the triangular wave of the example 2

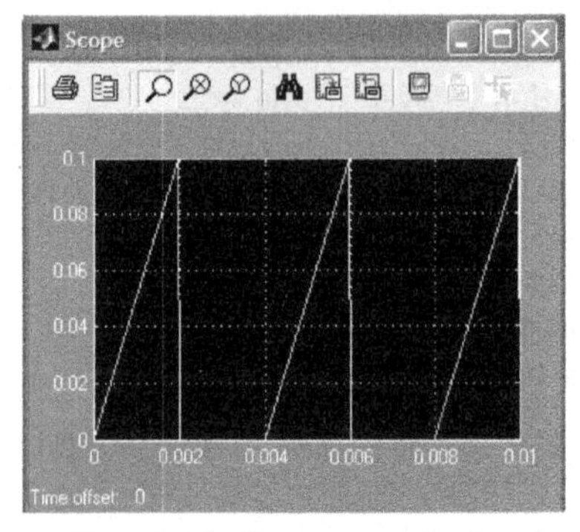

Figure 3.6(f) Scope output for the model of the figure 3.6(h)

$\begin{Bmatrix} \text{Frequency: 500} \\ \text{Amplitude: 0.05} \\ \text{Offset: 0} \end{Bmatrix}$ for the required wave in the prompt window. With that action, the updated wave appears in the design window.

Our wave design is finished and let us save it from the **Save** icon or from the **File** menu of the **Signal Builder** design window. We can close the **Signal Builder** window. Let us move onto the SIMULINK model file. SIMULINK does not have any information about the wave duration. Enter it by changing the solver **Stop time** to 0.01 (figure 1.7(g)). Run the model and see the **Scope** output as shown in figure 3.6(d) with the autoscale setting (figure 3.1(c)). Referring to the **Scope** output, the vertical edge of each wave does not seem to be vertical. This is because of the variable step or adaptive setting of the solver. If we make the solver setting fixed step and small within the duration, for sure the wave would appear as expected.

Example 2

We intend to generate a triangular wave which has the frequency $500\ Hz$, the amplitude swing from 0 to 0.1, and the duration 0.01secs.

This is basically the wave of the example 1 with little modification. The previous swing is from –0.05 to 0.05 but now we need from 0 to 0.1 so just adding an offset of 0.05 will have our simulation done. Referring to the design window of the figure 3.6(c), click the **Signal** down **Replace With** down **Triangle**, enter $\begin{Bmatrix} \text{Frequency: 500} \\ \text{Amplitude: 0.05} \\ \text{Offset: 0.05} \end{Bmatrix}$ in the prompt window, save the design, and run the SIMULINK model of the figure 3.6(b). The output is shown in figure 3.6(e) with the **Fixed Step** solver option and autoscale setting. For the **Fixed** step size 0.00001 (our chosen), click the **Simulation** in the model menu bar and then click **Configuration Parameters** to see the window in figure 1.7(g). As the **Solver options type** popup, select the **Fixed-step** from the popup and enter **0.00001** in the slot of **Fixed-step size** leaving other settings unchanged.

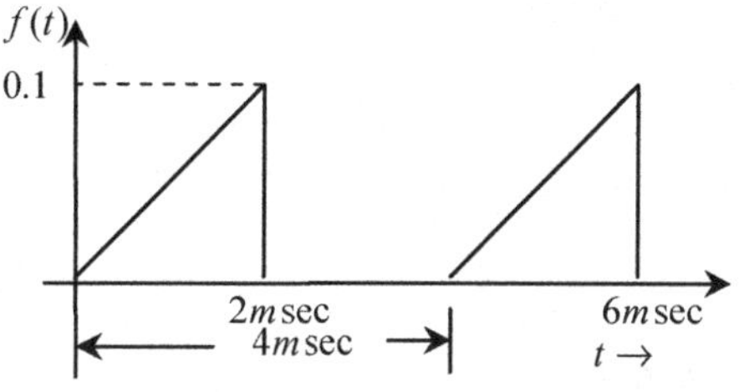

Figure 3.6(g) A triangular wave with some off interval

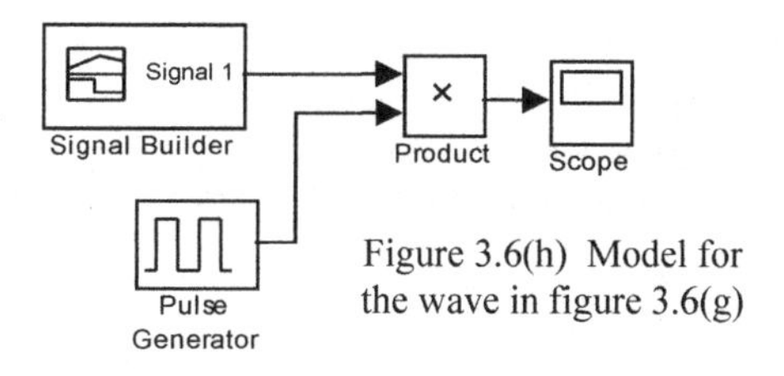

Figure 3.6(h) Model for the wave in figure 3.6(g)

Example 3

The periodic wave of the figure 3.6(g) is to be generated over $0 \le t \le 0.01$ secs.

The periodic wave has the time period $4\ m\text{sec}$ but the wave is off from 2 to $4\ m\text{sec}$ considering the first cycle. To simulate so, we first form a triangular wave (like example 2) of swing from 0 to 0.1 and the time period $2\ m\text{sec}$ (means frequency $500\ Hz$) and then set the alternate cycle to zero.

Let us imagine the first period of the rectangular pulse (similar to figure 3.5(a)) whose amplitude value is 1 from 0 to 2 $m\sec$ and 0 from 2 to 4 $m\sec$ and the pulse continues this way for the other cycles.

This wave generation will be accomplished if the rectangular pulse is multiplied with the triangular wave for what reason first we generate the wave of the example 2 employing the Signal Builder and connect the Pulse Generator as shown in the figure 3.6(h). The parameter window settings for the Pulse Generator should be $\begin{Bmatrix} \text{Amplitude: 1} \\ \text{Period(secs): 4e} - 3 \\ \text{Pulse Width (\% of period): 50} \end{Bmatrix}$. The Scope output is shown in the figure 3.6(f) with the autoscale setting.

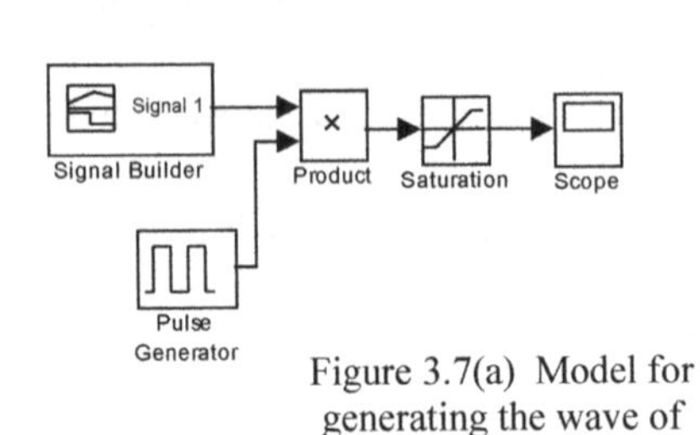

Figure 3.7(a) Model for generating the wave of the figure 3.7(c)

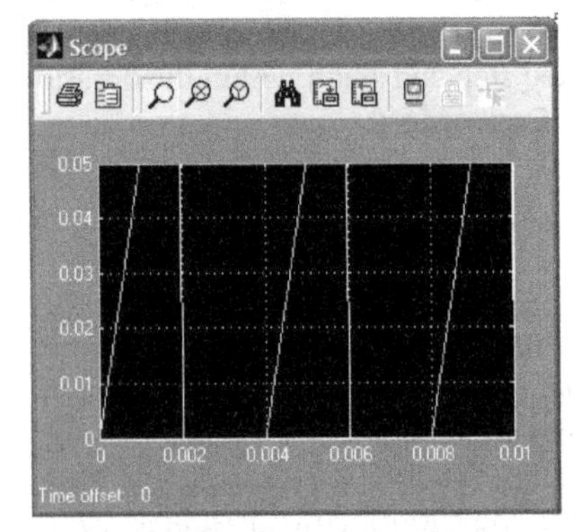

Figure 3.7(b) Scope output for the model of the figure 3.7(a)

Example 4

Suppose the wave of the figure 3.6(g) is cropped at 0.05 so that we only get the lower portion of the wave under the dotted line at 0.05 (shown in figure 3.7(c)). We intend to generate this wave.

The solution is just insert

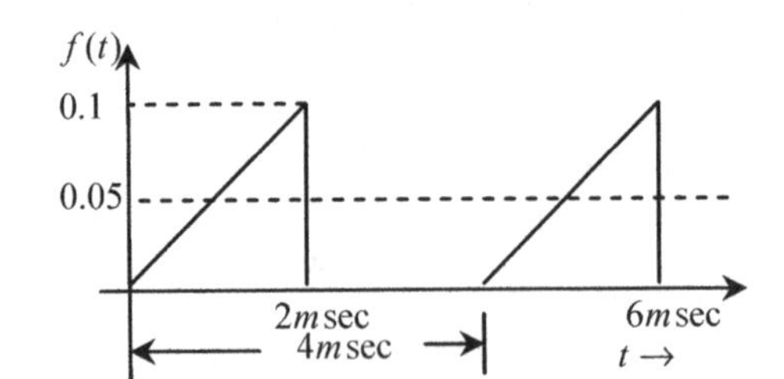

Figure 3.7(c) The triangular wave of the figure 3.6(g) is cropped at 0.05

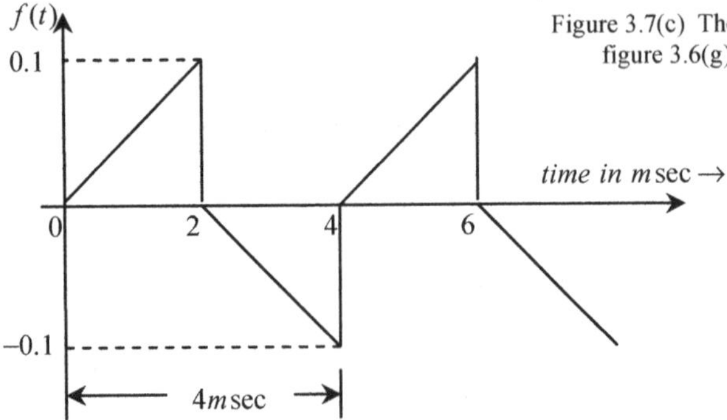

Figure 3.7(d) A symmetric positive and negative halves triangular wave

a Saturation block as indicated in the figure 3.7(a) with the settings $\left\{\begin{array}{l}\text{Upper limit}: 0.05 \\ \text{Lower limit}: 0\end{array}\right\}$ in the parameter window whose Scope output is the figure 3.7(b) with the autoscale setting.

Example 5

Figure 3.7(d) presents a wave in which the symmetric positive and negative halves are located and we wish to generate the wave.

Once we generate the wave (example 3) of the figure 3.6(g), it is easy to go through this wave. The positive half of this wave is exactly the same as that of the figure 3.6(g). We shift the wave of the figure 3.6(g) by $2\,m$sec to the right and then flip the wave about the horizontal axis that is how we form the negative halves. The time domain shifting operation is conducted by the block Transport Delay (appendix A for outlook and link) and the flipping about the t axis happens just multiplying any wave value by -1 (carried out by a Gain block of gain -1). We add these two halves by a Sum block to form the complete wave. Anyhow extending the model of the figure 3.6(h), we model the wave shape as shown in figure 3.7(e). Having brought the Transport Delay and Gain blocks in the model 3.6(h), doubleclick them to enter the Time Delay: 2e-3 and Gain: –1 in the parameter windows respectively. Figure 3.7(f) is the Scope result on the simulation following the autoscale setting.

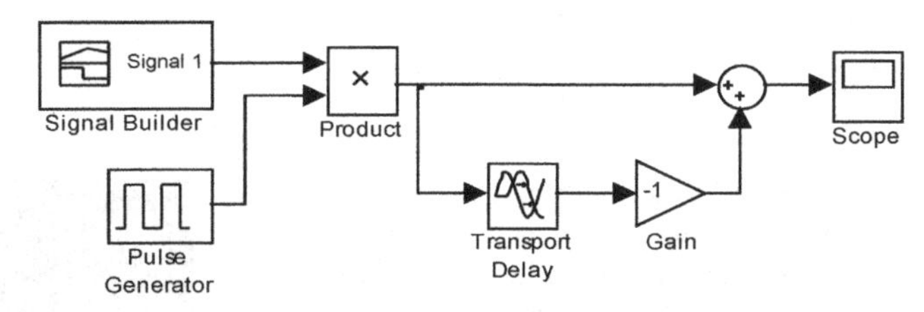

Figure 3.7(e) Model for generating the wave in figure 3.7(d)

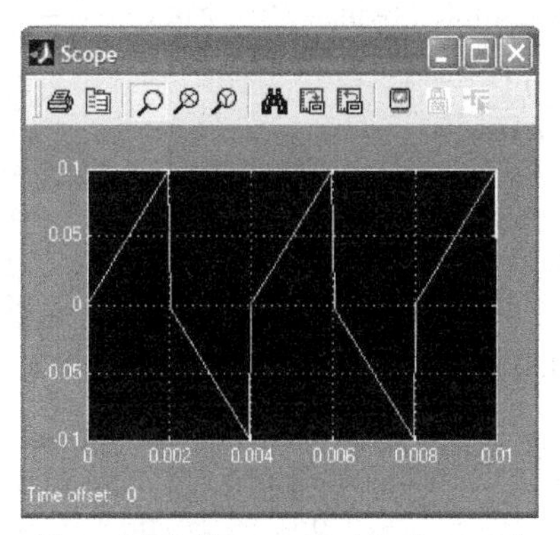

Figure 3.7(f) Scope output for the model in the figure 3.7(e)

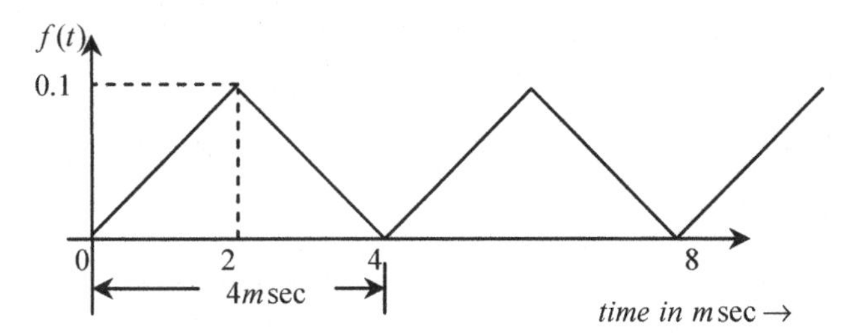

Figure 3.7(g) Negative half of wave in figure 3.7(d) is made +ve

Example 6

In this example we wish to simulate the wave of the figure 3.7(g) by a slightly different approach over $0 \le t \le 10 m\sec$.

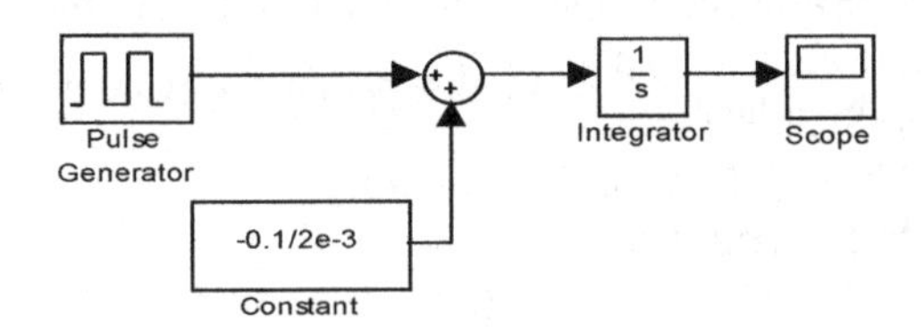

Figure 3.8(a) Model for generating the wave of the figure 3.7(g)

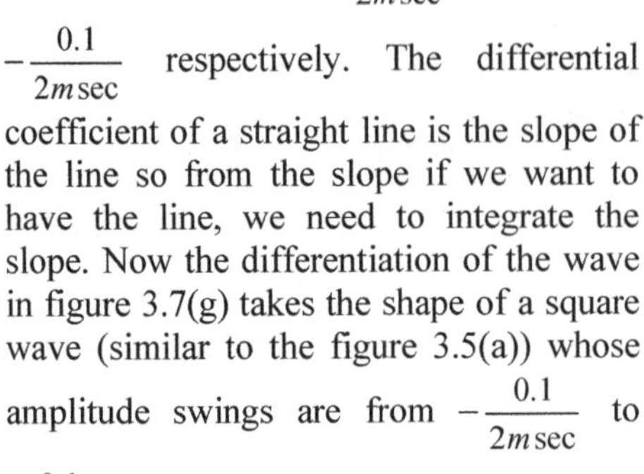

The first cycle of the wave has two ramps or straight line functions, the first and second of which have the slopes $\frac{0.1}{2m\sec}$ and $-\frac{0.1}{2m\sec}$ respectively. The differential coefficient of a straight line is the slope of the line so from the slope if we want to have the line, we need to integrate the slope. Now the differentiation of the wave in figure 3.7(g) takes the shape of a square wave (similar to the figure 3.5(a)) whose amplitude swings are from $-\frac{0.1}{2m\sec}$ to $\frac{0.1}{2m\sec}$.

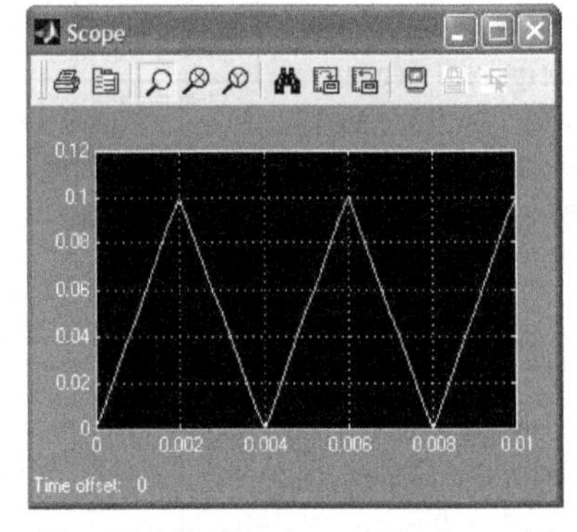

Figure 3.8(b) Scope output for the model of the figure 3.8(a)

The block Pulse Generator generates pulse from 0 to some value A. So if we intend to generate the swing from –A to A, we first generate the wave swing from 0 to 2A and then add a constant value –A to obtain the required swing. The slope $\frac{0.1}{2m\sec}$ has the code 0.1/2e-3 and $2\times\frac{0.1}{2m\sec}$ can be written as 2*0.1/2e-3. For the problem at hand, the Pulse Generator generates a pulse of swing 0 to 2*0.1/2e-3 and we add –0.1/2e-3 with the generator output to obtain the required swing from $-\frac{0.1}{2m\sec}$ to $\frac{0.1}{2m\sec}$. Afterwards we integrate (employing the block Integrator, appendix A for outlook and link) the resultant output to have the wave of the figure 3.7(g).

The model of the figure 3.8(a) illustrates the simulation strategy in which you need the settings for the Pulse Generator and Constant blocks as

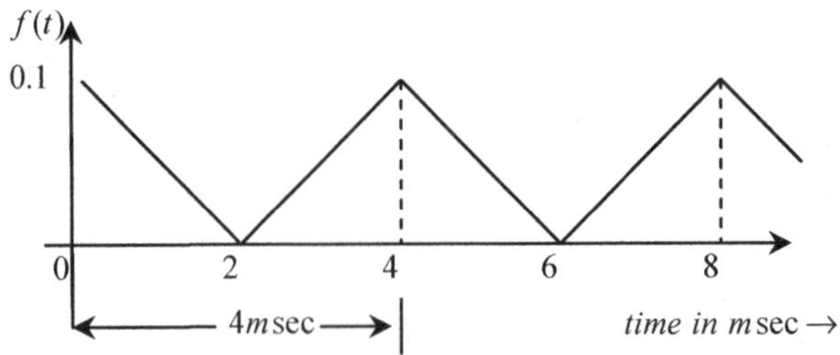

Figure 3.8(c) The wave of the figure 3.7(g) is shifted by 2msec

$\left\{\begin{array}{l}\text{Amplitude}: 2*0.1/2e-3 \\ \text{Period(secs)}: 4e-3\end{array}\right\}$ and **Constant Value: –0.1/2e-3** keeping the others as default in the parameter windows respectively. Also enlarge the **Constant** block to display its contents and enter the **Stop time** as **10e-3** for the interval. However the **Scope** output is presented in the figure 3.8(b) with the autoscale setting.

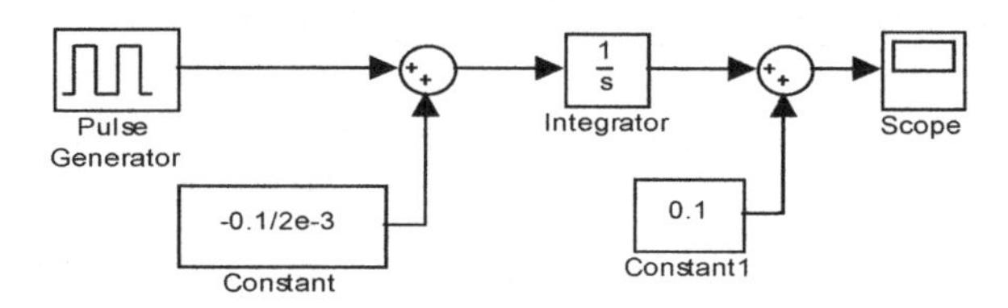

Figure 3.8(d) Model for generating the wave of the figure 3.8(c)

Example 7

The wave of the figure 3.8(c) is the shifted version of the wave in figure 3.7(g) by 2 $m$sec – which needs to be simulated.

The model in the figure 3.8(a) simulates the wave in figure 3.7(g) which requires slight modification to implement this wave. Doubleclick the **Pulse Generator** and change its **Phase Delay (secs)** to **2e-3** (for 2 $m$sec shifting) in the parameter window. In doing so the swing becomes –0.1 to 0. We just add a constant value of 0.1 to attain the swing from 0 to 0.1. The modified model is shown in the figure 3.8(d) whose **Scope** output is the figure 3.8(e) with the autoscale setting.

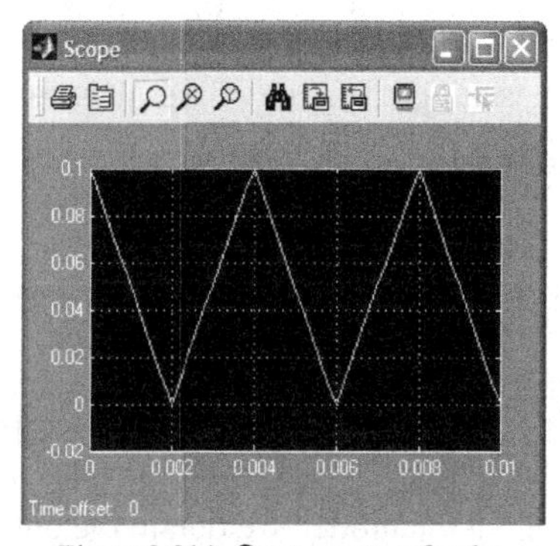

Figure 3.8(e) **Scope** output for the model of the figure 3.8(d)

3.7 Modeling triggered and user-defined nonperiodic signals

Any wave or signal can be turned on or off at some instant of its cycle depending on the user requirement – this is termed as triggering a wave.

Let us say we have a sine wave $y = A\sin 2\pi ft$ whose frequency and amplitude are f=50 Hz and A=0.8 respectively. The wave is off in one quarter of the period and in the rest three quarters of the period the wave is on. Or it can be rephrased as the wave is triggered at 25% of the period. Figure 3.9(a) illustrates the triggering of the sine wave over four cycles or 0.08 seconds – which we plan to simulate.

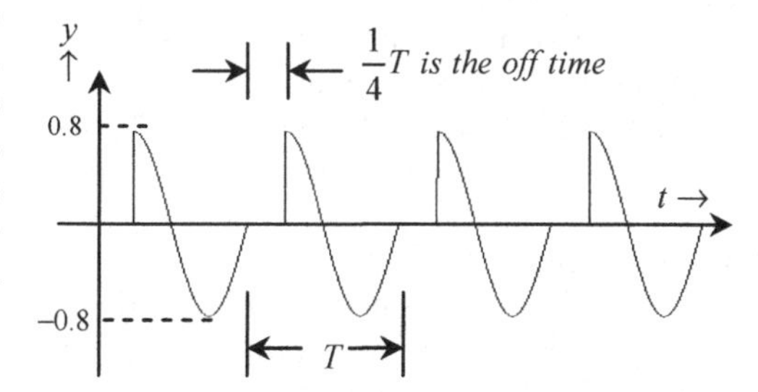

Figure 3.9(a) A sine wave is triggered at 25% of its period

To mention about the SIMULINK solution, we generate a rectangular pulse of the same period as that of the sine wave but with the amplitude swing from 0 to 1. The rectangular pulse (section 3.5) must have the duty cycle which is exactly the on time of the sine wave. Since the duty cycle of the pulse starts at zero, we shift the pulse to the right by exactly (1–duty cycle) where the duty cycle is expressed as the percentage of the time period.

So to simulate the current problem, let us bring one Sine Wave, one Pulse Generator, one Product, and one Scope blocks following the link of appendix A in a new simulink model file (subsection 1.2.2) and connect (subsection 1.2.3) them as presented in figure 3.9(b). The time period of the given wave is $\frac{1}{50Hz} = 0.02\,\text{sec}$ hence the duty cycle is either 75% of the period or 0.015secs. With that the off period is 25% of the period or 0.005secs. Enter the settings of the Sine Wave and Pulse Generator as $\left\{\begin{array}{l}\text{Amplitude: 0.8}\\ \text{Frequency(rad/sec)}: 2 * \text{pi} * 50\end{array}\right\}$ and $\left\{\begin{array}{l}\text{Period (secs): 0.02}\\ \text{Pulse Width (\% of period): 75}\\ \text{Phase Delay (secs): 0.005}\end{array}\right\}$ keeping the others as default in the parameter windows respectively and set the solver Stop time to 0.08 (figure 1.7(g)) for the four cycles. After running the model, we obtain the output from the Scope as depicted in the figure 3.9(c) with the autoscale setting (figure 3.1(c)).

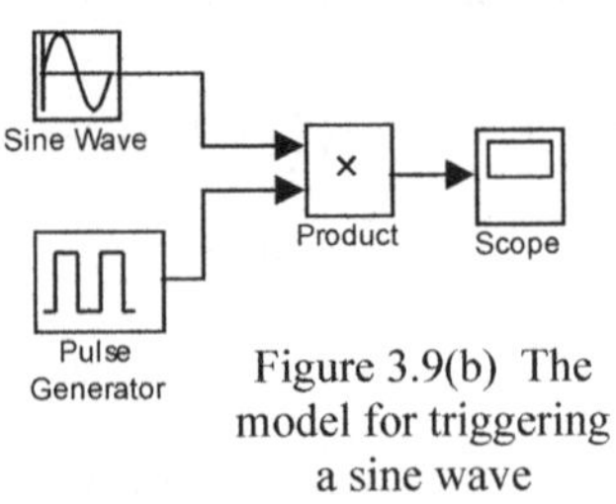

Figure 3.9(b) The model for triggering a sine wave

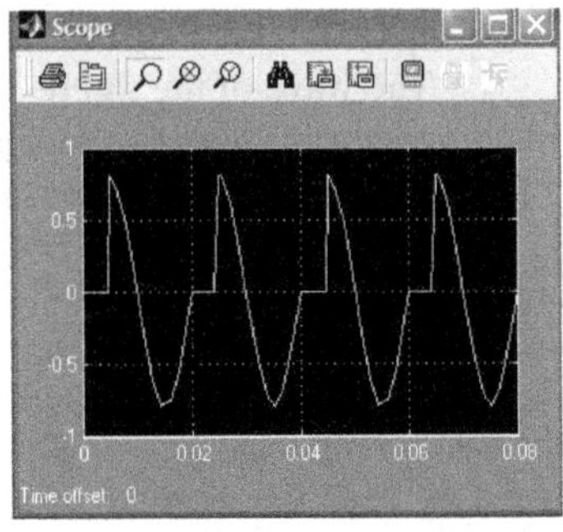

Figure 3.9(c) The Scope output for the triggered sine wave

This sort of triggering may take place for any other previously discussed waves.

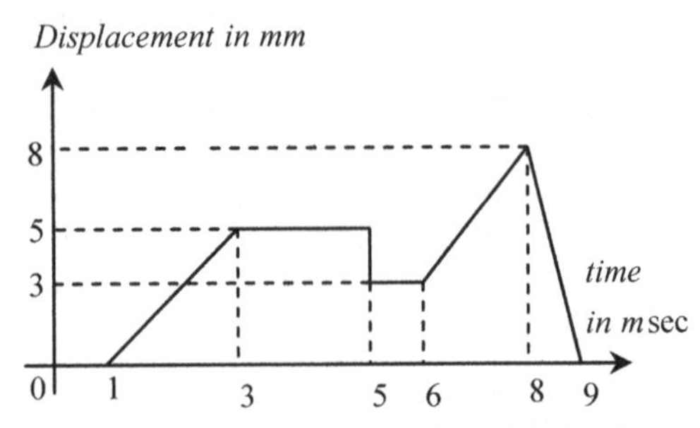

Figure 3.9(d) A displacement versus time function

Nonperiodic wave

All along we have been discussing the periodic waves of various types in previous sections. In the case of a nonperiodic wave, we just generate one cycle of the periodic wave by employing techniques illustrated so far. Nonetheless we come across finite duration signal of varying shapes.

If the signal is nonperiodic and defined by straight line segments and edges, the Signal Builder is the appropriate block to generate it which we introduced in section 3.6. Just to be specific by an example, we wish to design the finite displacement versus time variation of figure 3.9(c) by employing the Signal Builder block in SIMULINK.

Let us bring the block in a new SIMULINK model file and doubleclick it. The design window of the figure 3.6(c) appears with the default setting of a finite duration pulse. Referring to the figure 3.9(d), the variation has the vertical and horizontal axes ranges as 0 to 8mm and 1 to 9msec respectively but we design the function in the standard units – meter and second respectively. The line segments of the figure 3.9(d) have the coordinates (0msec,0mm), (1msec,0mm), (3msec,5mm), (5msec,5mm), (5msec,3mm), (6msec,3mm), (8msec,8mm), and (9msec,0mm). Collecting consecutive time and displacement coordinates, we have $[0 \;\; 1 \;\; 3 \;\; 5 \;\; 5 \;\; 6 \;\; 8 \;\; 9]\times10^{-3}$ sec and $[0 \;\; 0 \;\; 5 \;\; 5 \;\; 3 \;\; 3 \;\; 8 \;\; 0]\times10^{-3}$ meter respectively.

Concerning the design window of the figure 3.6(c), click the Signal down Replace with down Custom from the menu bar, enter $\begin{Bmatrix} \text{Time Values: [0 1 3 5 5 6 8 9]*1e}-3 \\ \text{Y Values: [0 0 5 5 3 3 8 0]*1e}-3 \end{Bmatrix}$ in the prompt window for entering the horizontal and vertical coordinates in standard units, click the Axes down Change time range in the menu bar, enter $\begin{Bmatrix} \text{Min time: 0} \\ \text{Max time: 9e}-3 \end{Bmatrix}$ in the prompt window for the horizontal range, and save the signal design by clicking the save icon of the design window menu bar. The design window displays the function of figure 3.9(d).

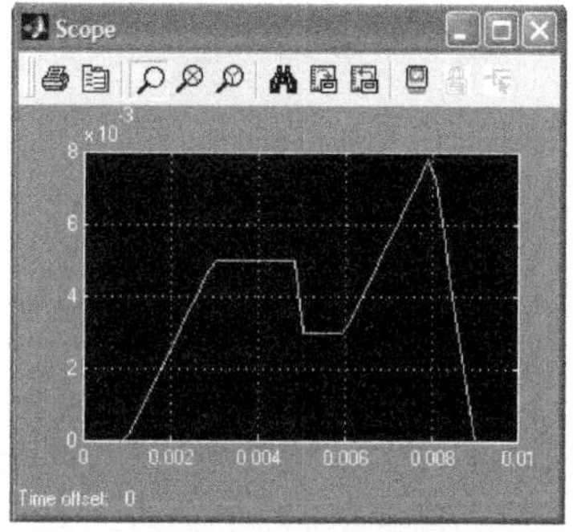

Figure 3.9(e) The Scope output for signal in figure 3.9(d) with the autoscale setting

Just to verify, let us connect the Signal Builder in conjunction with the Scope as shown in the figure 3.6(b), change the solver Stop time to 9e-3, and run the model. The Scope output with the autoscale setting as seen in figure 3.9(e) confirms the design.

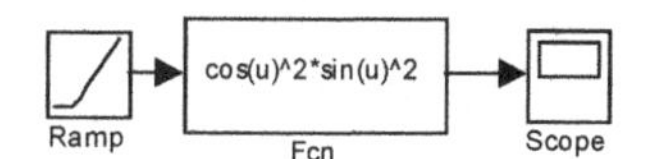

Figure 3.9(f) Expression based model of a signal

The reader might say what the displacement has to do with the voltage or current signal? Let us imagine a displacement sensor which produces voltage causing from the displacement. The sensor has some displacement to voltage proportionality constant often supplied by the manufacturer and which you can model by using a Gain block. Therefore the Signal Builder and Gain together represent the sensor voltage generation.

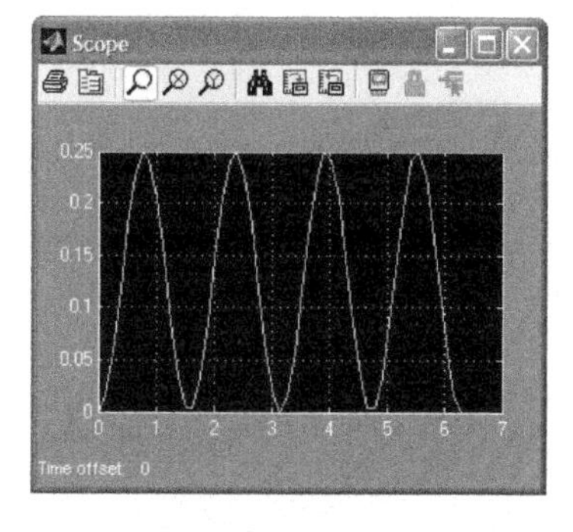

Figure 3.9(g) Scope output for the expression based signal

Expression based wave

All waves or signals discussed so far mostly employed built-in blocks. User can define his/her own function based signal. For example a voltage signal as expression is given by $v(t) = \cos^2 t \sin^2 t$ which we intend to generate over $0 \le t \le 2\pi$.

The block Fcn (appendix A) helps us simulate expression based signal. Recall that the Ramp block (section 3.3) simulates the independent variable t. We just enter the expressional vector code (appendix B) of the given signal in the Fcn parameter window but assuming that the independent variable is u.

Therefore we enter cos(u)^2*sin(u)^2 for $\cos^2 t \sin^2 t$ in the parameter window of the Fcn on doubleclicking it. Figure 3.9(f) depicts the model for such signal generation. Enlarge the Fcn block to see its contents and enter the Stop time as 2*pi for the interval. On running the model we see the output in the Scope like the figure 3.9(g) on autoscale setting for the given $v(t)$.

We intend to close the chapter with this modeling example.

Chapter 4

SIGNAL OPERATIONS

Introduction

Various signal operations are the part and parcel of signal study, signal processing, signal filtration, and in many signal related analysis. This chapter provides necessary tools for elementary signal operations which are linear in general. Through adequate examples and executions, both the continuous and discrete time signals are considered. Presented emphasis is on the following:

- Continuous to discrete conversion of signals
- Samplings, flippings, and convolutions for varying signal cases
- Quantization and other fundamental signal operations such as gain, integration, differentiation, etc

4.1 Continuous to discrete conversion

Today's signals are very much digital and the trend is becoming widespread by the day. Digital equipment such as computer, mobile, CD, DVD, etc are encompassing our daily life. Tremendous evolution of digital devices one day will make us forget about the analog or continuous systems. It is true that analog systems are phasing out gradually but without the understanding of continuous systems, understanding of discrete systems does not become perfect. The word analog (electrical engineering term) is synonym to continuous (mathematical science term). The words digital and discrete also follow similar construal.

Let us see the basic difference between a continuous and a discrete signals. Figure 4.1(a) shows the plot of continuous or analog ramp signal $f(t)=t$. The domain of t includes all real numbers – fraction or integer. Suppose we intend to store the $f(t)$ information as numbers for different t. How many values of t are required? It goes without saying that infinite number of points is required (figure 4.1(b)). Instead of taking infinite points, one can take substantial samples or points on the t axis to pick up the signal information in $f(t)$ that is how the idea of continuous to discrete is derived. How many points or samples we take is user-defined.

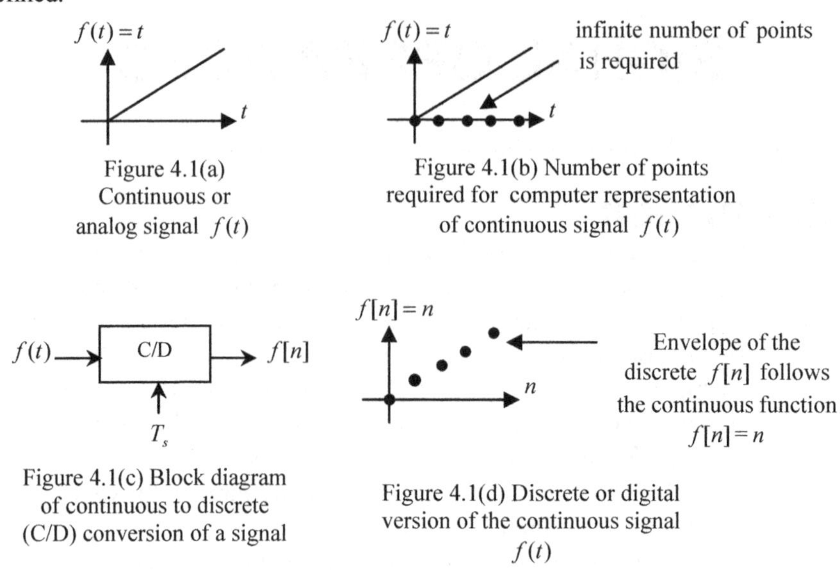

Figure 4.1(a) Continuous or analog signal $f(t)$

Figure 4.1(b) Number of points required for computer representation of continuous signal $f(t)$

Figure 4.1(c) Block diagram of continuous to discrete (C/D) conversion of a signal

Figure 4.1(d) Discrete or digital version of the continuous signal $f(t)$

The schematic representation of a continuous to analog conversion system is shown in the figure 4.1(c). Let us say t is changing integerwise from 0 to onward. Figure 4.1(d) shows the samples of $f(t)$ or discrete version of $f(t)$ which is symbolized by $f[n]$. The $f[n]$ values are not connected (so are the n's) that is why the term discrete is used. Of coarse the envelope of $f[n]$ follows the functional variation of $f(t)$. Figure 4.1(d) depicts the discrete version of the ramp signal of figure 4.1(a). Even though the $f[n]$ is related with the $f(t)$ by the relationship $f[n]=f(nT_s)$, just looking into $f[n]$ one can not say anything about the $f(t)$ because all we have is discrete numbers. The moment we calculate some $f(t)$ for some t, we consider the discrete one. Let us say we calculate the $f(t)$ for $0 \le t \le 2\sec$ with a step size 0.5 sec. Choosing a step size means we choose sampling period T_s of the block diagram of the figure 4.1(c). With the T_s=0.5sec, the sampling frequency becomes $f_s=\frac{1}{T_s}=2\,Hz$. Choosing another step size points to another sampling period or frequency. In mathematical term we say the step size as Δt which is exactly T_s of signal processing. Computer always represents the discrete one for finite memory reason. For a particular C/D conversion system, the f_s is constant therefore $f[n]$ is merely a function of integer index n. If one puts the $f[n]$ one after another for different n, the placement result is a row or column matrix. Let us see following two examples on C/D conversion.

❖❖ Example 1

Figure 4.1(e) shows the continuous to discrete conversion of a parabolic input signal. Our goal is to obtain $f[n]$ from $f(t)$. In accordance with the block diagram, we are forced to choose the sampling period T_s as $1\,m\text{sec}$. As a first step, we generate the t points as a row or column matrix at which $f(t)$'s are required so to say the t points are $0\,m\text{sec}$, $1\,m\text{sec}$, $2\,m\text{sec}$, ..., 0.5sec within the given interval. That happens exercising the command 0:1e-3:0.5 (subsection 1.1.2) in which $1\,m\text{sec} = 10^{-3}\text{ sec}$ has the code 1e-3. As a formal procedure, we do the following:

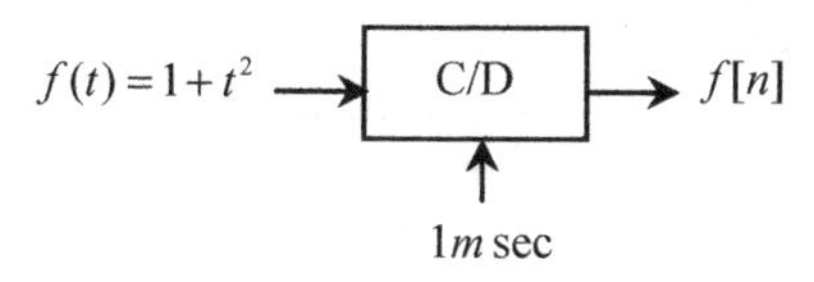

Figure 4.1(e) A parabolic continuous signal is sampled for digitization over $0 \le t \le 0.5\text{sec}$

>>t=0:1e-3:0.5; ↵ ← Workspace t holds all t points consecutively as a row matrix

The next step is to write the scalar code (appendix B) of the continuous parabolic signal $f(t)$ as follows:

>>f=1+t.^2; ↵ ← Workspace f holds $f[n]$ samples as a row matrix

The t and f in above implementation are user-chosen variable names. There is no n or t information in f. It is the user who has to keep the sampling information in mind.

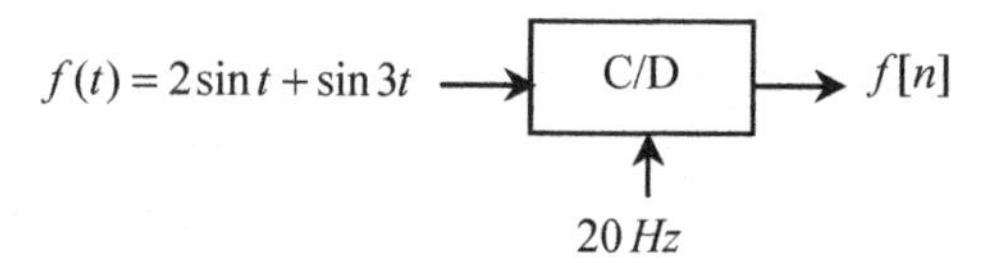

Figure 4.1(f) A two frequency continuous signal is sampled for digitization over $0 \le t \le 0.3\text{sec}$

❖❖ Example 2

A two frequency signal is to be digitized as shown in the figure 4.1(f). The procedure is similar to that of the example 1 with the exception that now we have sampling frequency instead of period. Having f_s=20 Hz, the T_s or step size becomes $\frac{1}{f_s}$=0.05sec therefore the conversion is as follows:

>>t=0:0.05:0.3; ↵ ← Workspace t holds all t points consecutively as a row matrix over $0 \le t \le 0.3\text{sec}$

>>f=2*sin(t)+sin(3*t); ↵ ← Scalar code of the two frequency signal is assigned to workspace f

Having executed as above, the discrete signal $f[n]$ is stored as a row matrix in the workspace f.

❖ Sample correspondence

The t and f are merely row matrices of identical size. The first element of t corresponds to the first element of f or t(1) to f(1), second element of t corresponds to the second element of f or t(2) to f(2) and so on.

4.2 Down and up samplings of signals

The down and up samplings of a signal mean contracting and dilating the time domain of the signal respectively. The operations are specially associated with the discrete signals. Both samplings happen according to user-defined factor and it is integer.

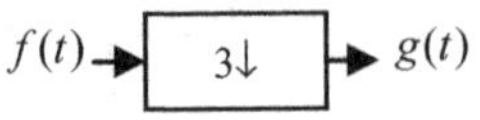

$f(t) \rightarrow \boxed{3\downarrow} \rightarrow g(t)$ $\qquad$ $f(t) \rightarrow \boxed{2\uparrow} \rightarrow g(t)$

Figure 4.2(a) Down sampling of $f(t)$ 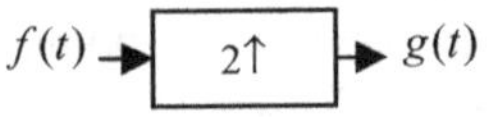 $\qquad$ Figure 4.2(b) Up sampling of $f(t)$

❖ Down and up samplings of a continuous signal

Computer never simulates a perfect continuous signal. The effect of sampling appears in the functional expression of the signal. If the signal is downsampled like the figure 4.2(a), downsampling by a factor of 3 (as an example) means $g(t) = f(3t)$. For example $f(t) = \sin t$ turns $g(t)$ as $\sin 3t$. In the case of upsampling by a factor of 2 (as an example) like the figure 4.2(b), the $g(t)$ becomes $f\left(\frac{t}{2}\right)$ or $\sin\frac{t}{2}$ when $f(t) = \sin t$. It means the down or up sampling is taken care of as the multiplier of t in the case of a continuous signal.

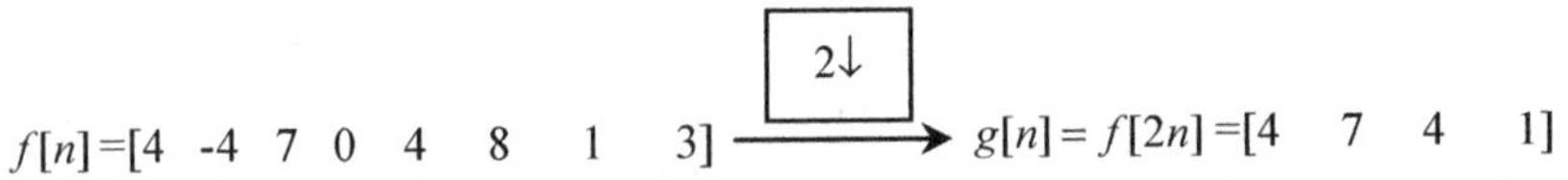

$$f[n] = [4 \;\; -4 \;\; 7 \;\; 0 \;\; 4 \;\; 8 \;\; 1 \;\; 3] \xrightarrow{\boxed{2\downarrow}} g[n] = f[2n] = [4 \;\; 7 \;\; 4 \;\; 1]$$

Figure 4.2(c) Down sampling of the discrete $f[n]$

$$f[n] = [4 \;\; -4 \;\; 7 \;\; 0 \;\; 4] \xrightarrow{\boxed{2\uparrow}} g[n] = f\left[\frac{n}{2}\right] = [4 \;\; 0 \;\; -4 \;\; 0 \;\; 7 \;\; 0 \;\; 0 \;\; 0 \;\; 4 \;\; 0]$$

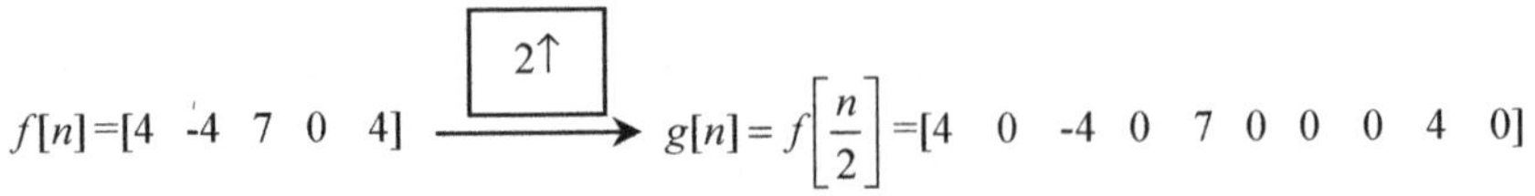

Figure 4.2(d) Up sampling of the discrete $f[n]$

❖ Down and up samplings of a discrete signal

We know that a discrete signal $f[n]$ takes the shape of a row matrix for different n. Downsampling the $f[n]$ by a factor of 2 (for example) means turning the $f[n]$ to $f[2n]$. The number of elements in $f[n]$ must be multiple of downsampling factor. For downsampling by a factor 2, the number of elements in

$f[n]$ can be 2, 4, 6, etc. Again for downsampling by a factor 3, the number of elements in $f[n]$ can be 3, 6, 9, etc and so on.

Figure 4.2(c) presents an example of downsampling. The discrete signal $f[n]$=[4 -4 7 0 4 8 1 3] turns to $g[n]=f[2n]$=[4 7 4 1] when downsampled by a factor of 2. We wish to implement this downsampling.

The MATLAB built-in function downsample implements this sort of operation with the syntax downsample(signal as a row matrix, downsampling factor). First we enter the $f[n]$ as a row matrix as follows:

```
>>f=[4 -4 7 0 4 8 1 3]; ↵     ← f holds the signal f[n], where f is user-chosen
>>g=downsample(f,2) ↵         ← Workspace g holds the downsampled discrete
                                 signal f[2n], g⇔ f[2n], g is user-chosen name
g =
         4   7   4   1
```

Skipping every alternate sample of $f[n]$ provides the $f[2n]$. If the factor were 3, the skipping would happen for every two alternate samples.

Upsampling is the reverse process of the downsampling. Figure 4.2(d) presents the upsampling of a discrete signal $f[n]$=[4 -4 7 0 4] by a factor of 2 which provides $g[n]=f\left[\frac{n}{2}\right]$=[4 0 -4 0 7 0 0 0 4 0]. We intend to implement this upsampling operation.

When upsampling is conducted by a factor of 2, it is the insertion of a single zero between the samples thereby forming $f\left[\frac{n}{2}\right]$. If the factor were 3, the insertion would occur with two zeroes. MATLAB built-in function upsample conducts upsampling with the syntax upsample(signal as a row matrix, upsampling factor) as follows:

```
>>f=[4 -4 7 0 4]; ↵        ← Workspace f holds the signal f[n], f⇔ f[n]
>>g=upsample(f,2) ↵        ← Workspace g holds the upsampled discrete signal
                              f[n/2], g⇔ f[n/2], g is user-chosen name
g =
         4   0  -4   0   7   0   0   0   4   0
```

Whether down or up sampling, the n information is lost during the sampling so it is the user's accountability to keep a mark on that. Let us see another example on the samplings.

Figure 4.2(e) shown discrete signal $f[n]=2^{-n}$ when downsampled over $0 \le n \le 11$, the $g[n]$ becomes [1 0.125 0.0156 0.002] which we intend to get.

2^{-n} → [3↓] → $g[n]$

Figure 4.2(e) The discrete signal 2^{-n} is downsampled by a factor 3

In this sort of problems we first generate the integer n values as a row matrix (subsection 1.1.2) and then write the scalar code (appendix B) to obtain the discrete 2^{-n} values. Let us go through the following in this regard:

```
>>n=0:11; ↵        ← n is a row matrix holding integers from 0 to 11
>>f=2.^(-n); ↵     ← f is a row matrix holding discrete 2^-n signal values
```

```
>>g=downsample(f,3) ↵    ← Workspace g holds the downsampled discrete
                           signal, g⇔ g[n] , g is user-chosen name

g =
        1.0000    0.1250    0.0156    0.0020
```

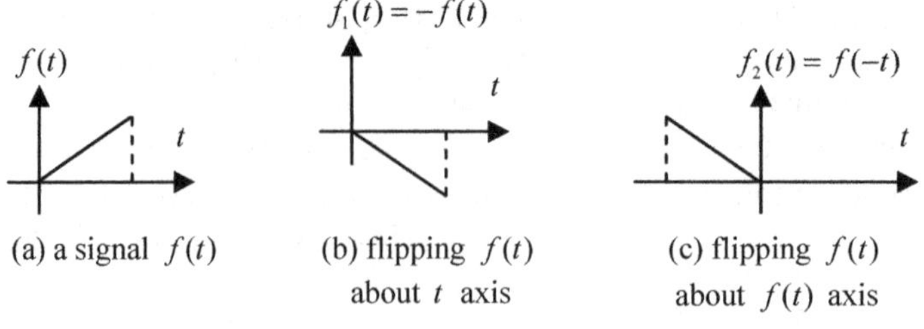

(a) a signal $f(t)$

(b) flipping $f(t)$ about t axis

(c) flipping $f(t)$ about $f(t)$ axis

Figure 4.3(a)-(c) Flipping a continuous signal $f(t)$

4.3 Flipping a signal

Flipping happens both in the case of continuous and discrete signals.

❖ Flipping a continuous signal

When a continuous signal $f(t)$ versus t is considered, the flipping can be about the $f(t)$ axis or about the t axis. Figure 4.3(a) shows a continuous signal $f(t)$ whose flipped versions about the t and $f(t)$ axes are $f_1(t)$ and $f_2(t)$ and which are graphed in the figures 4.3(b) and 4.3(c) respectively. Flipping about the t axis means taking the negative sign of the $f(t)$ but flipping about the $f(t)$ axis means the function $f(-t)$. In either case the change is utterly expressional.

Suppose we have a continuous signal $f(t) = t + 2\sin^2 t$. If it is flipped about the t and $f(t)$ axes, we have $f_1(t)$=$-(t + 2\sin^2 t)$=$-t - 2\sin^2 t$ and $f_2(t)$=$(-t)$+ $2\sin^2(-t)$=$-t + 2\sin^2 t$ respectively. The interval description for instance $0 \le t \le 6$ sec does not change for $f_1(t)$ but does for $f_2(t)$, which becomes $-6\text{sec} \le t \le 0$.

If we have to write the code (appendix B) for the signal, we have -t-2*sin(t)^2 and -t+2*sin(t)^2 for the vector code of $f_1(t)$ and $f_2(t)$ respectively. Also do we have -t-2*sin(t).^2 and -t+2*sin(t).^2 for the scalar code respectively.

If you are interested to see the graph of the flipped $-t + 2\sin^2 t$ over $-6\text{sec} \le t \le 0$, the command ezplot of appendix F can be exercised as ezplot('-t+2*sin(t)^2',[-6 0]).

❖ Flipping a discrete signal

In the discrete case like the signal $f[n]$ versus n of the figure 4.1(d), the flipping can be about the n and $f[n]$. We know that $f[n]$ are purely numbers for a discrete signal and turns to a row or column matrix for different n. So to say, flipping discrete signal $f[n]$ about the n means taking the negative of the numbers (figure 4.3(d)). For example f holds the discrete signal $f[n]$=[0 5 6 5 −6] in the workspace. The flipped signal $f_1[n]$ is obtained just by writing f1=-f; (where f1⇔ $f_1[n]$ and f1 is user-chosen).

Flipping about the $f[n]$ axis (that is for the $f_2[n]$ of the figure 4.3(e)) happens by employing the command fliplr(f) or flipud(f) if the signal is stored as a row or column matrix by the name f respectively.

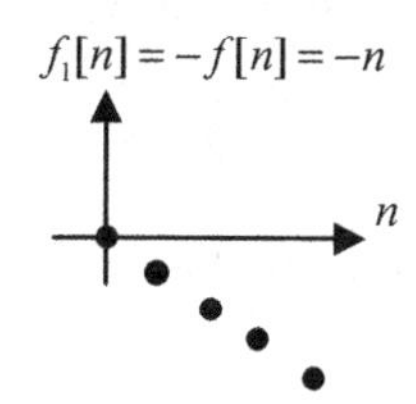

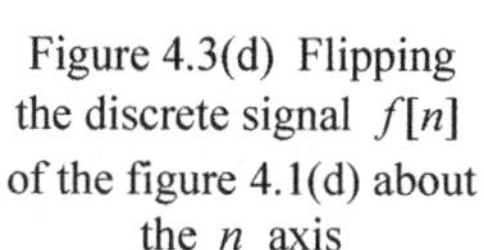
Figure 4.3(d) Flipping the discrete signal $f[n]$ of the figure 4.1(d) about the n axis

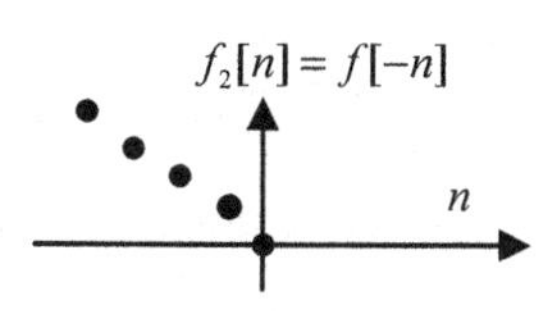

Figure 4.3(e) Flipping the discrete signal $f[n]$ of the figure 4.1(d) about the $f[n]$

Let us execute the row matrix case as follows:

```
>>f=[0 5 6 5 -6]; ↵             ← Assigning just cited f[n] to f
>>f2=fliplr(f) ↵                ← Workspace f2 means f2[n], f2 is user-chosen

f2 =

        -6   5   6   5   0
```

In the case of expression based discrete signal for example $f[n]=n^2$ over $2 \le n \le 8$ we first generate the integers as a row matrix over the interval by using n=2:8, then use the scalar code for sample values to be stored in f as a row matrix by writing f=n.^2, and after that apply f2=fliplr(f) where f2 is user-chosen variable. Note that the interval for the f2 or $f_2[n]$ becomes $-8 \le n \le -2$.

If the reader is interested to plot the flipped n^2, appendix F cited stem can be exercised as follows:

```
>>n=2:8;f=n.^2;f2=fliplr(f);stem(-fliplr(n),f2) ↵        ← graph not shown for space reason
```

4.4 Convolution of continuous signals

First of all computer never simulates perfect continuous signal and there is no built-in function for the convolution operation of continuous signals. Expression based convolution can be implemented by employing the int function of the appendix G.

The convolution of two continuous signals $f(t)$ and $g(t)$ is given by $f(t)*g(t)=\int_{p=-\infty}^{p=\infty} f(p)g(t-p)\,dp=\int_{p=-\infty}^{p=\infty} g(p)f(t-p)\,dp$ where p is a dummy variable (in most textbooks it is as τ) and the * is the convolution operator.

It is given that $f(t)*g(t)=\frac{t\sin t}{2}$ is the convolution of signals $f(t)=\sin t$ and $g(t)=\cos t$ for t >0 which we intend to implement.

By expression we have $f(t)*g(t)=\int_{p=0}^{p=t} \sin p\cos(t-p)\,dp$ whose implementation is shown below:

```
>>syms t p ↵   ← Declare variables t and p of the integrand as symbolic, t⇔t
                                      and p⇔p
>>y=sin(p)*cos(t-p); ↵ ← Integrand sin p cos(t − p) is assigned to workspace y
```

```
>>c=int(y,p,0,t) ↵ ← Integration with respect to p is carried out and the result is
                     assigned to c where c is user-chosen variable name
c =

1/2*sin(t)*t            ← c holds the code or string of t sin t / 2, c⇔ f(t) * g(t)
```

As another example let us calculate $f(t) * g(t) = \frac{3}{8} - \frac{3}{8}t^2 + \frac{1}{8}t^4 - \frac{3}{8}\cos^2 t$ where $f(t) = \sin^2 t$ and $g(t) = t^3$ for $t > 0$. For this we need to implement $f(t) * g(t) = \int_{p=0}^{p=t} (t-p)^3 \sin^2 p\, dp$ and MATLAB's exercise is as follows:

```
>>syms t p ↵           ← Declare variables t and p of the integrand as symbolic,
                          t⇔t and p⇔p
>>y=sin(p)^2*(t-p)^3; ↵ ← Integrand (t − p)^3 sin^2 p is assigned to workspace y
>>c=int(y,p,0,t); ↵     ← Integration with respect to p is carried out and the result
                          is assigned to c
>>pretty(simple(c)) ↵ ← c holds the string for f(t) * g(t), simplification on c is
                          conducted by simple

          2       4              2
  - 3/8 t  + 1/8 t  + 3/8 - 3/8 cos(t)
```

The pretty (appendix C.10) displays the mathematics readable form of the codes stored in c.

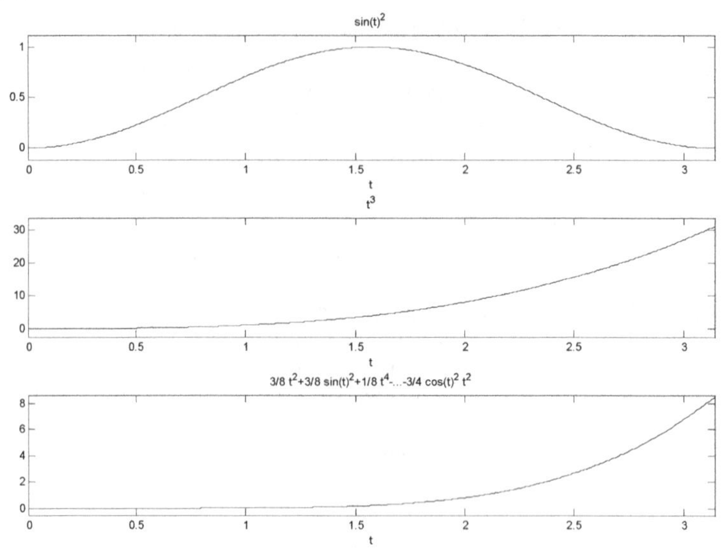

Figure 4.4(a) Convolution of two continuous signals

⬧⬧ Graphing the convolution result

If we have to plot the convolution result as well as the component signal, appendix F cited subplot can be helpful. Individual function plotting is executed by ezplot of the same appendix. Just now we have seen that $\frac{3}{8} - \frac{3}{8}t^2 + \frac{1}{8}t^4 - \frac{3}{8}\cos^2 t$ (which is held in c) is the convolution of $f(t) = \sin^2 t$ and $g(t) = t^3$. Interval selection is mandatory for the graphing and let it be $0 \le t \le \pi$ for all graphs. Let us go through the following for the graphing:

```
>>syms t ↵                                      ← Declare variable t as symbolic, t⇔t
>>subplot(311),ezplot(sin(t)^2,[0 pi]) ↵        ← Graphing only sin^2 t
>>subplot(312),ezplot(t^3,[0 pi]) ↵             ← Graphing only t^3
>>subplot(313),ezplot(c,[0 pi]) ↵   ← Graphing the codes stored in c, assuming
                                      that the c holds the f(t)*g(t) result
```

Figure 4.4(a) shows the plotting with order $\sin^2 t$, t^3, and $f(t)*g(t)$ from the top respectively.

In the figure we graphed all three signals over $0 \le t \le \pi$. As finite function suppose $f(t)$ and $g(t)$ exist over $a \le t \le b$ and $c \le t \le d$ respectively, the convolved signal $f(t)*g(t)$ total interval had better be $b-a+d-c$. The stretching of the interval depends where you start. Suppose we start at e, then the $f(t)*g(t)$ interval should be $e \le t \le b-a+d-c+e$. As a common practice we start at t=0 for all functions. For instance if $\sin^2 t$ and t^3 have the intervals $0 \le t \le \pi$ and $0 \le t \le 2\pi$ respectively, $f(t)*g(t)$ should have the interval $0 \le t \le 3\pi$.

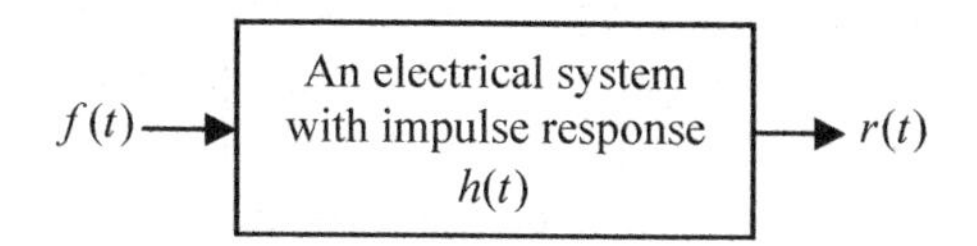

Figure 4.4(b) An electrical system has input $f(t)$ and output $r(t)$

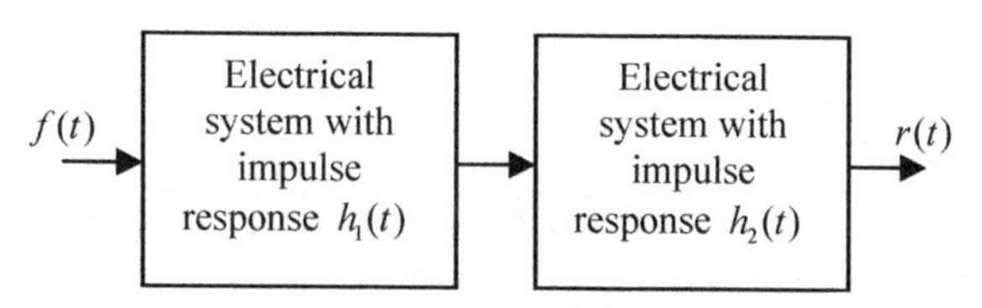

Figure 4.4(c) Two electrical systems connected in series

❖ **Usefulness of convolution**

Suppose the electrical system (can be a continuous time filter) of the figure 4.4(b) has the impulse response $h(t) = \sin^2 t$ and an input $f(t) = t^3$ is applied to the system. The output from the system is given by $h(t) * f(t)$ whose implementation we conducted just now.

Again, suppose we have two electrical systems connected in series like the figure 4.4(c). The systems have impulse responses $h_1(t) = \sin^2 t$ and $h_2(t) = t^3$. Their equivalent system function is determined by $h_1(t) * h_2(t)$ whose implementation we presented earlier.

4.5 Convolution of discrete signals

Formal definition of discrete convolution of two discrete signals $x[n]$ and $y[n]$ is given by $c[n] = x[n] * y[n] = \sum_{k=-\infty}^{k=\infty} x[k]y[n-k]$ where $c[n]$ is the discrete convolved signal. If lengths (mean numbers of samples) of the signals $x[n]$ and $y[n]$ are M and N respectively, the length of $c[n]$ is $M+N-1$. The summation $\sum_{k=-\infty}^{k=\infty} x[k]y[n-k]$ can be viewed as the polynomial multiplication of $x[n]$ and $y[n]$. We know that $x[n]$ or $y[n]$ is just a row or column matrix for different n and the n is always integer.

MATLAB built-in function conv (abbreviation for convolution) computes the convolution of two discrete signals $x[n]$ and $y[n]$ with the syntax conv($x[n]$ as a row matrix, $y[n]$ as a row matrix). The function does not provide any information about the index. Suppose $x[n]$ and $y[n]$ exist over $N_1 \le n \le N_2$ and $N_3 \le n \le N_4$ respectively, what should be the $c[n]$ interval? First the user has to decide where to start the index of $c[n]$ say p then the interval of $c[n]$ should be $p \le n \le N_2 - N_1 + N_4 - N_3 + p$. For example the $x[n]$ and $y[n]$ intervals are $-2 \le n \le 3$ and $0 \le n \le 3$ respectively, then the $c[n]$ interval is $0 \le n \le 8$ assuming the starting p is 0. Let us see the following examples on the convolution.

♦♦ Example 1

Given that $c[n]$=[−1 −2 −4 16 37 12 −43 −6 40] is the discrete convolution of the following tabulated discrete signal data which we intend to verify.

n	−2	−1	0	1	2	3
$x[n]$	1	2	7	−5	−6	8
$y[n]$			−1	0	3	5

First we assign the discrete signals $x[n]$ and $y[n]$ as a row matrix to x and y (user-chosen variable) respectively as follows:

```
>>x=[1 2 7 -5 -6 8]; ↵        ← x⇔ x[n]
>>y=[-1 0 3 5]; ↵             ← y⇔ y[n]
```

Then we call the conv with earlier mentioned syntax and put the return to c (user-chosen variable) as follows:

```
>>c=conv(x,y) ↵       ← c holds the convolved signal x[n] * y[n], c⇔ c[n]

c =
     -1  -2  -4  16  37  12  -43  -6  40
```

Should you intend to generate the index n (assuming that starts from −2), the interval of $c[n]$ becomes $-2 \le n \le 6$ and which we generate by writing -2:6 (subsection 1.1.2).

♦♦ Example 2

This example illustrates function based discrete convolution. When $x[n] = n$ over $0 \le n \le 3$ and $y[n] = n^2$ over $2 \le n \le 5$, the discrete convolution is $c[n] = x[n] * y[n]$=[0 4 17 46 84 98 75] which we wish to implement.

In this sort of problems we generate the n and sample generation of each signal takes place by the scalar code (appendix B) as follows:

```
>>n=0:3; ↵        ← n holds the integers 0, 1, 2, and 3 as a row matrix
>>x=n; ↵          ← Code of x[n] is assigned to x, x⇔ x[n]
>>n=2:5; ↵        ← n holds the integers 2, 3, 4, and 5 as a row matrix
>>y=n.^2; ↵       ← Code of y[n] is assigned to y, y⇔ y[n]
>>c=conv(x,y) ↵   ← Workspace c holds the discrete convolved signal x[n] * y[n],
                                c⇔ c[n]

c =
     0   4   17   46   84   98   75
```

Regarding the interval, the n should be from 0 to 6 if we assume the starting of $c[n]$ from n =0 and which is obtained by 0:6.

♦♦ Example 3

This example illustrates how we graph the discrete convolution. Let us consider the example 2 mentioned discrete signals. Appendix F cited **subplot** can be an option to plot the graphs of such discrete convolution. Each signal is plotted by using the **stem** of the same appendix. From example 2 we know that $x[n]$, $y[n]$, and $c[n]$ are available in the workspace x, y, and c respectively. Since n information is lost, we re-execute the n generation each time. Let us graph various signals as follows:

```
>>subplot(311),n=0:3; stem(n,x) ↵   ← Graphing x[n] versus n
>>subplot(312),n=2:5; stem(n,y) ↵   ← Graphing y[n] versus n
>>subplot(313),n=0:6; stem(n,c) ↵   ← Graphing c[n] versus n
```

Figure 4.4(d) shows the graphical response from MATLAB.

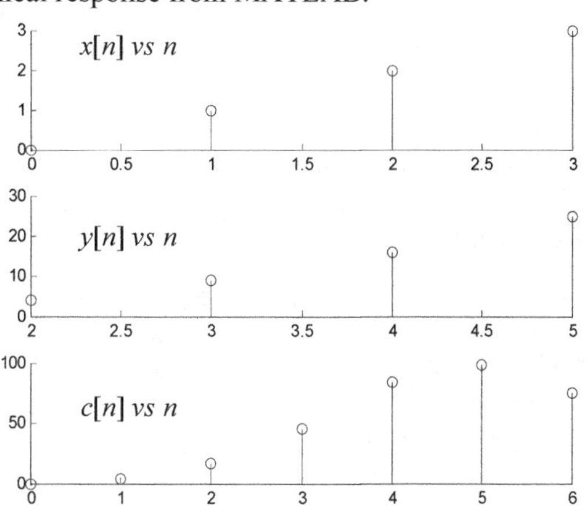

Figure 4.4(d) Plots of $x[n]$, $y[n]$, and $c[n]$ versus n (right side figure)

♦♦ Example 4

This example illustrates how the discrete convolution becomes useful in finding the discrete output.

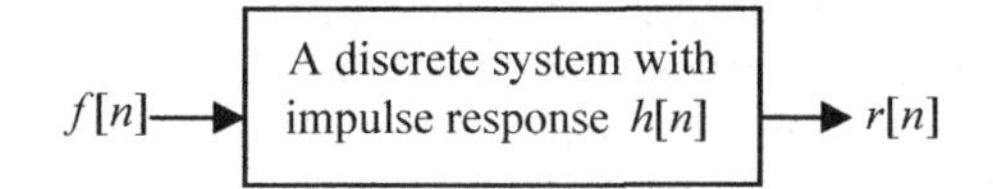

Figure 4.4(e) A discrete system has input $f[n]$ and output $r[n]$

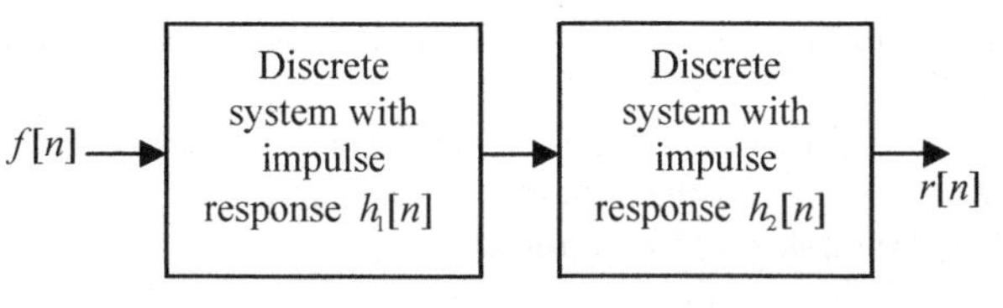

Figure 4.4(f) Two discrete systems connected in series

Figure 4.4(e) shows a discrete system (can be a discrete time filter) which has the impulse response $h[n] = n^2$ over $2 \le n \le 5$. An input $f[n] = n$ over $0 \le n \le 3$ is applied to the system. The output $r[n]$ is $f[n] * h[n]$ whose implementation is the example 2 with different dependent variable name.

♦♦ Example 5

Figure 4.4(f) shows two discrete time systems with impulse responses $h_1[n]$ and $h_2[n]$ which are seriesly placed. Their equivalent system function is $h_1[n] * h_2[n]$ which you can compute like the other four examples.

4.6 Basic signal operations

Having signal(s) available, definitive mathematical operations (usually schematized by a system block diagram) are conducted on the signals for various processing. Basic operations such as addition/subtraction, multiplication, powering, integration, etc are hidden in many electrical systems for example in communications. In this section we address few basic signal operations and their implementational link in MATLAB/SIMULINK.

♦♦ Gain of a signal

Figure 4.5(a) shows the gain c (any positive or negative constant) of the signal $f(t)$ wherefrom $r(t)=c\ f(t)$. Let us say $f(t)=5\sin 20\pi t$ and $c=-3$ hence $r(t)=-3\ f(t)$ i.e. the implementation works on the expression. As code (appendix B), we write first f=5*sin(20*pi*t) for $f(t)$ assignment to f and then r=-3*f for $r(t)$.

Again if $f[n]$ is a discrete signal and exists in the form of a row or column matrix by the name f, the gain operation is still conducted by the command -3*f.

When we need to model the gain in SIMULINK, appendix A cited Gain block simulates that. On doubleclicking the block, we find the slot for entering problem-defined gain.

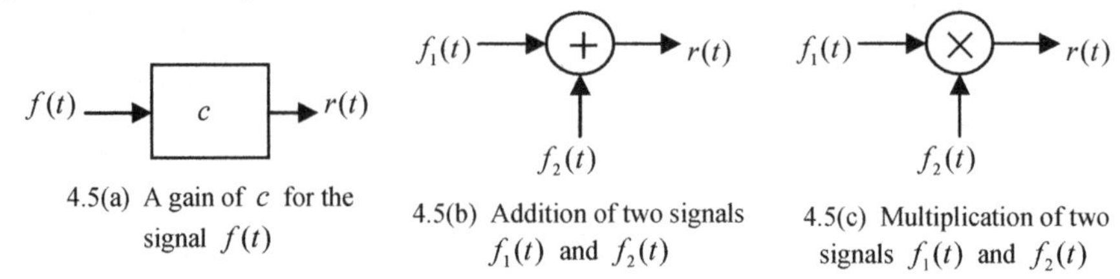

4.5(a) A gain of c for the signal $f(t)$

4.5(b) Addition of two signals $f_1(t)$ and $f_2(t)$

4.5(c) Multiplication of two signals $f_1(t)$ and $f_2(t)$

♦♦ Addition or subtraction of signals

The block diagram of the figure 4.5(b) presents the addition of two signals $f_1(t)$ and $f_2(t)$ and the output is simply $r(t)=f_1(t)+f_2(t)$. Suppose $f_1(t)=t$ and $f_2(t)=2\sin t$ from which we have $r(t)=t+2\sin t$. One implements the addition by + operator for the example f1=t; f2=2*sin(t); r=f1+f2; where r holds the resultant signal $r(t)$. Similarly we add three signals by r=f1+f2+f3. For the subtraction, we just bring the – operator for example subtraction of the f2 from f1 happens through f1-f2.

When we need to model the addition or subtraction in SIMULINK, appendix A cited Sum block simulates that.

♦♦ Multiplication of signals

In the multiplication of two continuous signals (figure 4.5(c)), we employ the * operator for example the addition-mentioned

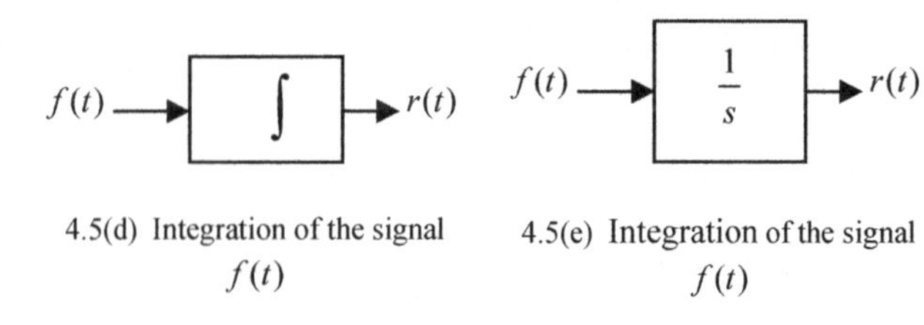

4.5(d) Integration of the signal $f(t)$

4.5(e) Integration of the signal $f(t)$

signals have the resultant signal r=f1*f2 or f1*f2*f3 for two or three input signals respectively.

When we need to model the multiplication in SIMULINK, appendix A cited Product block simulates that.

◆◆ Integration of a signal

Each of the block diagrams of figures 4.5(d) and 4.5(e) represents the integration of the continuous signal $f(t)$ that is $r(t)=\int f(t)dt$ where s is the Laplace transform variable. If the integration is expression based, we apply the appendix G cited int.

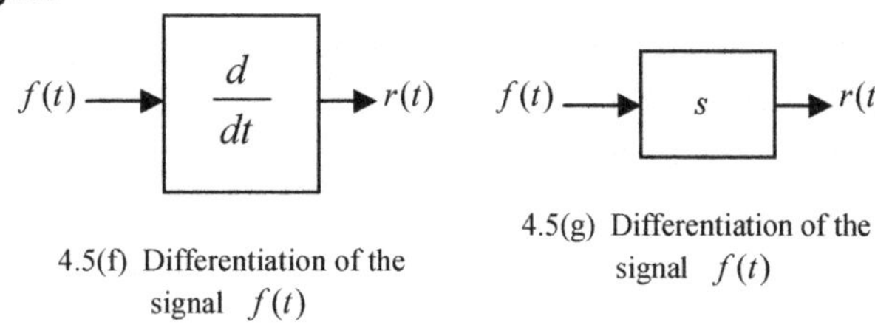

4.5(f) Differentiation of the signal $f(t)$

4.5(g) Differentiation of the signal $f(t)$

When we need to model the integration in SIMULINK, appendix A cited Integrator block simulates that.

◆◆ Differentiation of a signal

Each of the block diagrams of figures 4.5(f) and 4.5(g) represents the differentiation of the continuous signal $f(t)$ that is $r(t)=\frac{d\{f(t)\}}{dt}$. If the differentiation is expression based, we apply the command diff on the code of the $f(t)$ expression.

When we need to model the differentiation in SIMULINK, appendix A cited Derivative block simulates that.

4.7 Quantization of a signal

Quantization is a process which rounds or truncates a signal amplitude whether continuous or discrete to a user-defined finite number of levels. The quantization is solely associated with the signal amplitudes or values not with the time. If the signal value levels are equally spaced, the quantization is called uniform quantization. MATLAB function quant performs the uniform quantization. The function accepts two input arguments, the first and second of which are the signal as a row or column matrix and the quantization step size (called resolution Δ of the quantization process in signal processing term) respectively. Let us go through the following examples on the subject of quantization.

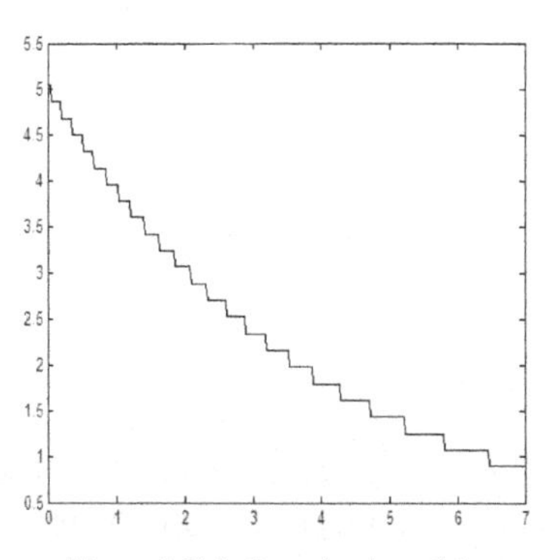

Figure 4.6(a) Quantization of the continuous signal of figure 2.1(f)

◆◆ Example 1

Given a single signal value $V=1.3$, quantize it by using $\Delta=0.2$. Computationally we divide the value of V by Δ and round the division result $\frac{V}{\Delta}$

towards the nearest integer M. The division result is checked according to the following: $\left\{\begin{matrix}\text{if the fractional part} \geq 0.5, \text{take that as } 1 \\ \text{if the fractional part} < 0.5, \text{take that as } 0\end{matrix}\right\}$. Once the integer M is found, the quantized signal value is given by $M\Delta$. For the numerical value, we have $\frac{V}{\Delta}$=6.5 or M=7 consequently the quantized single signal value should be $M\Delta$=1.4. Following is the implementation:

```
>>Q=quant(1.3,0.2) ↵   ← Workspace Q holds the quantized single signal value,
                          Q is user-chosen name

Q =

       1.4000
```

❖❖ Example 2

Example 1 mentioned Δ=0.2 is applied to the discrete signal $f[n]$=[−1.3 1.3 0 1.8] so that we finish up with the quantized signal $q[n]$=[−1.4 1.4 0 1.8] whose implementation is shown below:

```
>>f=[-1.3 1.3 0 1.8]; ↵ ← Workspace f holds the given discrete signal f[n] as a
                          row matrix
>>Q=quant(f,0.2) ↵   ← Workspace Q holds the quantized discrete signal q[n]

Q =

    -1.4000    1.4000    0    1.8000
```

❖❖ Example 3

Figure 2.1(f) presented continuous signal has the expression $f(t)=5e^{-\frac{t}{4}}$ over $0 \leq t \leq 7\text{sec}$ which is to be quantized at a Δ=0.18. One needs to choose some t step size (must be very smaller than Δ) to have the signal samples despite the signal is a continuous one and let it be 0.02. Our objective is to quantize and to plot the quantized signal in continuous sense. Let us conduct the following:

```
>>t=0:0.02:7; ↵    ← t holds the t sample points as a row matrix on chosen step
                     size over 0 ≤ t ≤ 7sec (subsection 1.1.2)
>>f=5*exp(-t/4); ↵ ← f holds the f(t) samples (section 2.3) as a row matrix at the
                     t points stored in t, scalar code of f(t)
>>Q=quant(f,0.18); ↵  ← Q holds quantized f(t) sample values as a row matrix
                     at the t points stored in t
>>plot(t,Q) ↵   ← Figure 4.6(a) shows the response, section 2.9 for the plot
```

❖❖ Example 4

In analog to digital conversion, the number of levels L rather than resolution Δ is invariably used. The L is frequently a power of 2 because quantized signals are mapped to the binary form. The relationship among quantization step, number of levels, and signal data is $\Delta = \frac{f(t)|_{\max} - f(t)|_{\min}}{L}$. Suppose example 3 mentioned exponential signal is to be quantized by 3-bit quantizer. From the exponential function properties, we have $f(t)|_{\max}=5$, $f(t)|_{\min}=0$, and $L=2^3=8$ whence Δ=0.625 and the relevant command would have been Q=quant(f,0.625);.

❖❖ Example 5

Quantization error and signal to noise power ration (SNR) are also of interest in the signal processing literature.

In example 3 (re-execute the commands of the example at the command prompt), the workspace variables f and Q hold the samples of the original and quantized signals respectively both as the identical size row matrix. The error signal (which is $e[n]=f[n]-q[n]$ in discrete sense and is also the same size row matrix) is just the f-Q. In discrete sense, the power in the original and error signals are given by $P_s=\frac{1}{N}\sum_{n=0}^{N-1}f^2[n]$ and $P_q=\frac{1}{N}\sum_{n=0}^{N-1}e^2[n]$ respectively (ignoring sampling period effect due to ratio taking, section 10.4) where N is the number of samples in $f[n]$ or the row matrix element number in f or Q. On account of the quantization, the signal to noise ratio in dB is given by SNR= $10\log_{10}\frac{P_s}{P_q}$ whose calculation is the following:

```
>>e=f-Q; ↵                      ← Workspace e holds the error signal e[n] as a row matrix
>>Ps=sum(f.^2)/length(f); ↵     ← Workspace Ps holds the computed Ps,
                                    appendices B and C.11
>>Pe=sum(e.^2)/length(e); ↵     ← Workspace Pe holds the computed Pq
>>R=10*log10(Ps/Pe) ↵           ← R holds computed SNR due to quantization

R =
        34.1734                 ← i.e. the SNR is 34.1734dB due to quantization
```

In above executions the e, Ps, Pe, and R are all user-chosen names.

4.8 Discrete to continuous conversion

Having gone through the section 4.1, the discrete to continuous (D/C) conversion is the reverse signal operation that is forming the $f(t)$ from discrete $f[n]$ whose block diagram representation is shown in the figure 4.6(b).

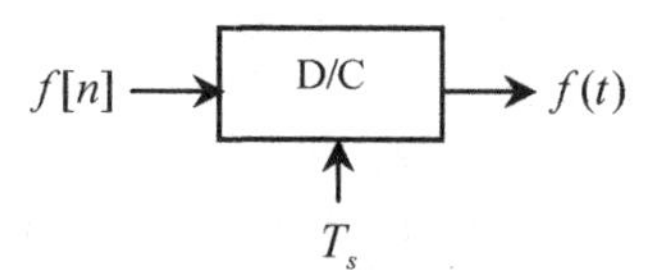

Figure 4.6(b) Block diagram for discrete to continuous (D/C) conversion of a discrete signal

Once again we can not have a perfect continuous system in computer. The best we can have is we can get the samples of the $f(t)$ from $f[n]$ knowing the step size or sampling period T_s or sampling frequency f_s.

The total sample number N in $f[n]$ and the value of T_s determine the duration of $f(t)$ over the t domain. If we assume the function $f(t)$ starting at $t=a$ sec, the interval of the $f(t)$ should be $a\le t\le a+(N-1)T_s$ (when T_s is given) or $a\le t\le a+\frac{N-1}{f_s}$ (when f_s is given).

```
function y=dtoc(f,t,Ts,a)
R=fix((t-a)/Ts);
if t<a
   y=0;
elseif t>(a+(length(f)-1)*Ts)
   y=0;
else
   y=f(R+1);
end
```

Figure 4.6(c) M-file for discrete to continuous (D/C) conversion

The main problem here we provide $f[n]$ and T_s and look for the $f(t)$ at any t. If the value of t is the multiple of T_s, direct $f[n]$ value can be taken else interpolation is required. The user has to decide the interpolation between the consecutive samples. We can assume that the samples follow zero order hold or linear interpolation.

Unfortunately there is no built-in function for the discrete to continuous conversion. Assuming zero-order hold (meaning signal value is constant until the next sample), we wrote a function file (appendix E) as presented in figure 4.6(c) which implements the D/C conversion. Type the codes of the figure 4.6(c) in a new M-file (subsection 1.1.2) and save the file by the name dtoc (any user-chosen name). The dtoc has four input arguments namely f, t, Ts, and a (correspond to $f[n]$ as a row matrix, t as a single value, T_s as a single value, and a as a single value respectively).

Let us see the following two examples.

Example 1

The discrete signal $f[n]$=[9 5 6 7 8 −9] with T_s=0.5sec and starting point of t as a=1sec should provide $f(t)$=8 at t=3.2sec considering zero-order hold. We wish to implement this. Its straightforward execution is as follows:

```
>>f=[9 5 6 7 8 -9]; ↵     ← Assigning the discrete signal f[n] to f as a row matrix
>>dtoc(f,3.2,0.5,1) ↵     ← Calling dtoc for t=3.2sec, Ts=0.5sec, and a=1sec

ans =
          8               ← It indicates f(t)=8 at t=3.2 sec
```

Since there are 6 samples in $f[n]$ i.e. N=6, T_s=0.5sec, and a=1sec, we can call any t within the interval $1 \le t \le 3.5$ sec.

Example 2

With $f[n]$=[9 5 6 7 8 −9] and T_s=0.4sec, we wish to generate the samples of the continuous $f(t)$ (considering zero-order hold) as a row matrix over $0 \le t \le 2$ sec with a step size 0.1sec.

The second step size 0.1sec or T_s is completely independent of the input sampling period T_s=0.4sec. All we need is call the function dtoc from the prompt by using the data accumulation technique of appendix C.8 as follows:

```
>>y=[9 5 6 7 8 -9]; ↵          ← Assigning the signal f[n] to y as a row matrix
>>f=[ ]; for t=0:0.1:2 f=[f dtoc(y,t,0.4,0)]; end ↵
```

Following the execution, the workspace variable f holds the samples of the continuous $f(t)$ with step size 0.1sec over $0 \le t \le 2$ sec. In the executions the y and f are user-chosen names.

With this example we wish to close the chapter.

Chapter 5
FOURIER SERIES AND TRANSFORM

Introduction

This chapter elucidates the implementation of Fourier analysis problems in MATLAB. Functional analysis through Fourier domain is very common in communications, biomedical engineering, all kinds of signal processing, and in many other disciplines. Fourier terminology always follows one common notion that is forward and inverse or in other words analysis-synthesis approach. Explicitly direct implementation for the Fourier series or transform is presented but efficient computational algorithm might be inherent to the terminology which is not discussed at all. Our implementations outline the following:

- What Fourier analysis is and different Fourier analyses for different signals
- Fourier series of a periodic signal in various forms along with the coefficient graphing
- Forward and inverse Fourier transforms of nonperiodic signals
- Discrete Fourier transform in the forward and inverse domains for discrete signals

5.1 What is Fourier analysis?

Fourier analysis is fundamentally the decomposition of any signal or function in terms of sinusoidal functions. The reason for the analysis is the sine wave has well-defined functional characteristics such as frequency. Most electrical system holds frequency related characteristics. The selection of Fourier method depends on the nature and properties of a signal. The sine wave has different forms of representation for example a cosine wave is a shifted sine wave. Complex exponential is also a sine wave with appropriate translation, rotation, or scaling. In a broad context the functions that fall into the Fourier analysis category are

$\begin{Bmatrix} perodic \\ nonperiodic \end{Bmatrix}$ and $\begin{Bmatrix} continuous \\ discrete \end{Bmatrix}$. In subsequent sections very briefly we highlight how one utilizes MATLAB/SIMULINK tools to implement different classes of Fourier terminology to various functions. All Fourier analysis has transform domain expression and its inverse. The transform domain expression can be real or complex. Inversion of the transform is just turning to the domain of the function we start with.

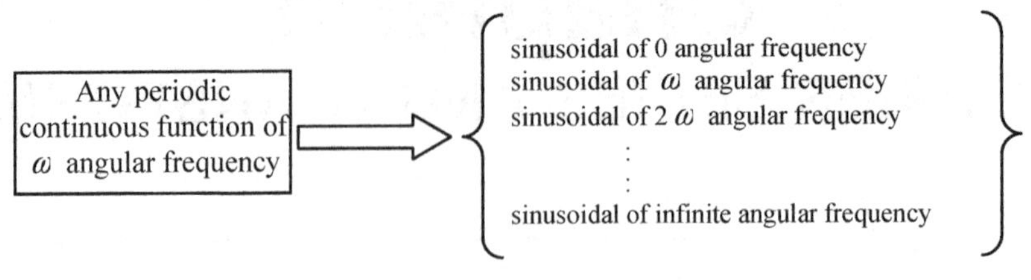

Figure 5.1(a) Basic idea behind the Fourier series

5.2 Fourier series of continuous periodic signals

Fourier series of a continuous periodic signal $f(t)$ with a period T expresses the signal in terms of the discrete sinusoidal functions. Figure 5.1(a) presents the basic concept behind the Fourier series analysis. Theoretically the continuous signal $f(t)$ is the sum of infinite continuous sinusoidal functions. The sinusoidal functions appear either in real or complex form however brief descriptions of various form Fourier series are as follows:

Real form 1: $f(t)=\frac{A_0}{2}+\sum_{n=1}^{\infty}\left(A_n\cos\frac{2\pi nt}{T}+B_n\sin\frac{2\pi nt}{T}\right)$ where the coefficients A_n and B_n are given by $A_n=\frac{2}{T}\int_T f(t)\cos\frac{2\pi nt}{T}dt$ and $B_n=\frac{2}{T}\int_T f(t)\sin\frac{2\pi nt}{T}dt$. Component sinusoidal frequencies are related with the n whose variations are n =0, 1, 2, ...etc for A_n and n =1, 2, ...etc for B_n. For each value of n, we have one sine function and the function is called a harmonic. The symbol $\int_T$ indicates the integration over one period T as the function $f(t)$ is defined. The term $\frac{A_0}{2}$ in the expansion is called the average value of the periodic function $f(t)$.

Real form 2: $f(t)=\frac{A_0}{2}+\sum_{n=1}^{\infty}C_n\cos\left(\frac{2\pi nt}{T}-\varphi_n\right)$ where $C_n=\sqrt{A_n^{\,2}+B_n^{\,2}}$ and $\varphi_n=\tan^{-1}\frac{B_n}{A_n}$. The meanings of the symbols are identical with those of the real form 1.

Complex form: $f(t)=\sum_{n=-\infty}^{\infty}C_n\, e^{j\frac{2\pi nt}{T}}$ where the coefficients are given by $C_n=\frac{1}{T}\int_T f(t)e^{-j\frac{2\pi nt}{T}}dt$ and the other symbols in the expression have the real form mentioned meanings (appendix C.12 for j).

The term $\frac{2\pi}{T}$ is the angular frequency ω in radian/sec if the T is in second. From the function description, one period of the function may appear from $-\frac{T}{2}$ to $\frac{T}{2}$, 0 to T, or other and the limits of the integration are set accordingly. We maintain the font equivalence like n⇔n in the following discussions. Computing in MATLAB and modeling in SIMULINK both are implementable.

5.2.1 Symbolic Fourier series coefficients

Finding Fourier series means finding the discrete coefficients A_n, B_n, C_n, and φ_n from the given continuous periodic function $f(t)$. We employ the int function (appendix G) to obtain the integration for various coefficients since no built-in function is available for the series computation to date. The syntax we need is int(integrand code, variable, lower limit, upper limit). In symbolic computation we do not use decimal or exponential data for example 0.2 is written as $\frac{2}{10}$. Also the exponential form 10^{-3} is written as $\frac{1}{1000}$. Prior to the integration we declare the integrand variables by using the command syms. Let us go through the following examples regarding the symbolic coefficient finding.

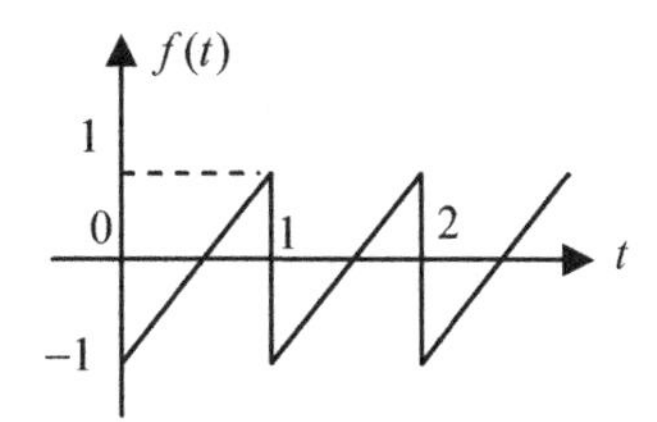

Figure 5.1(b) A sawtooth function

Example 1

It is given that Fourier series coefficients A_n=0 and $B_n=-\frac{2}{\pi n}$ are for the periodic wave of the figure 5.1(b) regarding the earlier mentioned real form 1 which we intend to verify.

From figure 5.1(b) the function has the period T=1 and the equation $f(t)=2t-1$ starting from t=0. The coefficients need the integrations $A_n=2\int_0^1(2t-1)\cos 2\pi nt\,dt$ and $B_n=2\int_0^1(2t-1)\sin 2\pi nt\,dt$. There are two related variables n and t and we declare them as follows:

```
>>syms n t ↵                    ← n⇔n and t⇔t
```

We perform the $A_n=2\int_0^1(2t-1)\cos 2\pi nt\,dt$ as follows:

```
>>An=2*int((2*t-1)*cos(2*pi*n*t),t,0,1); ↵ ← An⇔A_n where An is
                                              user-chosen variable name
```

Machine does not have any information about the n. We pass the information about the n into MATLAB by the assume function of maple (appendix C.5) as follows (in which the integer is reserve word and is applicable for the Fourier series context):

```
>>maple('assume(n,integer)'); ↵
```

In maple there is a function by the name simplify which simplifies any coded expression existing in a variable An in the workspace as follows:

```
>>An=maple('simplify',An) ↵
```

```
An =

0
```

In the last execution the left assignee An (indicated by An=) can be any user-chosen variable name. Just to maintain the consistency with A_n we again assigned the return from the An to An. However 0 is returned to An as expected. The computation of B_n is very similar to that of the A_n which is presented as follows:

```
>>Bn=2*int((2*t-1)*sin(2*pi*n*t),t,0,1) ↵   ← Bn⇔ B_n where Bn is
                                              user-chosen variable name
Bn =

-2/pi/n                                     ← The value of B_n
```

Once the integer condition declared in maple, it remains active. If you intend to remove the integer condition from the maple, exercise the command clear maplemex at the command prompt.

Example 2

Provided are the $C_n = \frac{2}{n\pi}$ and the $\varphi_n = \frac{\pi}{2}$ for the periodic wave of figure 5.1(b) as regards to the real form 2 Fourier series coefficients. We intend to verify these two coefficients.

The reader needs to execute the example 1 discussed commands completely. The last workspace variables An and Bn of the example 1 hold the codes for the coefficients A_n and B_n respectively. The C_n and φ_n just need writing the vector codes (appendix B) of $\sqrt{A_n^{\ 2} + B_n^{\ 2}}$ and $\tan^{-1}\frac{B_n}{A_n}$ which are sqrt(An^2+Bn^2) and atan(Bn/An) respectively and whose implementations are shown below:

```
>>Cn=sqrt(An^2+Bn^2) ↵

Cn =

2/pi*(1/n^2)^(1/2)                          ← The value of C_n
>>Pn=atan(Bn/An) ↵

Pn =

1/2*pi                                      ← The value of φ_n
```

The return to Cn (which is 2/pi*(1/n^2)^(1/2)) is the machine's way of representing $\frac{2}{n\pi}$. In just conducted executions we assigned the returns to the workspace Cn and Pn (can be any user-chosen name) which refer to C_n and φ_n respectively. The simplification on Cn is further conducted by using the command Cn=maple('simplify',Cn) if it is necessary which should return the code for $\frac{2}{n\pi}$.

Example 3

Complex Fourier series coefficient for the periodic function of figure 5.1(b) is $C_n = \frac{j}{\pi n}$ which we wish to check.

Plugging the periodic function parameters, one obtains $C_n = \int_0^1 (2t-1)e^{-j2\pi nt}dt$ with integer n. Applying ongoing function and symbology, we carry out the execution as follows (assuming that the integer condition of n is entered by maple like the example 1):

```
>>syms n t ↵
>>Cn=int((2*t-1)*exp(-j*2*pi*n*t),t,0,1) ↵

Cn =

i/pi/n                ← The value of complex Cn
```

The function int in general works for the complex functions. In the implementation the vector code for $(2t-1)e^{-j2\pi nt}$ is (2*t-1)*exp(-j*2*pi*n*t) where j or i is the imaginary number in MATLAB. We assigned the return to the workspace Cn which is this time our complex C_n.

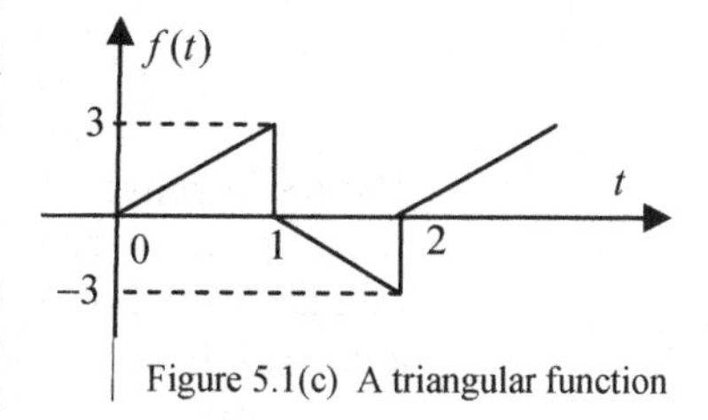

Figure 5.1(c) A triangular function

Example 4

In the last three examples we elaborately implemented the findings of Fourier series coefficients for all three forms of a single periodic function. We wish to do the same for the periodic function of the figure 5.1(c). It is given that various Fourier series coefficients are as follows:

$A_n = 6\frac{(-1)^n - 1}{n^2\pi^2}$ and $B_n = -3\frac{(-1)^n - 1}{\pi n}$ for the real form 1,

$C_n = \frac{-3((-1)^n - 1)\sqrt{4+n^2\pi^2}}{n^2\pi^2}$ and $\varphi_n = -\tan^{-1}\left(\frac{n\pi}{2}\right)$ for the real form 2, and

$C_n = \frac{3[-\pi nj - 2 + 2(-1)^n + (-1)^n \pi nj]}{2n^2\pi^2}$ for the complex form.

Our objective is to have these coefficients.

The function in one period is defined by $\begin{Bmatrix} f(t) = 3t & for\ 0 \le t \le 1 \\ f(t) = 3 - 3t & for\ 1 \le t \le 2 \end{Bmatrix}$.

Appendix C.6 mentioned maple function line helps us find the equation of a straight line passing through two points if the reader is interested to use that. Figure 5.1(c) says that the period T =2 starts from 0 and there are two linear functions and there should be two integrations for every Fourier series coefficient. We also assume that the reader has gone through the last three examples completely because we will be exercising the conducted symbology and function here.

for the real form 1:

Computations need us to implement $A_n = \int_0^1 3t \cos \pi nt \, dt + \int_1^2 (3-3t) \cos \pi nt \, dt$ and $B_n = \int_0^1 3t \sin \pi nt \, dt + \int_1^2 (3-3t) \sin \pi nt \, dt$ which we carry out as follows:

```
>>syms n t ↵
>>An=int(3*t*cos(pi*n*t),t,0,1)+int((3-3*t)*cos(pi*n*t),t,1,2); ↵
>>maple('assume(n,integer)'); An=maple('simplify',An); ↵
>>pretty(An) ↵
                n~
          (-1)   - 1
      6 -------------
               2   2
           pi    n~
>>Bn=int(3*t*sin(pi*n*t),t,0,1)+int((3-3*t)*sin(pi*n*t),t,1,2); ↵
>>Bn=maple('simplify',Bn); pretty(Bn) ↵
                 n~
            (-1)   - 1
        -3 ----------
              pi n~
```

The n~ in above execution indicates the real part of n where n is integer so n~ is equivalent to n in the returns from MATLAB as well as in the later executions. Executed code equivalences are as follows: n⇔n, t⇔t, 3*t*cos(pi*n*t)⇔$3t\cos \pi nt$, (3-3*t)*cos(pi*n*t)⇔$(3-3t)\cos \pi nt$, An⇔A_n, and Bn⇔B_n. The code of the expression for A_n is stored in the last An on which the command pretty shows the easy mathematics readable form. Similar displaying is also done on the Bn in above and in later executions.

for the real form 2:

Having the real form 1 executed, we need to implement $C_n = \sqrt{A_n^2 + B_n^2}$ and $\varphi_n = \tan^{-1}\frac{B_n}{A_n}$ for this computation as follows:

```
>>Cn=sqrt(An^2+Bn^2); ↵                                    ← Cn⇔Cn
>>pretty(simple(Cn)) ↵
                 n~                  2   2  1/2
           ((-1)   - 1) (4 + pi   n~   )
        -3 --------------------------------
                          2  2
                        pi  n~
>>Pn=atan(Bn/An) ↵                                         ← Pn⇔φn

Pn =
```

```
-atan(1/2*pi*n)
```

The command simple(Cn) is conducted in MATLAB not in maple unlike maple('simplify',Bn). Both the simple and simplify work for expression simplification but the simple tries to determine the shortest expression in machine term.

for the complex form:

For the complex form Fourier series coefficient we have to implement $C_n = \frac{1}{2}\int_0^1 3te^{-j\pi nt}dt + \frac{1}{2}\int_1^2 (3-3t)e^{-j\pi nt}dt$ as follows:

```
>>C1=1/2*int(3*t*exp(-j*pi*n*t),t,0,1); ↵
>>C2=1/2*int((3-3*t)*exp(-j*pi*n*t),t,1,2); ↵
>>Cn=C1+C2; Cn=maple('simplify',Cn); pretty(Cn) ↵
                                          n~        n~
              -pi n~ I - 2 + 2 (-1)   + (-1)   pi n~ I
        3/2 -------------------------------------------
                                2   2
                              pi  n~
```

When we have long integration it is better that we perform the integration part by part. The command C1=1/2*int(3*t*exp(-j*pi*n*t),t,0,1) in above implementation executes $\frac{1}{2}\int_0^1 3te^{-j\pi nt}dt$, so does C2=1/2*int((3-3*t)*exp(-j*pi*n*t),t,1,2) for $\frac{1}{2}\int_1^2 (3-3t)e^{-j\pi nt}dt$.

Both integration results are combined by writing Cn=C1+C2 where C1, C2, and Cn are user-chosen variable names. Some code equivalences are as follows: n~⇔ n, Cn⇔ C_n, 3*t*exp(-j*pi*n*t) ⇔ $3te^{-j\pi nt}$, I⇔ j, and (3-3*t)*exp(-j*pi*n*t)⇔ $(3-3t)e^{-j\pi nt}$. In maple the imaginary number is I (identical with the i or j of MATLAB) for this reason the return in the complex form is in terms of I.

5.2.2 Numeric Fourier series coefficients

In subsection 5.2.1 the implementation mainly highlights the symbolic or analytical form of the computation for any n or harmonic. Here we introduce the computational tactics for the numeric or decimal harmonics. Fourier series coefficient expressions for the A_n, B_n, C_n, or φ_n must be available in the workspace of MATLAB before the numeric value finding.

The command subs (abbreviation for the substitution) is helpful in the numeric value finding from coefficient expressions.

Let us consider the example 1 of the subsection 5.2.1 in which we have $B_n = -\frac{2}{n\pi}$ and the n varies from 1 to infinity. The $n=1$ should provide us $B_1 = -\frac{2}{\pi} =$ –0.6366 which we wish to determine.

Re-execute the commands of the example 1 until you get the Bn which holds the code for B_n and then execute the following:

Numeric coefficient: for a single n :

in symbolic form:

```
>>B1=subs(Bn,sym(1)) ↵

B1 =

-2/pi
```

in decimal form:

```
>>double(B1) ↵

ans =
      -0.6366
```

The subs in above execution has two input arguments, the first and second of which are the name of the assignee holding the code and the value of the n we are interested at respectively (for symbolic output we used the command sym). The return from the subs is assigned to B1 (can be any user-given name). For the decimal value, we exercise the command double on B1 which is also presented above on the right side in this page.

Again let us say we intend to find the values of B_n for n=1, 2, 3, and 4 and which should be $\begin{bmatrix} -\frac{2}{\pi} & -\frac{1}{\pi} & -\frac{2}{3\pi} & -\frac{1}{2\pi} \end{bmatrix}$ as a row matrix respectively and is conducted as follows:

Numeric coefficient: for multiple n :

```
>>B=subs(Bn,sym([1 2 3 4])) ↵          ← B is user-chosen variable name

B =

[ -2/pi,  -1/pi,  -2/3/pi, -1/2/pi]
```

The four values of n are written as the four element row matrix and over the matrix the sym is applied. If it is necessary, executing double(B) yields the four values in decimal form as a row matrix. The n values do not have to be consecutive for example writing the second argument as [4 2 3 1] returns the B_n value for n=4 first. The four n values could have been written as 1:4 according to MATLAB notation (subsection 1.1.2) and the command would have been B=subs(Bn,sym(1:4)). Now we present three more examples on finding the numerical Fourier series coefficient in the following.

Example 1

Find the average value of the periodic signal in figure 5.1(c) from A_n coefficient of the Fourier series.

By inspection the value should be zero or average value $\frac{A_0}{2}$ is obtained by setting n=0 in A_n of the series coefficient. Since A_n= $6\frac{(-1)^n - 1}{n^2\pi^2}$ (example 4 of the subsection 5.2.1), setting n=0 results a form which is indeterminate (here it is $\frac{0}{0}$). Machine can not handle undefined expression. If we use subs for n=0 by writing subs(An,sym(0)), we see the return as NaN indicating not a number. We redo the computation for A_n. Start pressing the up-arrow key one at a time from the keyboard and you find the expression for An. Replace n by 0 and execute the command for An again. After that An/2 provides the average value of the signal.

Example 2

It is given that the 3rd harmonic of the real form 2 Fourier series coefficients are $C_3 = \frac{2\sqrt{4+9\pi^2}}{3\pi^2}$ and $\varphi_3 = -\tan^{-1}\left(\frac{3\pi}{2}\right)$ for the periodic signal of the figure 5.1(c) which we wish to obtain. Also we need the 3rd, 5th, and 7th harmonics of the signal.

Example 4 of subsection 5.2.1 presents the finding of the real form 2 Fourier series coefficients C_n and φ_n which are stored in Cn and Pn respectively. Re-execute the commands until you get the Cn and Pn. Setting n=3 in the respective expression occurs as follows:

For C_3:

```
>>C3=subs(Cn,sym(3)); ↵
>>pretty(simple(C3)) ↵
                    2 1/2
          (4 + 9 pi  )
     2/3  ---------------
                 2
               pi
```

For φ_3:

```
>>P3=subs(Pn,sym(3)); ↵
>>pretty(P3) ↵
      -atan(3/2 pi)
```

The C3 and P3 (can be any user-given names) refer to C_3 and φ_3 respectively. The command simple on the C3 finds the best simplified expression for the codes stored in C3.

Finding the 3rd, 5th, and 7th harmonics happens by the command subs(Cn,sym([3 5 7])) in which the required harmonics are placed as a three element row matrix as the second input argument of the subs. The decimal form of the coefficient is exhibited by using the command double(subs(Cn,sym([3 5 7]))). Similar computational style applies to the phase coefficient φ_n as well.

Example 3

The 3rd harmonic of the complex form coefficient for the signal in figure 5.1(c) is $C_3 = \frac{-4-6j\pi}{6\pi^2}$ which we intend to obtain.

Example 4 of subsection 5.2.1 presents the implementation of C_n as Cn. Redo the implementation until you get the Cn. Substitution of the n=3 to the C_n is as follows:

```
>>C3=subs(Cn,sym(3)); ↵                  ← C3⇔C3
>>pretty(C3) ↵
          -6 I pi - 4
    1/6  ------------                    ← I⇔j
               2
             pi
```

A set of harmonics are also found as we have shown in example 2 but all coefficients will be complex. The command double for decimal finding equally applies here i.e. double(C3). The real, imaginary, magnitude, and phase angle separations from the complex coefficients take place by using the commands real, imag, abs, and angle on C3 respectively.

5.2.3 Graphing Fourier series coefficients

Graphing the Fourier series coefficient means graphing the coefficients A_n, B_n, C_n, or φ_n against n. Concerning their expressions as presented earlier, the values of the A_n, B_n, C_n, and φ_n are discrete and different for different n. MATLAB built-in function stem (appendix F) is the best option for discrete plotting of the Fourier series coefficients. From the expression of the coefficient, we first obtain numerical value by applying the functions of subsection 5.2.2 and then employ the stem. Let us see the following examples in this regard.

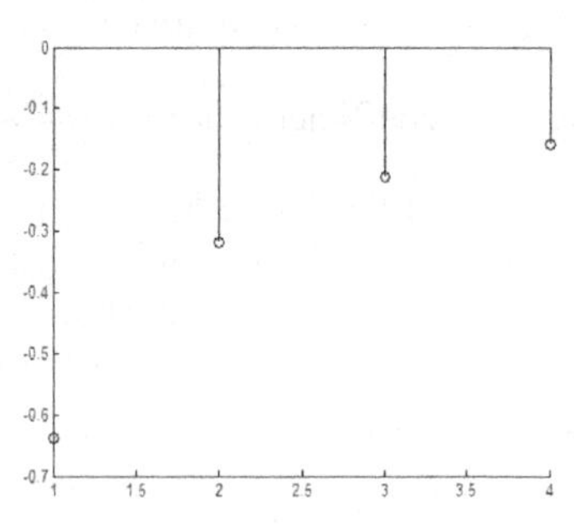

Figure 5.2(a) Plot of the four Fourier series coefficients

Example 1

Figure 5.2(a) is the plot of the real form 1 Fourier series coefficients B_n of the periodic signal in figure 5.1(b) for n=1, 2, 3, and 4 which we intend to implement.

The beginning example of subsection 5.2.2 presents that the workspace variable B holds the four harmonics $\begin{bmatrix} -\frac{2}{\pi} & -\frac{1}{\pi} & -\frac{2}{3\pi} & -\frac{1}{2\pi} \end{bmatrix}$ as a row matrix for n=1, 2, 3, and 4 respectively (redo the implementation until you get the B). After that we execute the following:

>>B=double(B); ↵ ← Symbolic values are converted to decimal, because the stem functions on decimal values and the result is again assigned to the workspace B

>>n=[1 2 3 4]; ↵ ← The four harmonics are assigned to n as a four element row matrix

>>stem(n,B) ↵

Execution of the last line command displays the plot of the figure 5.2(a). The horizontal and vertical axes of the figure refer to the n and Fourier series coefficient B_n values respectively.

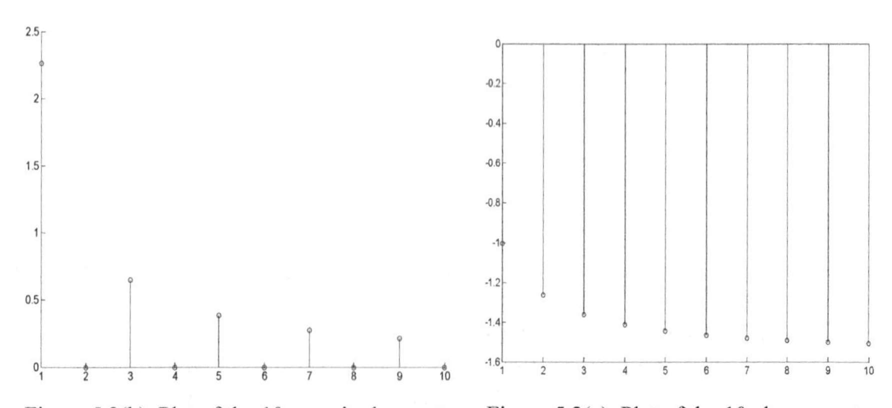

Figure 5.2(b) Plot of the 10 magnitude spectra Figure 5.2(c) Plot of the 10 phase spectra

Example 2

Figures 5.2(b) and 5.2(c) show the plots of the real form 2 Fourier series coefficients C_n and φ_n from $n=1$ to $n=10$ for the periodic signal in figure 5.1(c) respectively. Our objective is to get these two spectra.

Referring to the example 4 of subsection 5.2.1, the real form 2 magnitude and phase coefficients C_n and φ_n are stored in the workspace variables Cn and Pn respectively (redo the implementation until you obtain them). We generate the required n values as follows:

```
>>n=1:10; ↵        ← n holds the 10 integers as a row matrix, subsection 1.1.2
>>C10=subs(Cn,sym(1:10)); ↵ ← C10 holds the 10 magnitude spectrum values
                              as a row matrix followed by the substitution,
                              subsection 5.2.2 for substitution, C10 is
                              user-chosen name
>>stem(n,double(C10)) ↵     ← Graphing the spectrum first turning the
                              symbolic values stored in C10 to decimal
```

The last line command responds with the figure 5.2(b) displaying the spectra of the 10 magnitude coefficients. The figure says that the nature of the signal does not allow the even harmonics to exist. However the phase spectrum is obtained by employing similar exercise as follows:

```
>>P10=subs(Pn,sym(1:10)); ↵ ← P10 holds the 10 phase spectra values as a row
                              matrix followed by the substitution, the P10 is
                              user-chosen variable
>>stem(n,double(P10)) ↵ ← Graphing the phase spectra like the figure 5.2(c)
```

Critical situation:

Occasionally there is complicity while computing some spectrum without simplification for example ongoing φ_n. The φ_n could possess denominator like $(-1)^n-1$ and turn to zero for even harmonics thereby resulting an undefined expression. One can overcome this sort of difficulty by taking the limiting value at those harmonics. MATLAB function limit finds the limit of some expression symbolically with the syntax limit(function code, limit value). But the limit finding must take place one at a time. For the same phase spectra, a for-loop provides the limit finding for every harmonic as (appendix C.9) follows:

```
>>for k=1:10 P10(k)=limit(Pn,sym(k)); end ↵
>>stem(n,double(P10)) ↵          ← Graphs the figure 5.2(c)
```

In the first line of above execution the for-loop counter k changes from 1 to 10 which simulates the n variation. The function limit has two input arguments, the first and second of which are the expressional code (here it is the code for Pn) and the limit value (here it is $n=1,2,..,10$ but one at a time taken care of by the k) respectively. The sym(k) turns the number k to symbolic k. Each time the output of the limit is assigned to the kth element of the new workspace variable P10 by using the command P10(k) so that at end of the for-loop the P10 is also a row matrix.

5.2.4 Reconstruction from Fourier series coefficients

In this subsection our objective is to sum some harmonics so that we reconstruct approximately the original shape of the periodic signal $f(t)$. Theoretically speaking we must sum the Fourier series harmonics up to the infinity in order to reconstruct the function perfectly but a machine can not perform infinite sums or the harmonics might become insignificant after some value of n.

Considering the example 1 of the subsection 5.2.1, the periodic signal of figure 5.1(b) is expressed as $f(t)=\sum_{n=1}^{\infty}-\frac{2}{n\pi}\sin 2n\pi t$. We choose some finite integer instead of infinity say 5 and compute the sum $f_5(t)=\sum_{n=1}^{5}-\frac{2}{n\pi}\sin 2n\pi t$ where $f_5(t)$ indicates the construction of $f(t)$ from 5 harmonics or terms. There are 3 cycles in the signal of the figure 5.1(b). Let us concentrate on two cycles just to have less computational burden from what reason the t variation should be $0 \le t \le 2$ sec. Given that when the $f(t)$ and $f_5(t)$ are plotted together, figure 5.2(d) shows the plot. Our objective is to get this plot.

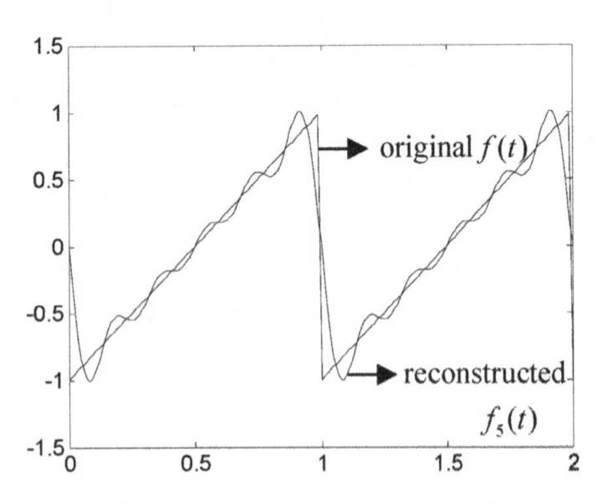

Figure 5.2(d) Reconstruction from Fourier series coefficients

We concentrate only on the sample point of t on the given continuous function as far as a computer performs discrete computation. The calculation or reconstruction takes place on the sample points on t as well. That necessitates some user-chosen step size on the t within the range $0 \le t \le 2$ sec and let it be 0.01sec.

For the computation we first generate a row matrix considering the time interval and step size by writing the command [0:0.01:2]. It means we are focusing on the t points at 0, 0.01, 0.02, ,2. Any particular harmonic n can be controlled by using a for-loop (appendix C.9). For each n, we then compute the expression $-\frac{2}{n\pi}\sin 2n\pi t$ on all t points using the scalar code (appendix B) and which is -2/n/pi*sin(2*n*pi*t). However the complete code for the computation is as follows:

```
>>t=[0:0.01:2]; ↵    ← The t variation as a row matrix is assigned to the workspace variable t
>>f5=0; ↵            ← Assigning 0 outside the for-loop to workspace variable f5, f5⇔ f5(t), f5 is user-chosen name
>>for n=1:5 f5=f5-2/n/pi*sin(2*n*pi*t); end ↵
```

A programming artifice is used here. Initially f5=0, the computed values of the function are added with the f5 for n=1. When n=2, the f5 holds the values for the n=1. The looping is continued until n=5 therefore f5 holds the sum for all harmonics at the end. Or in other words, two identical size row matrices are being added element by element for each harmonic or n. Consequently the workspace variable f5 holds the reconstructed signal samples of $f_5(t)$ as a row matrix by summing the 5 harmonics at those specific t points.

The original signal samples of the figure 5.1(b) are obtained by the example 6 quoted technique of section 2.6. The t vector existing in the workspace is equally applicable for the original signal. The code for the two cycle signal sample generation of the $f(t)$ in figure 5.1(b) is as follows:

```
>>f=sawtooth(2*pi*t); ↵
```
← f holds two cycle $f(t)$ samples of figure 5.1(b) as a row matrix at those specific t points

In order to plot the original signal and the reconstructed one together (appendix F), we exercise the following:

```
>>plot(t,f,t,f5) ↵
```

Figure 5.2(d) presents the outcome from the last line execution.

Understandably the reconstruction is not exact. The reason is we have chosen only 5 terms in lieu of infinite terms. To see the effect of 10 terms, the change we need is in the last counter index of the for-loop. Also we need to assign 0 to some other workspace variable say f10 to store the reconstruction data so we carry out the following:

```
>>f10=0; ↵
```

Start pressing the up-arrow key from the keyboard and modify the for-loop as follows:

```
>>for n=1:10 f10=f10-2/n/pi*sin(2*n*pi*t); end ↵
```
← f10 holds reconstruction from 10 harmonics

If you say I intend to plot all three cases together, the command you need is plot(t,f,t,f5,t,f10), graph is not shown for space reason. In order to differentiate the three plots, we execute the command legend('Original','5-terms','10-terms'). Thus the reader can explore the reconstruction on the Fourier series of any other function from different number harmonics.

Mean square error from reconstruction

Mean square error (mse) often provides a deterministic measure of deviation of the reconstructed signal from that of the actual one which is defined as $\text{mse}=\frac{1}{t\ span}\int_t [f(t)-f_5(t)]^2\,dt$ considering the 5-term. In the case of the discrete computation, the integration turns to summation however we calculate the mse for the 5-term as follows:

```
>>e=f-f5; ↵
```
← e holds the error as a row matrix at the specified points in the reconstruction, e⇔ $f(t)-f_5(t)$, e is any user-given name

```
>>mean(e.^2) ↵
```
← e.^2⇔ $[f(t)-f_5(t)]^2$

```
ans =

        0.0423
```
← mse from the 5-term reconstruction

The command e.^2 squares every element in the row matrix e and the mean finds the average of all elements in the row matrix e.^2. In order to see the mse for the 10 terms, all we need is execute e=f-f10; mean(e.^2) which should return 0.0256.

5.3 Fourier transform of nonperiodic signals

Fourier transform is applicable for a continuous and nonperiodic finite function $f(t)$. One can say that the Fourier transform is a Fourier series in which the

$$f(t)\xrightarrow{\text{forward Fourier transform}}F(\omega)\xrightarrow{\text{inverse Fourier transform}}f(t)$$

Figure 5.3(a) The concept behind the Fourier transform

period of the signal is infinity. The transform has two forms – forward and inverse. The forward transform is the analysis phase which turns the signal $f(t)$ from t or time domain to ω or angular frequency domain (let us call $F(\omega)$). On the contrary the inverse counterpart recovers the time domain signal $f(t)$ from the $F(\omega)$ in angular frequency domain which is also called the synthesis process. The finiteness of $f(t)$ is assured if the value of $\int_{t=-\infty}^{t=\infty} |f(t)|dt$ is finite. That is why the periodic signal does not have a Fourier transform. Also the function like t^t or e^{t^2} which increases continuously with t does not have the Fourier transform. Figure 5.3(a) presents the concept behind the Fourier transform. In the next two subsections we explain how the forward and inverse Fourier transforms of nonperiodic signals can be symbolically computed in MATLAB.

5.3.1 Forward Fourier transform

The forward Fourier transform converts a nonperiodic and continuous signal $f(t)$ to $F(\omega)$ through the formula $F(\omega)=\int_{t=-\infty}^{t=\infty} e^{-j\omega t} f(t)\,dt$. The transform $F(\omega)$ is a complex function of ω (omega) and continuous as well. MATLAB function that computes the forward Fourier transform is fourier. The common syntax of the function is fourier (given function $f(t)$ in vector string form – appendix B, independent variable of $f(t)$, wanted transform variable) but also works with the first input argument. The default return from the fourier is in terms of w (corresponds to the frequency variable ω) and in vector string form. The independent variable of $f(t)$ must be declared symbolically by using the command syms prior to the computation. The readable form of the forward transform string is viewed by the use of the command pretty (appendix C.10).

Let us see the Fourier transform of $f(t)=e^{-|t|}$ which should be $F(\omega)=\int_{t=-\infty}^{t=\infty} e^{-j\omega t} f(t)\,dt = \int_{t=-\infty}^{t=\infty} e^{-j\omega t} e^{-|t|}\,dt = \frac{2}{1+\omega^2}$. The vector code of the function $e^{-|t|}$ is exp(−abs(t)) and we compute the transform as follows:

```
>>syms t ↵                          ← Declaring the independent t of e^-|t| as symbolic, t⇔t
>>F=fourier(exp(-abs(t))) ↵ ← Forward Fourier transform is assigned to
                                      workspace variable F, F⇔ F(ω) , F is user-chosen name

F =

2/(1+w^2)                  ← Vector code of F(ω) , w⇔ ω
>>pretty(F) ↵              ← Displaying the transform string stored in F in readable form
      2
  ---------
         2
  1 + w
```

If the independent variable of the given function were x (use syms x before) i.e. $f(x)=e^{-|x|}$, the command would be F=fourier(exp(-abs(x))) and still the transform return is in terms of w. When the transform variable is required from w to z, one uses the command F=fourier(exp(-abs(x)),x,z) for which syms z should be conducted before.

For a function which does not have the transform for example e^{t^2} and whose code is exp(t^2), the response of the fourier is as follows:

```
>>F=fourier(exp(t^2)) ↵
```

```
F =

fourier(exp(t^2),t,w)
```

Above output indicates no close form results and just the definition of the transform.

Table 5.A Fourier transform and its MATLAB counterpart for some standard functions

Example 1 $f(t)=e^{-t^2}$ $F(\omega)=\sqrt{\pi}\,e^{-\frac{1}{4}\omega^2}$	Command for the example 1: >>syms t, F=fourier(exp(-t^2)); ↵ >>pretty(F) ↵ 1/2 2 pi exp(- 1/4 w)
Example 2 $f(t)=\dfrac{1}{1+t^2}$, $F(\omega)=$ $\pi[e^{\omega}u(-\omega)+e^{-\omega}u(\omega)]$	Command for the example 2: >>syms t ↵ >>F=fourier(1/(1+t^2)); ↵ >>pretty(F) ↵ exp(w) pi heaviside(-w)+exp(-w) pi heaviside(w)
Example 3 $f(t)=e^{-\lvert at\rvert}$ $F(\omega)=F[f(t)]=\dfrac{2a}{a^2+\omega^2}$ where $a>0$	Command for the example 3: >>maple('assume(a>0)'); ↵ >>F=maple('fourier(exp(-abs(a*t)),t,w)') ↵ F = 2*a/(a^2+w^2)
Example 4 $f(t)=\lvert 5t+7\rvert$ $F(\omega)=-\dfrac{10e^{j\frac{7}{5}\omega}}{\omega^2}$	Command for the example 4: >>syms t, F=fourier(abs(5*t+7)) ↵ F = -10*exp(7/5*i*w)/w^2

The unit step and impulse (or Dirac delta) functions are important functions to the context of Fourier transform. Figure 3.1(a) shows the plot of the unit step function $u(t)$ and whose MATLAB counterpart is heaviside(t). Forward Fourier transform of the unit step function is $F(\omega)=\pi\delta(\omega)-\frac{j}{\omega}$ where $\delta(\omega)$ is the Dirac delta or impulse function (figure 3.1(k)) as regards to ω and is obtained as follows by applying ongoing function and symbology:

```
>>syms t, F=fourier(heaviside(t)) ↵

F =
pi*dirac(w)-i/w
```

← Vector string of $\pi\delta(\omega)-\frac{j}{\omega}$ is held in F

Regarding the output, dirac(w) corresponds to $\delta(\omega)$. The unit step function can be shifted to the left or right for example $u(t)$ shifted to the right at t =2 is denoted by $u(t-2)$ and has the transform $F[u(t-2)]=e^{-2j\omega}(\pi\delta(\omega)-\frac{j}{\omega})$ which we obtain as follows (F is the Fourier transform operator):

```
>>F=fourier(heaviside(t-2)) ↵
```

← F⇔ $F[u(t-2)] \Leftrightarrow F(\omega)$

```
F =
exp(-2*i*w)*(pi*dirac(w)-i/w)
```

← Vector code of $e^{-2j\omega}(\pi\delta(\omega)-\frac{j}{\omega})$

Dirac delta or impulse function located at $t=t_0$ on t axis is denoted by $\delta(t-t_0)$ and has the Fourier transform $F(\omega)=e^{-j\omega t_0}$ for example $F[\delta(t-2)]=e^{-j2\omega}$. MATLAB representation for the $\delta(t)$ is dirac(t) and we find the transform for $\delta(t-2)$ as follows:

```
>>fourier(dirac(t-2)) ↵

ans =
exp(-2*i*w)
```
← Code of $e^{-j2\omega}$

Table 5.A presents more examples on the forward Fourier transform and their MATLAB commands maintaining ongoing function and symbology. In example 4 of the table the i means imaginary number j. The maple package (appendix C.5) also keeps the function (example 3 in the table) by the same name and syntax in addition constant declaration facility is provided with the maple one. Two different command lines are separated by a comma in the table.

5.3.2 Inverse Fourier transform

Given a complex frequency function $F(\omega)$, its inverse Fourier transform is defined as $f(t)=F^{-1}[F(\omega)]=\frac{1}{2\pi}\int_{\omega=-\infty}^{\omega=\infty}F(\omega)e^{j\omega t}dt$ where F^{-1} is the inverse Fourier transform operator. MATLAB counterpart of F^{-1} is ifourier (abbreviation for the inverse fourier). To have the inverse Fourier transform of $F(\omega)$, the command we conduct is ifourier($F(\omega)$ in vector string – appendix B, frequency variable ω, inverse transform variable t). Prior to applying the ifourier, declaration of the related variables in $F(\omega)$ as symbolic by employing the command syms is mandatory. The default return from the ifourier is in terms of x. Mathematics readable form can be viewed by the built-in command pretty.

Computational example is best for the concept. It is given that $F(\omega)=\frac{10}{6+5j\omega-\omega^2}$ is the Fourier transform of $f(t)=10[e^{-2t}-e^{-3t}]u(t)$. We find the $f(t)$ from the expression of $F(\omega)$ by using the ifourier as follows:

```
>>syms w t ↵
>>F=10/(6+5*i*w-w^2); ↵
>>f=ifourier(F,w,t); ↵
>>pretty(f) ↵
        10 heaviside(t) (exp(-2 t) - exp(-3 t))
```

- `>>syms w t` ← Defining ω and t, t⇔t and w⇔ω, wanted output in t
- `>>F=10/(6+5*i*w-w^2);` ← Vector code of $F(\omega)$ is assigned to F, F⇔$F(\omega)$, F is user-chosen name
- `>>f=ifourier(F,w,t);` ← Inverse transform of $F(\omega)$ is assigned to f, f⇔$f(t)$, f is user-chosen name
- `>>pretty(f)` ← Display the readable form of f
- `10 heaviside(t) (exp(-2 t) - exp(-3 t))` ← It indicates $10[e^{-2t}-e^{-3t}]u(t)$

In the following we have some forward transform frequency functions. The upper and lower case assignees represent the forward and inverse Fourier transforms respectively (for instance $X(\omega)$⇔forward Fourier transform and $x(t)$⇔inverse Fourier transform). One can easily implement the inverse Fourier transforms of these frequency functions by applying ongoing function and symbology as follows:

Example 1

Angular frequency function: $X(\omega)=\frac{5e^{j(2\omega-8)}}{6-(4-\omega)j}$ has its inverse transform: $x(t)=5\,e^{-6t-12+4jt}u(t+2)$:

Vector code representation of $X(\omega)$ is 5*exp(i*(2*w-8))/(6-(4-w)*i).
Command:

```
>>syms w t ↵                                   ← Defining ω and t as symbolic
>>X=5*exp(i*(2*w-8))/(6-(4-w)*i); ↵ ← Assigning the code of X(ω) to
                                                 workspace X
>>x=ifourier(X,w,t); ↵ ← Inverse transform of X(ω) is assigned to
                                   workspace x
>>pretty(x) ↵              ← Display the readable form of x
        5 heaviside(t + 2) exp(-12 - 6 t + 4 I t)    ← I means
                                                       imaginary number
```

Example 2

Angular frequency function: $H(\omega)=\dfrac{\sin\omega}{\omega}$ has its inverse transform: $h(t)=\dfrac{1}{2}[u(t+1)-\ u(t-1)]$:

Vector code representation of $H(\omega)$ is sin(w)/w.
Command:

```
>>syms w t ↵
>>h=ifourier(sin(w)/w,w,t); ↵ ← Inverse transform of H(ω) is assigned
                                  to workspace h, h⇔ h(t)
>>pretty(h) ↵                     ← Display the readable form of h
    - 1/2 heaviside(t - 1) + 1/2 heaviside(t + 1)
```

Example 3

Angular frequency function: $Y(\omega)=\dfrac{15\sin(4\omega-12)}{\omega-3}$ has its inverse transform: $y(t)=\dfrac{15}{2}\ e^{j3t}\ [\ u(t+4)-u(t-4)\ ]$

Vector string representation of $Y(\omega)$ is15*sin(4*w-12)/(w-3).
Command:

```
>>syms w t ↵
>>Y=15*sin(4*w-12)/(w-3); ↵ ← Assigning the code of Y(ω) to
                                workspace Y
>>y=ifourier(Y,w,t); ↵ ← Inverse transform of Y(ω) is assigned to
                                workspace y, y⇔ y(t)
>>pretty(simple(y)) ↵        ← Display the readable form of y following
                               simplification using the command simple
  15/2 exp(3 I t) (heaviside(t + 4) - heaviside(t - 4))          ← I⇔ j
```

Example 4

$H(\omega)=\begin{Bmatrix} 1 & if\ \omega>0 \\ -1 & if\ \omega<0 \end{Bmatrix}$ is an example when the frequency function is given in graphical form (figure 5.3(b)). Employing the unit step function, one writes $H(\omega)=u(\omega)-u(-\omega)$ whose inverse Fourier transform is $\dfrac{j}{t\pi}$ and following is the implementation:

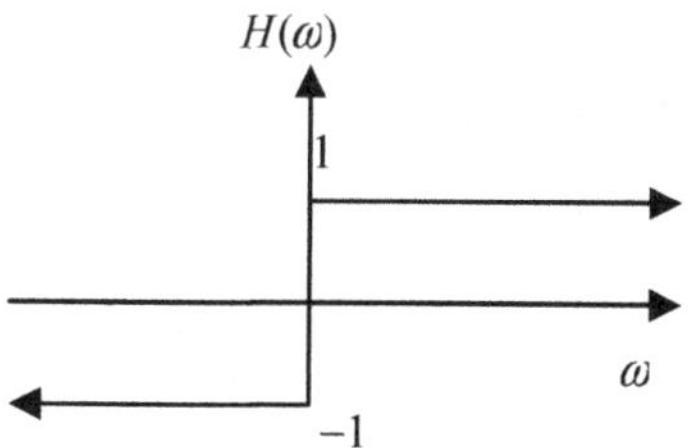

Figure 5.3(b) Plot of $H(\omega)$ vs ω

```
>>syms w t ↵
>>H=heaviside(w)-heaviside(-w); ↵ ← H⇔ H(ω)
>>h=ifourier(H,w,t); ↵              ← h⇔ h(t)
>>pretty(h) ↵                       ← Display the readable form of h, I⇔ j
        I
      -----
      t pi
```

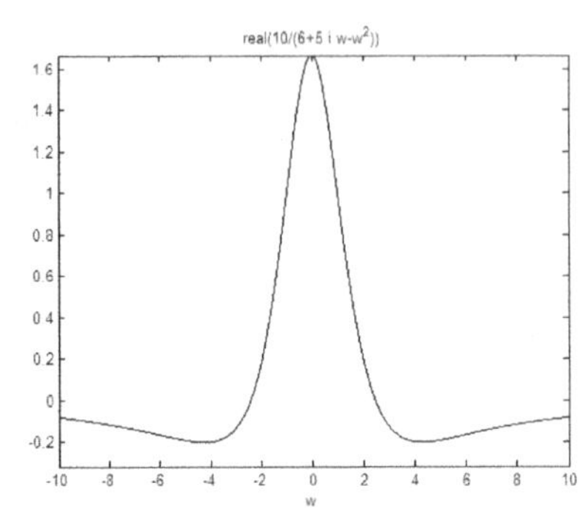

Figure 5.4(a) Plot of $\mathrm{Re}\{F(\omega)\}$ vs ω

Figure 5.4(b) Plot of $\mathrm{Im}\{F(\omega)\}$ vs ω

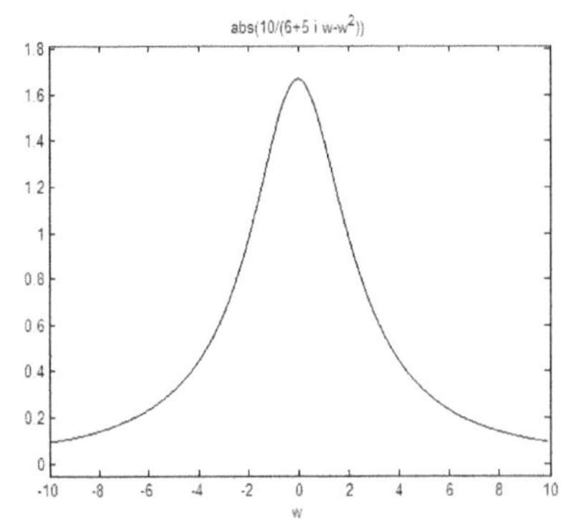

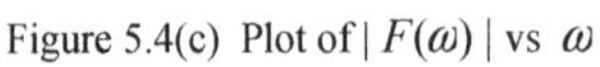

Figure 5.4(c) Plot of $|F(\omega)|$ vs ω

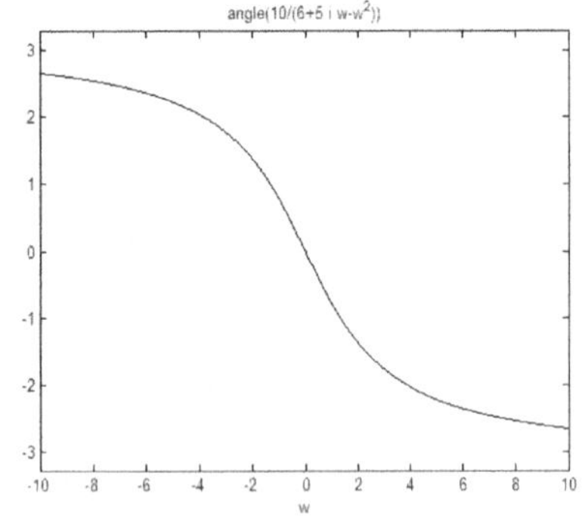

Figure 5.4(d) Plot of $\angle F(\omega)$ vs ω

5.3.3 Graphing the Fourier transform

The forward Fourier transform $F(\omega)$ is a continuous and complex function of ω. It possesses four components – real, imaginary, magnitude, and phase. The variation of these components with respect to ω is termed as the spectrum.

As an example let us find the four spectra of transform $F(\omega)=\dfrac{10}{6+5j\omega-\omega^2}$ over $-10\le\omega\le10$ rad/sec. It is important to mention that although the frequency ω is continuous, we choose substantial discrete points in the given ω interval to graph the $F(\omega)$ in continuous sense. There are two styles for graphing the transform – symbolic and numeric, both of which are addressed separately in the following.

Symbolic style:

In this approach we do not choose the step size. MATLAB automatically chooses the step size. The function that graphs the spectrum is the ezplot (appendix F). The $F(\omega)$ has the vector code 10/(6+5*i*w-w^2). Symbolically the real, imaginary, magnitude, and phase spectra are written as $\mathrm{Re}\{F(\omega)\}$, $\mathrm{Im}\{F(\omega)\}$, $|F(\omega)|$, and $\angle F(\omega)$ and their MATLAB counterparts are real, imag, abs, and angle respectively on that we graph these spectra as follows:

>>ezplot('real(10/(6+5*i*w-w^2))',[-10,10]) ↵ ← For graphing $\mathrm{Re}\{F(\omega)\}$ like figure 5.4(a)

>>ezplot('imag(10/(6+5*i*w-w^2))',[-10,10]) ↵ ← For graphing $\mathrm{Im}\{F(\omega)\}$ like figure 5.4(b)

>>ezplot('abs(10/(6+5*i*w-w^2))',[-10,10]) ↵ ← For graphing $|F(\omega)|$ like figure 5.4(c)

>>ezplot('angle(10/(6+5*i*w-w^2))',[-10,10]) ↵ ← For graphing $\angle F(\omega)$ like figure 5.4(d)

This style is better for quick plotting of a single spectrum.

Numeric style:

The numerical approach needs the scalar code writing of the $F(\omega)$ (appendix B) and the graphing function is the plot (appendix F). Also another important point is the user has to decide the step size or sample point of the ω variation. As a first step we generate a row or column vector data from the user-given ω sample point and range and then calculate $F(\omega)$ or other spectrum at those ω points. If the ω vector is a row or column matrix, so is the calculated spectrum (of identical length). However the graphings of the four spectra for the example transform are shown below:

>>w=-10:0.01:10; ↵ ← We chose the step size 0.01 for ω and w is a row matrix (subsection 1.1.2)

>>F=10./(6+5*i*w-w.^2); ↵ ← Scalar code of $F(\omega)$ is assigned to F and F holds the complex value of $F(\omega)$ as a row matrix for each element in w, F is user-chosen name

>>R=real(F); I=imag(F); A=abs(F); P=angle(F); ↵ ← Picking up the four components from $F(\omega)$ where the assignees are the following: R⇔ $\mathrm{Re}\{F(\omega)\}$, I⇔ $\mathrm{Im}\{F(\omega)\}$, A⇔ $|F(\omega)|$, and P⇔ $\angle F(\omega)$ but all of them is a row matrix of the same size as that of w, assignee names are user-chosen

>>plot(w,R) ↵ ← Plotting the real spectrum $\mathrm{Re}\{F(\omega)\}$ like figure 5.4(a), figure not shown for space reason

In a similar fashion the other three spectra can be plotted by using the commands plot(w,I), plot(w,A), and plot(w,P) respectively. The default phase angle return from the angle is from $-\pi$ to π. If the phase spectrum $\angle F(\omega)$ needs to be in degrees, the command P=180/pi*angle(F); can be exercised.

In the spectrum analysis sometimes we plot the magnitude spectrum $|F(\omega)|$ in decibel (dB) scale which is defined as $20\log_{10}|F(\omega)|$ or $10\log_{10}|F(\omega)|$ depending on $f(t)$ borne information. The decibel needs a reference value. Usually for the magnitude spectrum, it is the maximum value of the $|F(\omega)|$. The workspace variable

A is holding the $|F(\omega)|$ values. For sure the values stored in A are positive. Dividing the $|F(\omega)|$ by the maximum $|F(\omega)|$ normalizes the spectrum between 0 and 1. Theoretically speaking we implement $20\log_{10}\frac{|F(\omega)|}{|F(\omega)|_{max}}$ to obtain the decibel plot on the spectrum. However we find the maximum value in A by applying the command max (appendix C.4) as follows:

>>M=max(A); ↵ ← Workspace M holds the maximum value of $|F(\omega)|$,

M⇔ $|F(\omega)|_{max}$, M is user-chosen

>>D=20*log10(A/M); ↵ ←D holds normalized dB values of $|F(\omega)|$,

D⇔ $20\log_{10}\frac{|F(\omega)|}{|F(\omega)|_{max}}$, D is user-chosen

>>plot(w,D) ↵ ← graphing the decibel plot of $|F(\omega)|$ over $-10 \le \omega \le 10$ rad/sec

Figure 5.4(e) depicts the decibel magnitude spectrum without any labeling. The vertical axis of the graph now represents the decibel values for example 0 dB corresponds to the maximum value 1 in the normalized spectrum or the maximum in the magnitude spectrum of the figure 5.4(c). As you see, the spectrum spreads about the maximum due to the logarithmic scale.

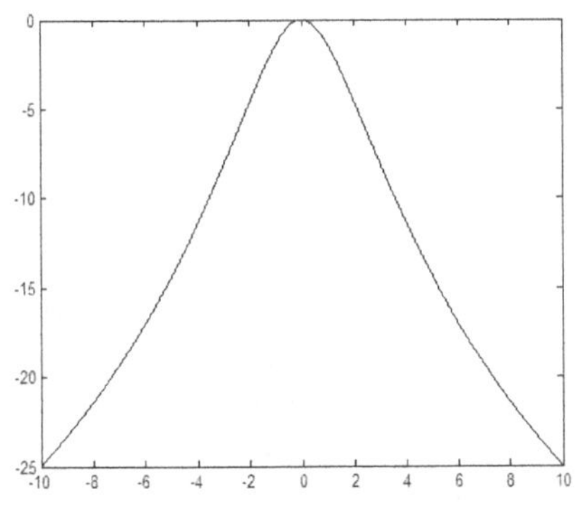

Figure 5.4(e) dB plot of the $|F(\omega)|$

When the decibel value becomes too much negative, it indicates the $|F(\omega)|$ is very small or close to 0. At exactly $|F(\omega)|$=0, the decibel value is minus infinity. For practical calculations we sometimes clip the unnecessary negative values. For example in the graph 5.4(e), the dB values are between 0 and –25. Let us say any dB value less than –20 should be clipped to –20. We know that the dB values are stored in D as a row matrix. We find the integer position indexes of such values in D through the find function (appendix C.3) as follows:

>>p=find(D<=-20); D(p)=-20; ↵

The integer indexes of less than or equal to –20 dB in D are found and assigned to workspace p (user-chosen name). Those index elements are set to –20 by using the command D(p)=-20;. Verify the calculation by exercising plot(w,D).

5.4 Discrete Fourier transform of discrete signals

The discrete Fourier transform (DFT) is merely relevant to a discrete function or sequence $f[n]$. The independent variable n is integer only. The values of $f[n]$ are the samples of a function but the $f[n]$ does not hold any sampling information. For

$$f[n] \xrightarrow{\text{forward discrete Fourier transform}} F[k] \xrightarrow{\text{inverse discrete Fourier transform}} f[n]$$

Figure 5.5(a) The concept behind the discrete Fourier transform different values of n, the $f[n]$ is just a row or column matrix.

Forward transform:

The forward discrete Fourier transform $F[k]$ of the discrete signal $f[n]$ is defined as $F[k]=\sum_{n=1}^{N} f[n]e^{-j2\pi(k-1)\frac{(n-1)}{N}}$ where N is the number of samples in $f[n]$. The transform $F[k]$ is discrete and possesses N samples as well. The discrete frequency variable k is integer.

Inverse transform:

The recovery of $f[n]$ from $F[k]$ takes place through the inverse discrete Fourier transform (IDFT) which is defined as $f[n]=\frac{1}{N}\sum_{n=1}^{N} f[n]e^{j2\pi(k-1)\frac{(n-1)}{N}}$ where the symbols have just mentioned meanings.

The index n or k can vary from 1 to N in the presented expressions. But it can vary from 0 to $N-1$ (most textbook follows this convention) in that case the modified expressions become $F[k]=\sum_{n=0}^{N-1} f[n]e^{-j2\pi k\frac{n}{N}}$ and $f[n]=\frac{1}{N}\sum_{n=0}^{N-1} f[n]e^{j2\pi k\frac{n}{N}}$ for DFT and IDFT respectively. Usage of either expression does not change the signal or its transform contents at all.

The $F[k]$ is in general complex. Figure 5.5(a) presents the concept behind the discrete Fourier transform. When we handle the $f[n]$ of length N being the power of 2, the transform is called the fast Fourier transform. In MATLAB the DFT and IDFT are simulated by using the built-in functions fft (abbreviation for the fast Fourier transform but also handles the sequence whose length is other than power of 2) and ifft (abbreviation for the inverse fast Fourier transform) respectively.

Example on the transform:

Let us consider the discrete signal $f[n]$=[1 2 7 –5 –6 8]. There are 6 samples in the $f[n]$. The choice of index n is the user's. Let us say n varies from 1 to 6, so does k. The expression $F[k]=\sum_{n=1}^{N} f[n]e^{-j2\pi(k-1)\frac{(n-1)}{N}}$ yields the computation of $F[k]$ as [7 $10.5-j6.0622$ $-9.5+j16.4545$ -3 $-9.5-j16.4545$ $10.5+j6.0622$]. Our objective is to obtain the complex sequence $F[k]$ from $f[n]$ and we do so as follows:

>>f=[1 2 7 -5 -6 8]; ↵ ← $f[n]$ is assigned to the workspace f as a row matrix, f is user-chosen name

>>F=fft(f) ↵ ← DFT on $f[n]$ is taken and assigned to the workspace F, F⇔ $F[k]$, F is user-chosen name

```
F =
    7.0000  10.5000 - 6.0622i  -9.5000 +16.4545i  -3.0000  -9.5000 -16.4545i  10.5000 + 6.0622i
```

Starting from the computed values of $F[k]$, the expression $f[n]=\frac{1}{N}\sum_{n=0}^{N-1} f[n]e^{j2\pi k\frac{n}{N}}$ brings the signal $f[n]$ back. Having the discrete forward transform values available in the workspace F, one can verify that as follows:

>>f=ifft(F) ↵ ← IDFT on $F[k]$ is taken and assigned to the workspace f, f⇔ $f[n]$, f is user-chosen name

```
f =
      1.0000   2.0000   7.0000  -5.0000  -6.0000   8.0000
```

Component separation:

The plot of the $F[k]$ versus integer k is termed as the discrete Fourier spectrum. Since the $F[k]$ is complex, there are four discrete component spectra namely real, imaginary, magnitude, and phase whose symbolic representations are $\text{Re}\{F[k]\}$, $\text{Im}\{F[k]\}$, $|F[k]|$, and $\angle F[k]$ and whose MATLAB counterparts are real, imag, abs, and angle respectively. For instance the discrete phase angle spectrum (in radians) $\angle F[k]$ can be picked up from $F[k]$ by employing the command angle(F).

More examples:

Let us see two more examples on DFT computation in the following.

Example 1

The beginning example presents the $f[n]$ to be formed from some observational data. A discrete signal may follow some functional pattern for example sine.

Let us say the discrete sine signal is $f[n]=5\sin\frac{2\pi n}{5}$ where n is integer and n exists from 0 to 5 thus making $f[n]$ discrete. We intend to find the discrete magnitude and phase spectra of the finite duration discrete sine signal.

Applying the $F[k]=\sum_{n=0}^{N-1} f[n]e^{-j2\pi k\frac{n}{N}}$, one obtains $|F[k]|=$ [0 12.4495 5.0904 3.6327 5.0904 12.4495] and $\angle F[k]$=[180^0 -60^0 150^0 180^0 -150^0 60^0] for which the implementation is as follows:

>>n=0:5; ↵ ← Placing the index n as a row matrix to the workspace n, n is user-chosen assignee

>>f=5*sin(2*pi*n/5); ↵ ← Scalar code (appendix B) of $5\sin\frac{2\pi n}{5}$ is assigned to f, f⇔discrete $f[n]$

>>F=fft(f); ↵ ← DFT on $f[n]$ is taken and workspace F holds $F[k]$

>>A=abs(F) ↵ ← A holds $|F[k]|$ as a row matrix, A is user-chosen assignee

```
A =
      0.0000   12.4495   5.0904   3.6327   5.0904   12.4495
```

>>P=180/pi*angle(F) ↵ ← P holds $\angle F[k]$ in degrees, default return of angle is in radian from $-\pi$ to π, P is user-chosen assignee

```
P =
      180.0000 -60.0000 150.0000 180.0000 -150.0000   60.0000
```

Example 2

Suppose we wish to find the same discrete spectra as in example 1 but for the discrete function $f[n]=2^{-n}$ on the same n duration. Only change in previous commands do we need is in the scalar code of $f[n]$ which is 2.^(-n).

5.4.1 Graphing the discrete Fourier transform

Graphing the discrete Fourier transform happens by the use of the discrete function plotter **stem** (appendix F). Referring to the beginning example (the $f[n]$ with six samples) of $F[k]$ computation, the workspace variable F holds the $F[k]$ for the data based $f[n]$. We wish to plot $\text{Re}\{F[k]\}$ versus k for this $f[n]$ i.e. real discrete spectrum of $f[n]$.

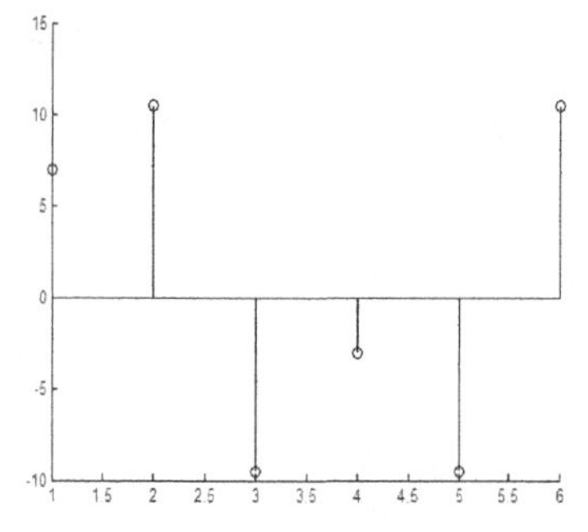

Figure 5.5(b) Plot of the discrete real spectrum $\text{Re}\{F[k]\}$ versus k

Let us obtain the discrete real spectrum from the complex values stored in F as follows:

```
>>R=real(F) ↵

R =
        7.0000   10.5000   -9.5000   -3.0000   -9.5000   10.5000
```

In above execution the R holds $\text{Re}\{F[k]\}$ values as a row matrix where R is a user-chosen assignee. In order to plot the real spectrum, the integer index k needs to be known. The k changes from 1 to 6 therefore the graphing is as follows:

```
>>k=1:6; ↵        ← k holds the index integers as a row matrix
>>stem(k,R) ↵     ← first and second input arguments are index and real spectrum
                    both as a row matrix
```

Above execution results in the figure 5.5(b). One can execute the commands stem(k,imag(F)), stem(k,abs(F)), and stem(k,angle(F)) for the discrete spectra $\text{Im}\{F[k]\}$, $|F[k]|$, and $\angle F[k]$ respectively.

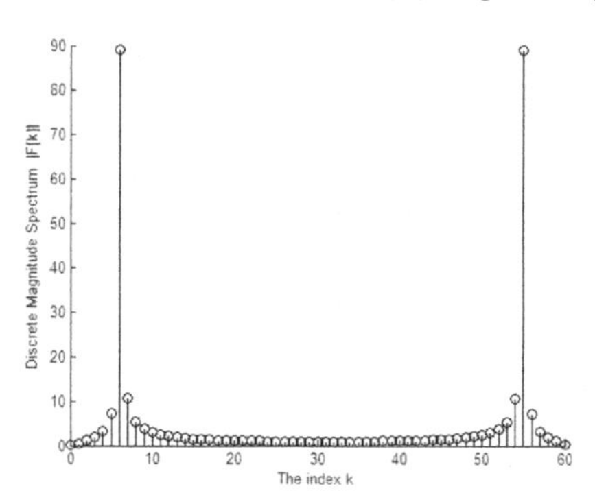

Figure 5.5(c) Discrete magnitude spectrum of a pure discrete sine wave

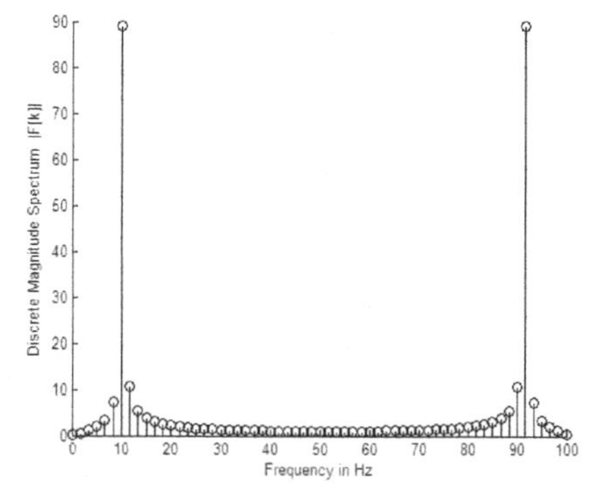

Figure 5.5(d) Discrete magnitude spectrum of the pure discrete sine wave plotted against frequency

5.4.2 DFT implications on discrete sine signal

The sine function is the most addressed one in the electrical engineering transform-based systems because of its widespread involvement. In discrete analysis, always we are focused on the sample. Behind the discrete terminology, a physical system is often continuous. Recall that $f[n]$ is just the samples taken at different index or integer n. Samples do not reveal actual system until we relate them with the sampling information. We know that the discrete function $f[n]$ is generated after sampling the function $f(t)$ with a specific frequency, called sampling frequency f_S. Or in other words the function $f(t)$ is sampled with a step size T_s where $T_s = \frac{1}{f_s}$. Now we demonstrate what allied implication holds the sampling frequency regarding the discrete Fourier transform.

Let us consider a single frequency sine function of $\begin{Bmatrix} amplitude \pm 3 \\ frequency\ 10Hz \end{Bmatrix}$ when t is in second and which has the continuous expression $x(t) = 3\sin 2\pi 10t$ and exists over $0 \le t \le 0.6$sec. The user has to decide the sampling frequency to discretize this continuous function.

Let us say the sampling frequency is f_S=100 Hz with that the sampling period or the step size must be $T_s = \frac{1}{f_s}$=0.01sec which implies that we must choose the t step as 0.01sec while discretizing the $x(t)$. The moment we choose the t step size, we take the samples of the continuous function. Let us conduct the following:

>>t=0:0.01:0.6; ↵ ← t holds a row matrix whose elements are points on t at which the samples to be taken

>>x=3*sin(2*pi*10*t); ↵ ← x holds the discrete sine wave as a row matrix in which t information is lost

The x corresponds to the DFT theory mentioned $f[n]$. In order to decide the N, the number of elements in x is to be found as follows:

>>numel(x) ↵ ← The function numel finds the number of elements in the matrix x

```
ans =
        61
```

It indicates that we must take N=61 in the DFT computation thereby changing the index n or k from 0 to 60.

An interesting fact is revealed from the following implementation. Let us graph (subsection 5.4.1) the discrete magnitude spectrum $|F[k]|$ against the index k as follows:

>>k=0:60; ↵ ← k holds the DFT mentioned indexes from 0 to 60 as a row matrix

>>A=abs(fft(x)); ↵ ← A holds the discrete magnitude spectrum $|F[k]|$ for ongoing discrete sine function

>>stem(k,A) ↵ ← Graphing the magnitude spectrum

Figure 5.5(c) presents the discrete magnitude spectrum for the sine wave. The DFT has the half index symmetry that is $|F[k]|$ is symmetric about $\frac{N-1}{2}$ or k=30. For k=0 to 30, there is only one strong discrete peak which corresponds to the frequency of the continuous sine function $x(t)$ we started with.

Knowing the sampling frequency, it is also convenient to display the discrete magnitude plot in terms of the discrete frequency f (not $f[n]$) rather than the sample index k. The relationship is given by $f = \frac{f_S}{N-1}k$. For ongoing example, we have $\begin{Bmatrix} f_s = 100Hz \\ N = 61 \\ k = 0 \; to \; 60 \end{Bmatrix}$ thereby providing $f = \frac{100}{60}k$ hence the f changes from 0 to 100 Hz with a step $\frac{100}{60}$ Hz. Now we plot the horizontal axis of the figure 5.5(c) in terms of the discrete frequency as follows:

```
>>f=0:100/60:100; ↵        ← f holds the discrete frequencies as a row matrix
>>stem(f,A) ↵              ← The first input argument of the stem is now the
                             discrete frequency row matrix
```

Above execution results the graph of the figure 5.5(d) in which you see the horizontal axis in terms of frequency in Hertz. In the plot exactly at f =10 does the peak magnitude appear – which is the signal frequency of $x(t)$ we started with.

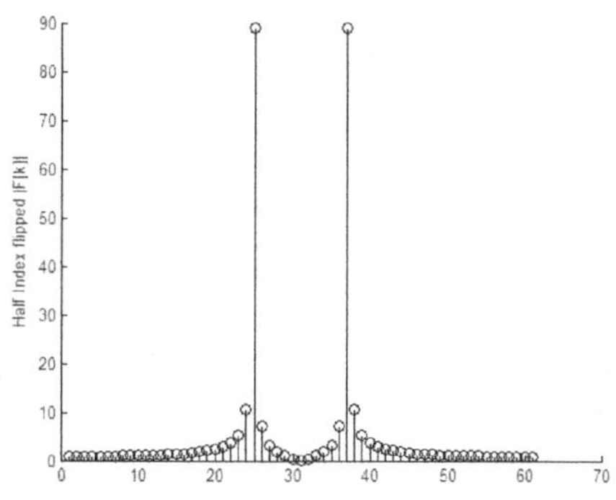

Figure 5.5(e) Half indexedly flipped discrete magnitude spectrum of a pure discrete sine wave

⬍⬍ Inference from the study

Suppose we do not have any information about the sinusoidal frequency present in a given discrete function. We can start with some sampling frequency and look for the strongest peak as done in the figure 5.5(d). If it does not appear, we can try with another sampling frequency. Thus the hidden frequency present in a discrete function can be detected.

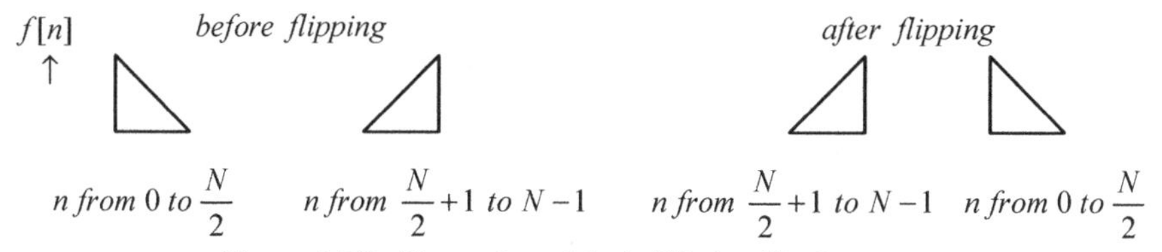

Figure 5.5(f) Illustration of the half index flipping

5.4.3 Half index flipping of the DFT

The discrete magnitude spectrum $|F[k]|$ of $f[n]$ is related with information content in the continuous counterpart $f(t)$ that is why it receives ample attention in the Fourier literature. From ongoing discussion, the $|F[k]|$ has half index symmetry when the total number of samples in it is N. Not only the magnitude one but the other spectra may also have even or odd symmetry about the half index.

If a signal is composed of many frequencies, the significant frequency components of $|F[k]|$ are located at the smaller and larger values of k (for example k =0, 1, 2,... and $k = N-1$, $N-2$,...). Flipping the $|F[k]|$ about the half index gathers significant frequencies in the middle of the k axis which exhibits more perceptibleness in frequency in the $|F[k]|$ versus k.

Considering a discrete function $f[n]$ whose integer index n varies from 0 to $N-1$ and where N is the number of samples in the $f[n]$. Flipping of the $f[n]$

about its half index $\frac{N}{2}$ is graphically interpreted in figure 5.5(f). The MATLAB built-in function fftshift helps us obtain half index flipping as illustrated in the figure 5.5(f).

Let us say we have the odd and even sample number discrete signals $x[n]$ = [10 2 1 1 2] and $y[n]$=[10 2 1 1 2 10] respectively. Following the half index flipping, one should obtain the flipped signals as [1 2 10 2 1] and [1 2 10 10 2 1] for $x[n]$ and $y[n]$ respectively whose implementations are shown below:

Flipping for the odd number sample,

```
>>x=[10 2 1 1 2]; ↵
>>F=fftshift(x) ↵

F =

          1   2  10   2   1
```

Flipping for the even number sample,

```
>>y=[10 2 1 1 2 10]; ↵
>>F=fftshift(y) ↵

F =

          1   2  10   10   2   1
```

In either case in last execution the F holds the flipped result as a row matrix where F is user-chosen name. We know that any discrete one dimensional function takes the form of a row or column matrix. Working on a row or column matrix in a sense is working on the discrete function. Concerning the fftshift if the number of samples in a discrete function is odd, the half index flipping happens about the index $k=\frac{N-1}{2}$ assuming k changes from 0 to $N-1$. Elements whose indexes are from $k=\frac{N+1}{2}$ to $k=N-1$ are placed in front of the other half. If N is even, elements whose indexes are from $k=\frac{N}{2}$ to $k=N-1$ are placed in front of the other. Two times half index flipping brings back the discrete signal we start with provided that the number of samples in the signal is even. That is to say, the fftshift(fftshift(y)) returns y. Let us see one example on the half index flipping.

Example

From the subsection 5.4.2, the last workspace variable A (re-execute the commands until you get the A) holds the discrete magnitude spectrum for the single frequency sine signal whose plot is presented in the figure 5.5(c) and in which you find the two strong peaks are located close to k=0 and k=60. The half index flipping operation should bring these two peaks in the middle of the k axis for which we conduct the following:

```
>>F=fftshift(A); ↵ ← The return is assigned to F, where F is user-chosen
>>stem(F) ↵        ← Graphs the figure 5.5(e)
```

The workspace F holds the half indexedly flipped $|F[k]|$. As the figure 5.5(e) displays, the two strong peaks are brought in the middle. Note that we used only the flipped spectrum as the input argument of the stem. The fftshift does not hold the k information. It is the user who keeps the information about the index k. By default the stem numbers the horizontal indexes from 1 to the number of samples present in A (which is here 61). If the information about the k is necessary, use fftshift on k and assign the fftshift(k) to some workspace variable.

We close the Fourier discussion with this article.

Chapter 6
LAPLACE TRANSFORM

Introduction

In signal and system course Laplace transform receives special attention. Many electrical systems are preferably defined in Laplace domain rather than in time domain. Analogous to Fourier transform the transform also shares the notion of forward-inverse terminology. Some of the Fourier transform limitations are overcome by using the Laplace transform. MATLAB is also equipped with the appropriate functions to handle the problems related to the transform. We are planning to show how the code writing of MATLAB turns clumsy mathematical manipulation to simplistic symbolic obtaining for:

- Forward transform of a continuous signal with varieties
- System function finding from governing equations
- Inverse Laplace transform from system function

6.1 Laplace transform

Laplace transform is also integral transform like the Fourier transform. The finiteness condition of the Fourier transform (chapter 5) imposes the limitation that the function must converge as t increases in the positive and negative directions thereby turning it unsuitable for all functions. If the $f(t)$ is not convergent, multiplying the function by the factor e^{-st} forces the function to be convergent yet not all. However Laplace transform also has the forward and inverse counterparts. Figure 6.1(a) depicts the analysis – synthesis notion behind the transform.

$$\begin{array}{l} continuous\ f(t) \xrightarrow{forward\ Laplace\ transform} F(s) \xrightarrow{inverse\ Laplace\ transform} f(t) \\ and\ \ t \geq 0 \end{array}$$

Figure 6.1(a) The concept behind the Laplace transform

6.2 Forward Laplace transform of continuous functions

Forward Laplace transform $F(s)$ of a continuous function $f(t)$ is defined as $L\,[f(t)] = F(s) = \int_{t=0}^{t=\infty} f(t)e^{-st}dt$ where $t > 0$, L is the forward transform operator, and s is the transform variable. MATLAB function that computes the transform is laplace. The common syntax of the function is laplace($f(t)$ in vector code form – appendix B, independent variable t, wanted transform variable s). Since the computation happens through symbolic toolbox, declaration of related variables in $f(t)$ by using the syms is compulsory before applying the laplace. The default return from the laplace is a function of s. Readable form from the return of laplace is seen by using the command pretty.

Let us begin with the forward Laplace transform of e^{at} where $t > 0$. We have $L\,[\,e^{at}\,] = F(s) = \int_{t=0}^{t=\infty} e^{at}\,e^{-st}dt = \dfrac{1}{s-a}$. Our objective is to obtain $\dfrac{1}{s-a}$ from e^{at} for which we conduct the following:

```
>>syms a t ↵                  ← Declaring the related variables a and t of e^at
                                      as symbolic, a⇔a and t⇔t
>>F=laplace(exp(a*t)) ↵       ← Vector code of e^at is exp(a*t) and F holds the
                                      code of F(s), F is any user-chosen assignee
F =

1/(s-a)
>>pretty(F) ↵                 ← Display the readable form stored in F
          1
       ------
        s - a
```

We conducted above commands in three lines, we could have executed all three commands in one line but separated by a comma like syms a t, F= laplace(exp(a*t)); pretty(F) (for the use of ;, see subsection 1.1.2). The transform by making specific choice of independent variable (x) and transform variable (z) is implemented by using the commands syms a x z first and laplace(exp(a*x),x,z) afterwards which returns $\dfrac{1}{z-a}$.

Employing the same function, symbology, and notation, table 6.A presents the Laplace transforms for a number of continuous functions. Function like e^{2t^2} or t^t does not provide a close form function of s therefore the return becomes laplace(exp(2*t^2),t,s) meaning just the definition of the transform. Sometimes a mathematical expression may exist in different forms in that case the command

Table 6.A Laplace transforms of various standard functions and their MATLAB implementations

Mathematical form	MATLAB commands
$L\ [\sin bt] = \dfrac{b}{s^2 + b^2}$	>>syms b t, F=laplace(sin(b*t)); ↵ >>pretty(F) ↵ b ------------ 2 2 s + b
$L\ [\sinh bt] = \dfrac{b}{s^2 - b^2}$	>>syms b t, F=laplace(sinh(b*t)); ↵ >>pretty(F) ↵ b ------------ 2 2 s - b
$L\ [\cos bt] = \dfrac{s}{s^2 + b^2}$	>>syms b t, F=laplace(cos(b*t)); ↵ >>pretty(F) ↵ s ------------- 2 2 s + b
$L\ [\cosh bt] = \dfrac{s}{s^2 - b^2}$	>>syms b t, F=laplace(cosh(b*t)); ↵ >>pretty(F) ↵ s ------------- 2 2 s - b
$L\ [e^{at} \cos bt] = \dfrac{s-a}{(s-a)^2 + b^2}$	>>syms a b t ↵ >>F=laplace(exp(a*t)*cos(b*t)); ↵ >>pretty(F) ↵ s - a ----------------- 2 2 (s - a) + b
$L\ [e^{at} \sin bt] = \dfrac{b}{(s-a)^2 + b^2}$	>>syms a b t ↵ >>F=laplace(exp(a*t)*sin(b*t)); ↵ >>pretty(F) ↵ b ----------------- 2 2 (s - a) + b
$L\ [u(t)] = \dfrac{1}{s}$	>>syms t, F=laplace(heaviside(t)); ↵ >>pretty(F) ↵ 1/s
$L\ [\delta(t)] = 1$	>>syms t, F=laplace(dirac(t)); ↵ >>pretty(F) ↵ 1
$L\,[\,Si(t)\,] = \dfrac{1}{s}\cot^{-1} s$ where $Si(t) = \int_{x=0}^{x=t} \dfrac{\sin x}{x} dx$ is the sine integral	>>syms t ↵ >>F=laplace(sinint(t)); ↵ >>pretty(F) ↵ acot(s) --------- s

Continuation of the table 6.A:

Mathematical form	MATLAB commands
$L[Ci(t)] = -\frac{\ln(s^2+1)}{2s}$ where $Ci(t) = \int_{x=\infty}^{x=t} \frac{\cos x}{x} dx$ is the cosine integral	>>syms t ↵ >>F=laplace(cosint(t)); ↵ >>pretty(F) ↵ 2 log(s + 1) - 1/2 ------------- s
$L[-7+2t+t^4] = \frac{-7s^4+2s^3+24}{s^5}$	>>syms t, F=laplace(-7+2*t+t^4); ↵ >>pretty(simplify(F)) ↵ 4 3 7 s - 2 s - 24 - ------------------ 5 s
$L[t\sin at] = \frac{2sa}{(s^2+a^2)^2}$	>>syms a t, F=laplace(t*sin(a*t)); ↵ >>pretty(F) ↵ s a 2 -------------- 2 2 2 (s + a)
$L\left[\frac{\sinh mt}{t}\right] = \frac{1}{2}\ln\frac{s+m}{s-m}$	>>syms m t, F=laplace(sinh(m*t)/t); ↵ >>pretty(F) ↵ s + m 1/2 log(---------) s - m
$L[t^{\frac{3}{2}}e^{t-1}] = \frac{3e^{-1}\sqrt{\pi}}{4(s-1)^{\frac{5}{2}}}$	>>syms t, F=laplace(t^(3/2)*exp(t-1)); ↵ >>pretty(F) ↵ 1/2 exp(-1) pi 3/4 --------------- 5/2 (s - 1)
$L\left[\frac{1}{\sqrt{\pi(t+a)}}\right] = \frac{e^{as}erfc(\sqrt{as})}{\sqrt{s}}$ where $erfc(x)$ is complementary error function	>>syms a t, F=laplace(1/sqrt(pi*(t+a))); ↵ >>pretty(F) ↵ 1/2 exp(a s) erfc((a s)) ----------------------------- 1/2 s
$L[J_0(at)] = \frac{1}{\sqrt{s^2+a^2}}$ where $J_0(x)$ is the Bessel function of the first kind of order 0	>>syms a t, F=laplace(besselj(0,a*t)); ↵ >>pretty(F) ↵ 1 -------------------- 2 2 1/2 (s + a)

simplify can be applied to simplify the expression as conducted for $L[-7+2t+t^4]$ in the table. The mathematical functions $u(t)$, $\delta(t)$, $J_0(at)$, $erfc(x)$, $Ci(t)$, and $Si(t)$ have the MATLAB symbolic counterparts heaviside(t), dirac(t), besselj(0,a*t), erfc(x), cosint(t), and sinint(t) as applied in the table respectively. Shifted unit step and Dirac delta functions are also computable in a similar fashion for example $L[u(t-2)] = \frac{e^{-2s}}{s}$ or $L[\delta(t-2)] = e^{-2s}$.

6.3 Laplace transform of graphical functions

When Laplace transform on graphical functions is needed, we first write the expression of the graphical function in terms of the standard functions of last section and then apply the transform. We illustrate following three examples in this regard.

Example 1

Given that the finite pulse $f(t)$ of the figure 6.1(b) has the $F(s)=\frac{1}{s}-\frac{e^{-3s}}{s}$ which we intend to obtain.

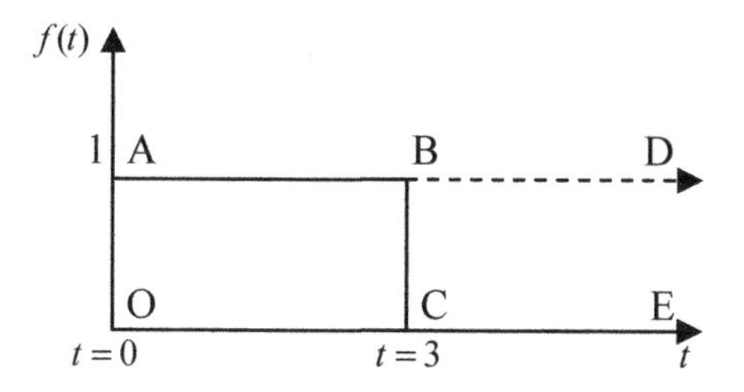

Figure 6.1(b) A finite rectangular pulse

The pulse $f(t)$ of figure 6.1(b) is expressed in terms of the unit step function $u(t)$ and its shifted version i.e. $f(t)$=OABC= OADE−ECBD=$u(t)-u(t-3)$ therefore we apply the section 6.2 cited terminology and function as follows:

```
>>syms t ↵                                  ← Declaring the t as symbolic, t⇔t
>>f=heaviside(t)-heaviside(t-3); ↵          ← Defining the f(t), f⇔f(t)
>>F=laplace(f); ↵                  ← Transform on f, F⇔F(s), F holds code of F(s)
>>pretty(F) ↵                      ← Displaying the readable form of F
          exp(-3 s)
    1/s - -----------
              s
```

Example 2

$F(s)=\frac{2}{s}-\frac{4e^{-2s}}{s}+\frac{2e^{-4s}}{s}$ is the Laplace transform of the square pulse $g(t)$ in figure 6.1(c) which we plan to verify.

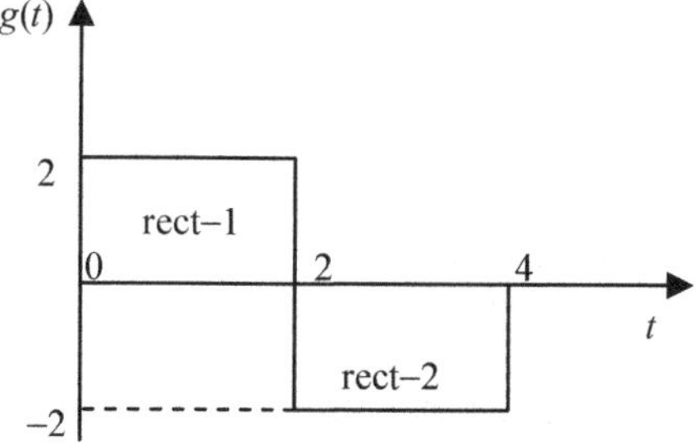

Figure 6.1(c) A square pulse

The $g(t)$ is composed of rect−1 and rect−2 as indicated in the figure 6.(c). By equation we have $g(t)$=rect−1+rect−2= $2[u(t)-u(t-2)]-2[u(t-2)-u(t-4)]=2u(t)-4u(t-2)+2u(t-4)$ therefore as we have been exercising:

```
>>syms t ↵                                          ← Declaring the t as symbolic, t⇔t
>>g=2*heaviside(t)-4*heaviside(t-2)+2*heaviside(t-4); ↵   ← Defining g(t),
                                                                    g⇔g(t)
>>F=laplace(g); ↵                  ← Transform on g, F⇔F(s), F holds code of F(s)
>>pretty(F) ↵                      ← Displaying the readable form of F

            exp(-2 s)     exp(-4 s)
    2/s - 4 ----------- + 2 -----------
                s             s
```

Example 3

The triangular pulse of figure 6.1(d) has $H(s) = \frac{1}{s^2} - \frac{2e^{-3s}}{s^2} + \frac{e^{-6s}}{s^2}$ which we wish to verify.

The $h(t)$ of the figure 6.1(d) is constructed as $h(t)=$ triangle 0B3+triangle 3B6. Equations of the straight lines 0BA and B6C are $h(t)=t$ and $h(t)=-t+6$ respectively. The triangle 0B3 is formed by $t[u(t)-u(t-3)]$ and the 3B6 is formed by $(6-t)[u(t-3)-u(t-6)]$ so $h(t)= t[u(t)-u(t-3)]+(6-t)[u(t-3)-u(t-6)]$ $= t\,u(t)+(6-2t)u(t-3)-(6-t)u(t-6)$. We have it executed as follows by exercising ongoing symbology and function:

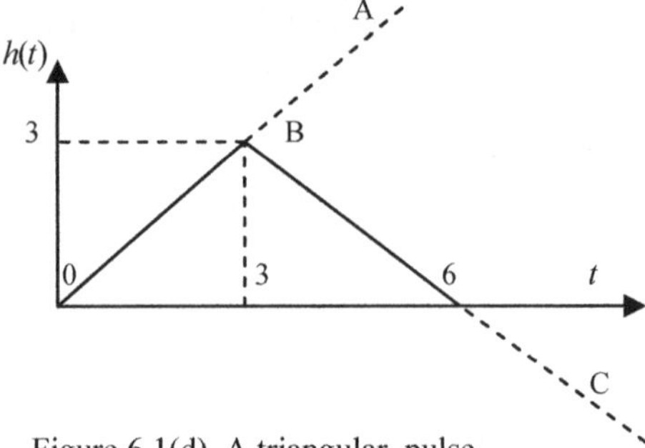

Figure 6.1(d) A triangular pulse

```
>>syms t ↵                  ← Declaring the t as symbolic, t⇔t
>>h=t*heaviside(t)+(6-2*t)*heaviside(t-3)-(6-t)*heaviside(t-6); ↵ ← Defining
                                                         h(t), h⇔h(t)
>>H=laplace(h); ↵           ← Transform on h, H⇔H(s), H holds code of H(s)
>>pretty(H) ↵               ← Displaying the readable form of H
      1        exp(-3 s)    exp(-6 s)
   ------- - 2 ----------- + -----------
      2            2             2
     s            s             s
```

These techniques can be extended for other finite functions in a similar fashion for example that follow the sinusoid or exponential variation over some interval.

6.4 Laplace transform of differential coefficients

Laplace transforms of different derivatives of a continuous signal $y(t)$ are given as follows: $L\left(\frac{dy}{dt}\right)=sY(s)-y(0)$, $L\left(\frac{d^2y}{dt^2}\right)=s^2Y(s)-sy(0)-y'(0)=s[sY(s)-y(0)]-y'(0)$, $L\left(\frac{d^3y}{dt^3}\right)=s^3Y(s)-s^2y(0)-sy'(0)-y''(0)=s[s[sY(s)-y(0)]-y'(0)]-y''(0)$ in general $L\left(\frac{d^ny}{dt^n}\right)=s^nY(s)-s^{n-1}y(0)-s^{n-2}y'(0)-.............-sy^{n-2}(0)-y^{n-1}(0)$ where $Y(s)$ is the Laplace transform of $y(t)$. $Y(s)$ is equivalent to laplace(y(t),t,s) in MATLAB terminology. The dependent variable y is a function of t and is entered by exactly writing the command y=sym('y(t)'). The nth derivative of y that is

$\frac{d^n y(t)}{dt^n}$ is written as diff(y,n) where the command diff is the differential operator $\frac{d}{dt}$ and in which the first and second input arguments of diff are the dependent variable and order of the derivative respectively.

To exemplify the computation, Laplace transform of the 4th order derivative is given by $L\left[\frac{d^4 y(t)}{dt^4}\right] = s[s[s[sY(s) - y(0)] - y'(0)] - y''(0)] - y'''(0)$ and we implement that as follows:

```
>>y=sym('y(t)'); ↵            ← y holds y(t)
>>F=laplace(diff(y,4)) ↵ ← Workspace F holds the whole string for the transform of
                              the derivative, F is any user-supplied assignee
F =

s*(s*(s*(s*laplace(y(t),t,s)-y(0))-D(y)(0))-@@(D,2)(y)(0))-@@(D,3)(y)(0)
```

By virtue of the commands simplify and pretty, you can even view the readable form on execution of pretty(simplify(F)) at the command prompt.

We did not describe the entering syntax of the initial conditions so far. Their MATLAB analogue strings are $y(t)\big|_{t=0} = y(0) \Leftrightarrow$ y(0), $\frac{dy(t)}{dt}\bigg|_{t=0} = y'(0) \Leftrightarrow$ D(y)(0), $\frac{d^2 y(t)}{dt^2}\bigg|_{t=0} = y''(0) \Leftrightarrow$ `@@`(D,2)(y)(0), $\frac{d^3 y(t)}{dt^3}\bigg|_{t=0} = y'''(0) \Leftrightarrow$ `@@`(D,3)(y)(0), and so are the other order derivatives. The characters in the string of any particular derivative are consecutive, there is no blank space between the characters. Note that the higher order derivatives more than one hold `@@` in which the ` is a character from the keyboard and is different from the transpose operator. In the string of F, the character ` is not displayed but it must be present when the user wants to access the derivative string.

Let us find the Laplace transform of $7\frac{d^3 y}{dt^3} - 5\frac{dy}{dt} + 3y$ when $y(0) = 9$, $y'(0) = 3$, and $y''(0) = 2$. Following the substitution and simplification, one should get $L\left(7\frac{d^3 y}{dt^3} - 5\frac{dy}{dt} + 3y\right) = [7s^3 - 5s + 3]Y(s) - 63s^2 - 21s + 31$. We intend to obtain this in MATLAB and the implementation is as follows:

```
>>y=sym('y(t)'); ↵              ← y holds the dependent variable of derivatives, y(t)
>>F=laplace(7*diff(y,3)-5*diff(y,1)+3*y); ↵ ← F holds the whole string for the
                                              transform of the derivative
```

In just conducted execution the derivative $7\frac{d^3 y}{dt^3}$ is coded by 7*diff(y,3), so is the others. From earlier discussion we know that the string laplace(y(t),t,s) is equivalent to $Y(s)$ and is a part of the string stored in F. We intend to replace the string (because it is too long) by single Y but we are dealing with the symbolic toolbox functions so the Y needs to be mentioned as symbolic that is why we declare the Y as follows:

```
>>syms Y ↵
```

With the help of the built-in function subs now we substitute various strings as follows:

```
>>S=subs(F,{'laplace(y(t),t,s)','y(0)','D(y)(0)','`@@`(D,2)(y)(0)'},{Y,9,3,2}) ↵

S =

7*s*(s*(s*Y-9)-3)+31-5*s*Y+3*Y
              ↑
```

rearranged $[7s^3-5s+3]Y(s)-63s^2-21s+31$ we are after

The subs used for substitution inherits different options. The subs employs here three input arguments – the first, second, and third of which are the string to be substituted which is here the transform string F, parts of the string which are to be replaced under single inverted comma, and the new values of the various parts in the string respectively. The second and third input arguments of the subs are placed within the second brace {.. }. Inside the second brace individual string components are separated by commas. The placement of the strings in the second and third input arguments of the subs must take place in order. For example here the order we have is $Y(s)\rightarrow y(0)\rightarrow y'(0)\rightarrow y''(0)$ both in the second and the third input arguments of the subs.

The applied string equivalence is as follows: laplace(y(t),t,s)⇔Y⇔ $Y(s)$, y(0)⇔ $y(0)$, D(y)(0)⇔ $y'(0)$, and `@@`(D,2)(y)(0)⇔ $y''(0)$. However the workspace S holds the transform string following the initial value substitution where the S is a user-chosen variable name.

Finding system function from a differential equation

Occasionally setting the transformed derivative expression to zero, only is needed the $Y(s)$ as a function of the transform variable s.

For example we found the transform string for $7\frac{d^3y}{dt^3}-5\frac{dy}{dt}+3y$ just now as $[7s^3-5s+3]Y(s)-63s^2-21s+31$, setting that to 0 provides $Y(s)=\frac{63s^2+21s-31}{7s^3-5s+3}$ which we intend to obtain.

That is to say we have to solve the expression stored in S for Y with the aid of function solve (appendix D) as follows:

```
>>T=solve(S,Y); ↵     ← T holds the string for solution, T is any user-chosen name
>>pretty(T) ↵         ← Display the readable form of the string stored in T

          2
    63 s  + 21 s - 31
    -------------------        ← The Y(s) we are looking for
          3
       7 s  - 5 s + 3
```

In above implementation the first and second input arguments of the solve are the equation expression and the variable for which it is to be solved respectively.

6.5 Laplace transform of integrals

Laplace transform of an integral type signal for example $L\left[\int_{x=0}^{x=t} f(x)\,dx\right] = \frac{F(s)}{s}$ is also obtainable where $F(s)$ is the Laplace transform of the continuous signal $f(t)$. Here the integration variable and upper limit variable both are declared by the built-in command syms before the transform as follows:

```
>>syms x t ↵
```
← Declaring the x and t as symbolic, x⇔x and t⇔t

"The function $f(x)$ is a function of x" is entered by exactly writing the command sym('f(x)') and which we assign to f (can be any user-chosen name) again as follows:

```
>>f=sym('f(x)'); ↵
```
← Declaring the $f(x)$ as symbolic and f⇔ $f(x)$

Appendix G cited int codes the $\int_{x=0}^{x=t} f(x)dx$ as int(f,x,0,t) so the transform is taken by the function of section 6.2 as follows:

```
>>L=laplace(int(f,x,0,t)) ↵
Warning: Explicit integral could not be found.
> In sym.int at 58
L =

laplace(f(t),t,s)/s
```

In above implementation the workspace L holds the $L\left[\int_{x=0}^{x=t} f(x)\,dx\right]$ where L is any user-chosen name. A warning message appears because no function is assigned to $f(x)$ which we ignore. Anyhow the output is laplace(f(t),t,s)/s i.e. $\frac{F(s)}{s}$ with the definition $F(s)$ ⇔laplace(f(t),t,s).

6.6 Laplace transform of integrodifferential equations

An integrodifferential equation is composed of both the differential $\frac{d}{dt}$ and integral $\int .... dt$ operators. Sections 6.4 and 6.5 illustrate the implementation of Laplace transform on the differential and the integral operators respectively. We combine these two techniques to find the system function from an integrodifferential equation.

There is some mathematical intricacy before finding the transform. To solve an integrodifferential equation, the equation is rearranged in two parts – the first is the nonintegral part and the second is the integral part of the equation but rearranging the right side of the equation to zero for instance the integrodifferential equation $y' + \int_{x=0}^{x=t} y(x)dx = 2u(t)$ is written as $[y' - 2u(t)] + \int_{x=0}^{x=t} y(x)dx = 0$.

Laplace transforms of the two parts are taken separately and then added to form the complete integrodifferential equation transform.

From an integrodifferential equation we are mainly interested to find the transfer function of a system which is obtained by setting the transformed equation to zero.

Let us see the following two examples in this regard.

Example A

It is given that $Y(s) = \frac{2(2s+1)}{s^2+1}$ is the system function for integrodifferential equation $y' + \int_{x=0}^{x=t} y(x)dx = 2u(t)$ subject to $y(0) = 4$ which we intend to obtain.

The nonintegral and integral parts of the equation are written as $[y' - 2u(t)] + \int_{x=0}^{x=t} y(x)dx = 0$ on that account following is the implementation by applying aforementioned function and symbology.

For the nonintegral part:

>>syms t ↵ ← Declaration of independent variable t in the part $[y' - 2u(t)]$ as symbolic

>>y=sym('y(t)'); ↵ ← Declaration of the dependent $y(t)$ in the part $[y' - 2u(t)]$ as symbolic

>>P1=laplace(diff(y)-2*heaviside(t)); ↵ ← Workspace P1 holds the transform of the $[y' - 2u(t)]$ where P1 is user-chosen variable

For the integral part:

>>y=sym('y(x)'); syms x ↵ ← Defining $y(x)$ as function of x for the integral part as well as x as symbolic

>>P2=laplace(int(y,x,0,t)); ↵ ← Workspace P2 holds the transform of the integral part ignoring the warning where P2 is user-chosen variable

Forming the whole equation:

>>E=P1+P2; ↵ ← The complete integrodifferential equation is formed from P1+P2 and assigned to E where E is user-chosen variable

>>E=subs(E,{'y(0)'},{4}); ↵ ← Initial condition $y(0) = 4$ is inserted in the string of E and assigned to E again

>>syms Y ↵ ← Declaring Y as symbolic for replacing the string laplace(y(t),t,s) stored in E by Y

>>E=subs(E,{'laplace(y(t),t,s)'},{Y}); ↵ ← String laplace(y(t),t,s) is replaced by Y and again assigned to E

>>Y=solve(E,Y); ↵ ← $Y(s)$ is obtained by forming an equation E=0 and again assigned to Y, Y⇔ $Y(s)$

>>pretty(Y) ↵ ← Displaying the readable form of Y from its string

```
       2 s + 1
   2 -----------
         2
        s  + 1
```

Example B

Given $Y(s) = \frac{2s(12+e^{-2s})}{8s^3+12s^2+6s+1}$ is the system function for integrodifferential equation $8y'' + 12y' + 6y + \int_{x=0}^{x=t} y(x)dx = 2\delta(t-2)$ with $y(0) = 0$ and $y'(0) = 3$ which we wish to determine.

As we did in the example A the nonintegral and the integral parts of the equation are written as $[8y''+12y'+6y-2\delta(t-2)]+\int_{x=0}^{x=t} y(x)dx=0$ therefore we proceed as follows:

For the nonintegral part:

>>syms t ↵ ← Declaration of independent variable t in the part $[8y''+12y'+6y-2\delta(t-2)]$ as symbolic

>>y=sym('y(t)'); ↵ ← Declaration of the dependent $y(t)$ in the part $[8y''+12y'+6y-2\delta(t-2)]$ as symbolic

>>P1=laplace(8*diff(y,2)+12*diff(y)+6*y-2*dirac(t-2)); ↵ ←Workspace P1 holds the transform on $[8y''+12y'+6y-2\delta(t-2)]$

For the integral part:

>>y=sym('y(x)'); syms x ↵ ← Defining $y(x)$ as function of x for the integral part as well as x as symbolic

>>P2=laplace(int(y,x,0,t)); ↵ ← Workspace P2 holds the transform of the integral part ignoring the warning where P2 is user-chosen variable

Forming the whole equation:

>>E=P1+P2; ↵ ← The complete integrodifferential equation is formed from P1+P2 and assigned to E

>>E=subs(E,{'y(0)','D(y)(0)'},{0,3}); ↵ ← Initial conditions $y(0)=0$ and $y'(0)=3$ are inserted in the string of E and assigned to E again

>>syms Y ↵ ← Declaring Y as symbolic for replacing the string laplace(y(t),t,s) stored in E by Y

>>E=subs(E,{'laplace(y(t),t,s)'},{Y}); ↵ ← String laplace(y(t),t,s) is replaced by Y and again assigned to E

>>Y=solve(E,Y); ↵ ← $Y(s)$ is obtained by forming an equation E=0 and assigned to Y, Y⇔ $Y(s)$

>>pretty(Y) ↵ ← Displaying the readable form of Y from its string

```
        s (12 + exp(-2 s) )
   2  --------------------------
           3      2
        8 s  + 12 s  + 6 s + 1
```

6.7 System function from a system of differential equations

In previous sections we have seen the number of dependent variables related to differential equations being 1. When we have more than 1 dependent variable, we can still employ the function laplace of section 6.4 for the system function but the function is applicable to each equation. The reader is expected to go through sections 6.2-6 before solving these types of problems because all the techniques are combined here. Appendix D cited solve finds each system function in terms of s. Let us see the following examples in this regard.

Example A

It is provided that $X(s)=\frac{s}{s^2-7}$ and $Y(s)=\frac{2s+7}{s^2-7}$ are the system functions for the system of differential equations $\begin{Bmatrix} x'=-2x+y, & x(0)=1 \\ y'=3x+2y, & y(0)=2 \end{Bmatrix}$ which we intend to obtain.

We know that $X(s)=L[x(t)]$ and $Y(s)=L[y(t)]$ and the rearranged equation is $\begin{Bmatrix} x'+2x-y=0 \\ y'-3x-2y=0 \end{Bmatrix}$. First, defining each dependent variable (i.e. $x(t)$ and $y(t)$) takes place as follows (section 6.4):

```
>>x=sym('x(t)'); ↵          ← Declaration of x(t), x⇔x(t)
>>y=sym('y(t)'); ↵          ← Declaration of y(t), y⇔y(t)
```

Then Laplace transforms on left sides of the organized equations are assigned to E1 and E2 (where E1 and E2 are user-chosen variable names) as follows respectively:

```
>>E1=laplace(diff(x)+2*x-y); ↵
>>E2=laplace(diff(y)-3*x-2*y); ↵
```

Next we insert the initial conditions and put the outputs to the E1 and E2 again as follows respectively:

```
>>E1=subs(E1,{'x(0)'},{'1'}); ↵        ← Substitution of x(0)=1
>>E2=subs(E2,{'y(0)'},{'2'}); ↵        ← Substitution of y(0)=2
```

After that we replace the strings laplace(x(t),t,s) by X and laplace(y(t),t,s) by Y and assign the outputs again to E1 and E2 respectively as follows:

```
>>E1=subs(E1,{'laplace(x(t),t,s)','laplace(y(t),t,s)'},{'X','Y'}); ↵
>>E2=subs(E2,{'laplace(x(t),t,s)','laplace(y(t),t,s)'},{'X','Y'}); ↵
```

Then we have the $X(s)$ and $Y(s)$ related two equations by forming the equations E1=0 and E2=0 and solve the two equations by employing the solve to get the $X(s)$ and $Y(s)$ as follows:

```
>>R=solve(E1,E2,'X','Y') ↵           ← R is any user-chosen variable which
                                        holds the return from the solve
R =                                   ← i.e. R is a structure array having two members X and Y
        X: [1x1 sym]
        Y: [1x1 sym]
```

From the structure array R we pick up the expressions for the $X(s)$ and $Y(s)$ one at a time as follows:

```
>>X=R.X; ↵          ← Assigning the first member of R to X, X is user-chosen
>>Y=R.Y; ↵          ← Assigning the second member of R to Y, Y is user-chosen
>>pretty(X) ↵            ← Displaying the contents of X, which is X(s)
           s
      -----------
        2
       s  -  7
>>pretty(Y) ↵            ← Displaying the contents of Y, which is Y(s)
        7 + 2 s
      -------------
         2
        s  -  7
```

Example B

It is given that $X(s)=\frac{2s^5+17s^4+31s^3+36s^2+12s+10}{s^6+6s^5+14s^4+16s^3+8s^2}$ and $Y(s)=\frac{s^5-5s^4-11s^3-18s^2-6s-6}{s^6+6s^5+14s^4+16s^3+8s^2}$ are the two system functions for the system of differential equations $\begin{cases} x' = x + 3y + t, & x(0) = 2 \\ y' = -3x - 5y + \sin t\, e^{-t}, & y(0) = 1 \end{cases}$. Our objective is to obtain the system functions starting from the differential equations.

In example A there is no t related expression in the given differential equations that is why we did not declare the t as symbolic. In this example we have t related expression so the declaration is as follows:

>>syms t ↵ ← Declaration of t, t⇔t

Dependent variable declaration:

>>x=sym('x(t)'); ↵ ← Declaration of $x(t)$, x⇔ $x(t)$

>>y=sym('y(t)'); ↵ ← Declaration of $y(t)$, y⇔ $y(t)$

Laplace transform on each given differential equation, right side of which is 0:

>>E1=laplace(diff(x)-x-3*y-t); ↵ ← E1 holds the first transformed equation

>>E2=laplace(diff(y)+3*x+5*y-sin(t)*exp(-t)); ↵ ← E2 holds the second transformed equation

Initial condition substitution:

>>E1=subs(E1,{'x(0)'},{'2'}); ↵ ← Inserting initial condition of equation 1

>>E2=subs(E2,{'y(0)'},{'1'}); ↵ ← Inserting initial condition of equation 2

Long string replacement for convenience (i.e. laplace(x(t),t,s) by X and laplace(y(t),t,s) by Y):

>>E1=subs(E1,{'laplace(x(t),t,s)','laplace(y(t),t,s)'},{'X','Y'}); ↵

>>E2=subs(E2,{'laplace(x(t),t,s)','laplace(y(t),t,s)'},{'X','Y'}); ↵

Solving for $X(s)$ and $Y(s)$ from E1=0 and E2=0:

>>R=solve(E1,E2,'X','Y') ↵

R = ← R is a structure array having two members X and Y

X: [1x1 sym]

Y: [1x1 sym]

>>X=R.X; ↵ ← Assign the first member of R to X

>>Y=R.Y; ↵ ← Assign the second member of R to Y

>>pretty(X) ↵ ← Display the contents of X, X⇔ $X(s)$

```
     4      5       2        3
17 s  + 2 s  + 36 s  + 31 s  + 12 s + 10
------------------------------------------
      2   4      3        2
     s  (s  + 6 s  + 14 s  + 16 s + 8)
```

>>pretty(Y) ↵ ← Display the contents of Y, Y⇔ $Y(s)$

```
      5        3      4      2
     s  - 11 s  - 5 s  -18 s  - 6 s - 6
  ---------------------------------------
      2   4      3        2
     s  (s  + 6 s  + 14 s  + 16 s + 8)
```

Example C

In the first two examples the systems of equations are expressible in terms of the first order derivatives. Mixed derivatives can also be solved similarly. It is given that $X(s)=\dfrac{60s^4+25s^3-56s^2+42s+42}{15s^4(s+1)}$, $Y(s)=\dfrac{5s^4-8s^2+6s+6}{5s^4(s+1)}$, and $Z(s)=\dfrac{2s+3}{s(s+1)}$ are the three s domain system functions for the system defined by $3\dfrac{dx}{dt}=7\dfrac{dy}{dt}$, $5\dfrac{d^2y}{dt^2}+3\dfrac{dz}{dt}=6t$, $\dfrac{dz}{dt}=e^{-t}$, $x(0)=4$, $y(0)=1$, $y'(0)=-1$, and $z(0)=2$ which we intend to determine.

Maintaining ongoing function and symbology we go through the following:

Independent variable declaration:

>>syms t ↵ ← Declaration of t, t⇔t

Dependent variable declaration (three variables $x(t)$, $y(t)$, and $z(t)$):

>>x=sym('x(t)'); ↵ ← Declaration of $x(t)$, x⇔ $x(t)$

>>y=sym('y(t)'); ↵ ← Declaration of $y(t)$, y⇔ $y(t)$

>>z=sym('z(t)'); ↵ ← Declaration of $z(t)$, z⇔ $z(t)$

Laplace transform on each given differential equation, right side of which is 0:

>>E1=laplace(3*diff(x)-7*diff(y)); ↵ ← E1 holds the first transformed equation

>>E2=laplace(5*diff(y,2)+3*diff(z)-6*t); ↵ ← E2 holds the second transformed equation

>>E3=laplace(diff(z)-exp(-t)); ↵ ← E3 holds the third transformed equation

Initial condition substitution:

Special attention is needed here. We insert only the string which is related with the given equation. The first differential equation needs only the $x(0)$ and $y(0)$. The second differential equation does only the $y'(0)$, $y(0)$, and $z(0)$. The third differential equation requires only $z(0)$. Insertion of initial conditions happens accordingly:

>>E1=subs(E1,{'x(0)','y(0)'},{'4','1'}); ↵ ← Inserting initial condition of equation 1

>>E2=subs(E2,{'y(0)','D(y)(0)','z(0)'},{'1', '-1', '2'}); ↵ ← Inserting initial condition of equation 2

>>E3=subs(E3,{'z(0)'},{'2'}); ↵ ← Inserting initial condition of equation 3

Long string replacement for convenience (i.e. laplace(x(t),t,s) by X, laplace(y(t),t,s) by Y, and laplace(z(t),t,s) by Z):

Like the initial condition attention must be paid here during the string substitution. In the given first, second, and third differential equations we have $X(s)$ - $Y(s)$, $Y(s)$ - $Z(s)$, and $Z(s)$ and the replacement occurs likewise as follows:

>>E1=subs(E1,{'laplace(x(t),t,s)','laplace(y(t),t,s)'},{'X','Y'}); ↵

>>E2=subs(E2,{'laplace(y(t),t,s)','laplace(z(t),t,s)'},{'Y','Z'}); ↵

>>E3=subs(E3,{'laplace(z(t),t,s)'},{'Z'}); ↵

Solving for $X(s)$, $Y(s)$, and $Z(s)$ from E1=0, E2=0, and E3=0:

>>R=solve(E1,E2,E3,'X','Y','Z') ↵

```
R =                                  ← R is a structure array having
                                       three members X, Y, and Z
        X: [1x1 sym]
        Y: [1x1 sym]
        Z: [1x1 sym]
```

Extracting the three members from the solution held in R:

```
>>X=R.X; Y=R.Y; Z=R.Z; ↵
>>pretty(X) ↵                 ← Display the contents of X, X⇔ X(s)
                  3       4                   2
            25 s  + 60 s  + 42 + 42 s - 56 s
   1/15 -----------------------------------------
                         4
                        s  (1 + s)
>>pretty(Y) ↵                 ← Display the contents of Y, Y⇔ Y(s)
                    4     2
        6 + 6 s + 5 s  - 8 s
   1/5 -----------------------
                  4
                 s  (1 + s)
>>pretty(Z) ↵                 ← Display the contents of Z, Z⇔ Z(s)
          3 + 2 s
          -----------
          s (1 + s)
```

Example D

Integrodifferential equations can also be handled in a similar fashion. The s domain system functions $X(s)=-\frac{2(84s^2+s-2)}{s(24s^2-1)}$ and $Y(s)=\frac{72s^4+12s^3+39s^2-1}{s^3(24s^2-1)}$ are obtained from the system of differential equations $4\frac{dy}{dt}+\int_{p=0}^{p=t}x(p)dp=2u(t)$, $\frac{dy}{dt}+6\frac{dx}{dt}=t$, $x(0)=-7$, and $y(0)=3$. We aim at determining the system functions from the integrodifferential equations in MATLAB.

The first given equation has the independent variable t and the dummy variable p. To us the $x(t)$ or $x(p)$ bears the same meaning but machine understands differently. For this reason in the first given equation we declare the dependent variable as $x(p)$ but in the second given equation we declare the dependent variable as $x(t)$:

Independent variable declaration:

```
>>syms t p ↵                  ← Declaration of t and p, t⇔t, p⇔p
```

First given equation dependent variable declaration (i.e. $x(p)$ and $y(t)$):

```
>>x=sym('x(p)'); ↵            ← Declaration of x(p), x⇔ x(p)
>>y=sym('y(t)'); ↵            ← Declaration of y(t), y⇔ y(t)
```

E1 holds the transform on the first equation (turning the right side to 0) as follows:

```
>>E1=laplace(4*diff(y)-2*heaviside(t)+int(x,p,0,t)); ↵
```

From the last execution the reader is supposed to experience some warning which we ignore.

Second given equation dependent variable declaration (i.e. $x(t)$ and $y(t)$):

>>x=sym('x(t)'); ↵ ← Declaration of $x(t)$, X⇔ $x(t)$, $y(t)$ already done

E2 holds the transform on the second equation (turning the right side to 0) as follows:

>>E2=laplace(diff(y)+6*diff(x)-t); ↵

Initial condition substitution:

The given first and second equations need only the $y(0)$ and $x(0)$ - $y(0)$ respectively therefore we enter them as follows:

>>E1=subs(E1,{'y(0)'},{'3'}); ↵ ← Inserting initial condition to equation 1

>>E2=subs(E2,{'x(0)','y(0)'},{'-7','3'}); ↵ ← Inserting initial condition to equation 2

Long string replacement for convenience (i.e. laplace(x(t),t,s) by X and laplace(y(t),t,s) by Y):

Both in the given first and second equations we have $X(s)$ - $Y(s)$ so the substitution is as follows:

>>E1=subs(E1,{'laplace(x(t),t,s)','laplace(y(t),t,s)'},{'X','Y'}); ↵
>>E2=subs(E2,{'laplace(x(t),t,s)','laplace(y(t),t,s)'},{'X','Y'}); ↵

Solving for $X(s)$ and $Y(s)$ from E1=0 and E2=0:

>>R=solve(E1,E2,'X','Y') ↵ ← R is a structure array having two members X and Y

```
R =
      X: [1x1 sym]
      Y: [1x1 sym]
```

Extracting the two members from the solution held in R:

>>X=R.X; Y=R.Y; ↵

>>pretty(X) ↵ ← Display the contents of X, X⇔ $X(s)$

```
             2
      84 s  - 2 + s
  -2 ---------------
               2
      s (24 s  - 1)
```

>>pretty(Y) ↵ ← Display the contents of Y, Y⇔ $Y(s)$

```
       3        2        4
  12 s  + 39 s  + 72 s  - 1
  -----------------------------
          3       2
         s  (24 s  - 1)
```

M-file approach

So far we explained all necessary steps to obtain the system function at the command prompt. Since we have been assigning different returns to one variable (for example E1 to E1 again and again), going back by one statement might show some error. This problem is overcome by using different assignee variables or writing all statements in an M-file (subsection 1.1.2).

Open a new M-file. Considering the example A of this section, we put all statements of the example in the opened M-file as shown in the figure 6.1(e). Save the file by the name test (can be any user-chosen name) and call it as follows:

>>test ↵

After that execute pretty(X) or pretty(Y) to see the $X(s)$ or $Y(s)$ like the example A at the command prompt.

```
x=sym('x(t)');
y=sym('y(t)');
E1=laplace(diff(x)+2*x-y);
E2=laplace(diff(y)-3*x-2*y);
E1=subs(E1,{'x(0)'},{'1'});
E2=subs(E2,{'y(0)'},{'2'});
E1=subs(E1,{'laplace(x(t),t,s)','laplace(y(t),t,s)'},{'X','Y'});
E2=subs(E2,{'laplace(x(t),t,s)','laplace(y(t),t,s)'},{'X','Y'});
R=solve(E1,E2,'X','Y');
X=R.X;
Y=R.Y;
```

Figure 6.1(e) All program statements of the example A

6.8 Inverse Laplace transform

If $F(s)$ is a function of Laplace transform variable s, the inverse Laplace transform of $F(s)$ is defined as $f(t)=L^{-1}[F(s)]=\frac{1}{2\pi j}\int_{s=c-j\infty}^{s=c+j\infty} e^{st}F(s)\,ds$ where L^{-1} is the inverse transform operator and c is a real number. The number c is selected in such a way that all singularities of $F(s)$ are to the left of the line $s=c$ in the s plane. This is the formal definition of the inverse Laplace transform but as a working convenience, the $F(s)$ is maneuvered to take the form of commonly known functions whose inverse Laplace transforms are known.

MATLAB function that performs the inverse Laplace transform is ilaplace. To take the inverse Laplace transform of $F(s)$, we apply the command ilaplace(code of $F(s)$ in vector string form – appendix B, independent variable of $F(s)$, return variable for the inverse transform $f(t)$). Also we declare the related variable by using the command syms before applying the ilaplace. Readable form of the inverse string is seen by using the command pretty.

Let us see the implementation for $F(s)=\frac{1}{s^2+1}$. From the table 6.A, we read $f(t)=L^{-1}\left[\frac{1}{s^2+1}\right]=\sin t$ for which we carry out the following:

>>syms s t ↵ ← Independent variables of $F(s)$ and $f(t)$ are declared as symbolic, t⇔t and s⇔s

>>F=1/(s^2+1); ↵ ← Vector code of $F(s)$ is assigned to workspace F, F⇔$F(s)$, F is any user-chosen name

>>f=ilaplace(F,s,t) ↵ ← Inverse Laplace transform is returned to workspace f, f⇔$f(t)$, f is any user-chosen name

f =

sin(t) ← Code for the $\sin t$

Since the default return of ilaplace is also in terms of t, the command ilaplace(1/(s^2+1)) would bring about the same outcome. Maintaining the symbology just discussed, three more examples are illustrated in the following.

❖❖ Example A

A polynomial of s for example $F(s)=3s-7s^2$ has the inverse transform $f(t)=3\frac{d[\delta(t)]}{dt}-7\frac{d^2[\delta(t)]}{dt^2}$:

```
>>syms s, F=3*s-7*s^2; ↵   ← code of 3s – 7s^2 is assigned to F
>>f=ilaplace(F) ↵          ← f holds the f(t)

f =

3*dirac(1,t)-7*dirac(2,t)
```

The dirac(2,t) in the return indicates $\frac{d^2[\delta(t)]}{dt^2}$ where $\delta(t)$ is the Dirac delta function (section 6.2) and the first input argument of the dirac is the derivative order.

❖❖ Example B

Rational form function $F(s)=\frac{2s^3+3}{s^4-6s^3+32s}$ has the inverse $f(t)=\frac{3}{32}+\frac{13}{72}e^{-2t}+\frac{497}{288}e^{4t}+\frac{131}{24}te^{4t}$:

```
>>syms s, F=(2*s^3+3)/(s^4-6*s^3+32*s); ↵  ← Code of F(s) is assigned to
                                              the workspace F
>>f=ilaplace(F); ↵    ← f contains the inverse transform string for f(t)
>>pretty(f) ↵         ← Displaying the readable form of the string stored in f

           /131     497\              13
  3/32 + |----- t + -----| exp(4 t) + --- exp(-2 t)      ← Rearranged f(t)
           \ 24     288/              72
```

❖❖ Example C

$F(s)=e^{-3s}\frac{s+2}{s^2+2s+5}$ has the inverse transform $f(t)=u(t-3)\,[\,e^{-t+3}\cos(2t-6)+\frac{1}{2}e^{-t+3}\sin(2t-6)\,]$:

```
>>syms s, F=exp(-3*s)*(s+2)/(s^2+2*s+5); ↵  ← Code of F(s) is assigned to
                                               the workspace F
>>f=ilaplace(F); ↵    ← f contains the inverse transform string for f(t)
>>pretty(f) ↵         ← Displaying the readable form of the string stored in f

  (exp(-t + 3) cos(2 t - 6) + 1/2 exp(-t + 3) sin(2 t - 6)) heaviside(t - 3)
```

In above return the heaviside(t - 3) is equivalent to $u(t-3)$.

That brings an end to this chapter.

Chapter 7
Z TRANSFORM

Introduction

Z transform is the discrete counterpart of the Laplace transform. Fourier and Laplace transforms are suitable for continuous time signals and systems whereas Z transform is solely for the discrete time signals and systems. In Fourier transform the convergence is achieved by considering complex exponential as basis function contrarily Z transform finds its region of convergence in terms of power series basis. Our topic outline is the following:

- Techniques to exercise forward-inverse terminology of the Z transform
- Z transform for standard signals as well as for finite sequences
- Z transform system function for discrete system and difference equation solving

7.1 Z transform

Z transform is applicable for the discrete signals. Like the Fourier and Laplace transforms, the Z transform also shares the strategy of the forward-inverse as shown in the figure 7.1(a). The signal $f[n]$ is discrete but the transform $F(z)$ is continuous. Depending on the integer index n variation, the Z transform can be bilateral or unilateral. In the former the n changes from minus infinity to plus infinity whereas the n changes from 0 to plus infinity in the latter. In MATLAB we implement the unilateral one as addressed in the following section.

$$discrete\ f[n] \xrightarrow{forward\ Z\ transform} F(z) \xrightarrow{inverse\ Z\ transform} f[n]$$

Figure 7.1(a) The concept behind the Z transform

7.2 Forward Z transform of discrete signals

Unilateral forward Z transform $F(z)$ of the discrete signal $f[n]$ is defined as $Z\ \{f[n]\} = F(z) = \sum_{n=0}^{n=\infty} f[n] z^{-n}$ where $n \geq 0$ and n is integer, and Z is the forward transform operator. MATLAB function ztrans (abbreviation for the Z transform) provides many symbolically known forward transforms. To have the unilateral Z transform of the discrete signal $f[n]$ (assuming that the envelope of $f[n]$ follows some functional variation), we employ the command ztrans(vector code of $f[n]$ – appendix B, independent variable of $f[n]$, wanted transform variable). Since the computation is merely symbolic, declaration of the associated variables in the signal expression by using the syms is mandatory. Following the transform, simplification of the expression is conducted by the command simple or simplify and mathematics readable form is seen by the command pretty (appendix C.10).

Let us implement the transform with $f[n] = e^{na}$. Its unilateral Z transform is $F(z) = Z\ \{[e^{an}]\} = \sum_{n=0}^{n=\infty} e^{an} z^{-n} = (1 - e^a z^{-1})^{-1} = \frac{z}{z - e^a}$. Starting from e^{na}, we intend to obtain $\frac{z}{z - e^a}$ – that is the problem statement for which we conduct the following:

```
>>syms n a z ↵ ← Declaring related and wanted transform variables as symbolic,
                                   a⇔a , n⇔n , z⇔z
>>F=ztrans(exp(n*a),n,z); ↵ ← F holds the Z transform on the code of e^na ,
                                   F⇔F(z) , F is user-chosen variable
>>pretty(simple(F)) ↵      ← Displaying the readable form of the string for F(z)
         z                         stored in F
   -------------
   z - exp(a)
```

We could have implemented all commands in one line like syms n a z, F= ztrans(exp(n*a),n,z); pretty(simple(F)). The command ztrans(exp(n*a)) also brings about the same result because of the default return. If the independent and transform variables were p and w respectively, the command would be ztrans(exp(p*a),p,w).

Discrete function like $f[n] = e^{-n^2}$ does not have Z transform, we see then ztrans(exp(-n^2),n,z) as the return which is the code for the definition $Z\ \{[e^{-n^2}]\}$. Applying the same symbols and functions, table 7.A presents the mathematical and MATLAB correspondence of some unilateral forward Z transforms.

Table 7.A Unilateral forward Z transforms of some discrete standard functions and their MATLAB counterparts

Mathematical form	MATLAB commands
$Z\{\sin[na]\}=\frac{z\sin a}{z^2-2z\cos a+1}$	>>syms a n, F=ztrans(sin(n*a)); ↵ >>pretty(F) ↵ sin(a) z ---------------------------- 2 - 2 z cos(a) + z + 1
$Z\{\cos[na]\}=$ $\frac{z(z-\cos a)}{z^2-2z\cos a+1}$	>>syms a n, F=ztrans(cos(n*a)); ↵ >>pretty(F) ↵ z (-cos(a) + z) -------------------------- 2 - 2 z cos(a) + z + 1
$Z\{[a^n]\}=\frac{z}{z-a}$	>>syms a n, F=ztrans(a^n); ↵ >>pretty(simple(F)) ↵ z - ---------- -z + a
$Z\{u[n]\}=\frac{z}{z-1}$ The $u[n]$ is represented by heaviside(n)	>>syms n, F=ztrans(heaviside(n)); ↵ >>pretty(F) ↵ z ------- z - 1
$Z\{\delta[n]\}=1$ The $\delta[n]$ is represented by sym('charfcn[0](n)') not by dirac(n)	>>d=sym('charfcn[0](n)'); ↵ ← d⇔ $\delta[n]$ >>F=ztrans(d); ↵ >>pretty(F) ↵ 1
$Z\{[n^2a^n]\}=\frac{za(z+a)}{(z-a)^3}$	>>syms a n, F=ztrans(n^2*a^n); ↵ >>pretty(F) ↵ z a (z + a) - ------------- 3 (-z + a)
$Z\{[r^n\sin na]\}=$ $\frac{rz\sin a}{z^2-2zr\cos a+r^2}$	>>syms a n r, F=ztrans(r^n*sin(n*a)); ↵ >>pretty(simple(F)) ↵ r sin(a) z ------------------------------ 2 2 -2 z r cos(a) + z + r
$Z\{a^{n-3}\,u[n-3]\}=\frac{1}{z^2(z-a)}$	>>syms a n ↵ >>F=ztrans(a^(n-3)*heaviside(n-3)); ↵ >>pretty(simple(F)) ↵ 1 ------------ 2 (z - a) z

7.3 Forward Z transform of finite sequences

The discrete signal $f[n]$ is also called a sequence. Most examples of $f[n]$ set in the table 7.A conceive n in the range $0 \le n \le \infty$. Discrete signals may exist only for few values of n then the $f[n]$ is called a finite sequence.

Special attention must be paid to interval while finding the transform. For example $0 \le n \le 4$ means 4 is inclusive and there should be five samples in the interval. We describe any function within the interval in terms of the unit step sequence by writing $u[n]-u[n-4]$. Since the variation is integer-wise, we lose the sample at $n=4$. If we are strict to have five samples for $0 \le n \le 4$, we must use $u[n]-u[n-5]$. Anyhow we present three examples on the finite sequence transform in the following.

♦♦ Example 1

It is given that $F(z)=\dfrac{1}{4z^7}\left[4z^7+2\times 2^{\frac{3}{4}}z^6+2\times 2^{\frac{1}{2}}z^5+2\times 2^{\frac{1}{4}}z^4+\ 2z^3+2^{\frac{3}{4}}z^2+ 2^{\frac{1}{2}}z+2^{\frac{1}{4}}\right]$ is the Z transform of the finite sequence $f[n]=\begin{Bmatrix} 2^{-\frac{n}{4}} & for\ 0 \le n < 8 \\ 0 & elsewhere \end{Bmatrix}$ which we wish to obtain.

By dint of the unit step sequence $u[n]$, the $f[n]$ is expressed as $f[n]=[2^{-\frac{n}{4}}]\{u[n]-u[n-8]\}$ on that we apply the section 7.2 cited function as follows:

```
>>syms n ↵                    ← Declaring related n as symbolic, n⇔n
>>f=2^(-n/4)*(heaviside(n)-heaviside(n-8)); ↵ ← Vector code of f[n] is
                                       assigned to f, f⇔f[n], f is user-chosen
>>F=ztrans(f); ↵          ← F holds the transform on f[n], F⇔F(z), F is user-chosen
>>pretty(simple(F)) ↵ ← Displaying readable form of F(z) following
                                simplification on the string stored in F

            7     3/4  6      1/2   5        1/4   4        3    3/4  2     1/2          1/4
 1/4 (4 z  + 2 2     z  + 2 2     z  + 2 2     z  + 2 z  + 2     z  + 2     z + 2

                                                                   /  7
                                                                 ) /  z
                                                                  /
```

♦♦ Example 2

$F(z)=\dfrac{6z^7-4z^6+3z^5+2z^3+6z^2+8z+2}{z^{10}}$ is the Z transform of the finite sequence $\begin{Bmatrix} f[n] & \rightarrow & 6 & -4 & 3 & 0 & 2 & 6 & 8 & 2 \\ n & \rightarrow & 3 & 4 & 5 & 6 & 7 & 8 & 9 & 10 \end{Bmatrix}$ which we intend to implement.

Given $f[n]$ does not follow any specific function and that $f[n]$ exists over $3 \le n \le 10$. At $n=3$, $f[n]$=6 is represented as $f[3]=6\delta[n-3]$. As a functional form, we write $f[n]=6\delta[n-3]-4\delta[n-4]+3\delta[n-5]+0\delta[n-6]+2\delta[n-7]+6\delta[n-8]+8\delta[n-9]+2\delta[n-10]$. The computation here is completely symbolic that is why first we enter all functional values of $f[n]$ without delta to a row matrix y (user-chosen name) but turning the elements to symbolic by the command sym as follows:

```
>>y=sym([6 -4 3 0 2 6 8 2]); ↵
```

Table 7.A says that sym('charfcn[0](n)') represents $\delta[n]$ and the general delta sample $\delta[n-m]$ is written as sym('charfcn[m](n)') so we enter it as follows:

```
>>d=sym('charfcn[m](n)'); ↵      ← d⇔δ[n − m] where d is user-chosen
```

The m is changing in various delta functions but not the n in different samples of $f[n]$ so we define the m as symbolic as follows:

```
>>syms m ↵                       ← m⇔m
```

By appendix C.8-9 cited data accumulation technique, we create only delta functions of $f[n]$ from $\delta[n-3]$ to $\delta[n-10]$ and assign those to s as follows:

```
>>s=[ ]; ↵
>>for k=3:10 s=[s subs(d,m,k)]; end ↵      ← s and k are user-chosen variables
```

In last command line the for-loop counter k provides control on m and we substitute different m sequentially to $\delta[n-m]$ (which is stored in d) by the command subs(d,m,k). Now the delta functions (held in row matrix s) and functional values (held in y) need to be multiplied to form $f[n]$ elements as a row matrix which happens by the scalar code (appendix B) as follows:

```
>>g=s.*y; ↵
```

In just conducted execution the g is a user-chosen variable. But again the g is a row matrix in which we have the delta elements of $f[n]$. In order to sum them all, we apply the sum (appendix C.11) as follows:

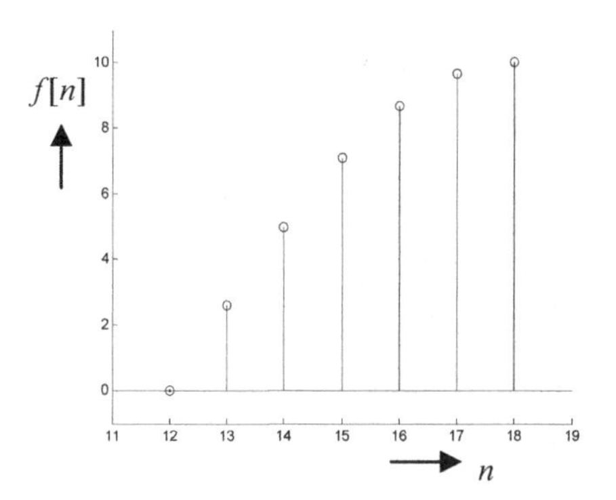

Figure 7.1(b) A discrete sine function

```
>>f=sum(g); ↵                  ← f is forming the f[n] where f is user-chosen
>>F=ztrans(f); ↵               ← F⇔F(z)
>>pretty(simplify(F)) ↵        ← Displaying readable form of F(z) after simplification
                                  on the string stored in F

       7      6      5      3      2
    6 z  - 4 z  + 3 z  + 2 z  + 6 z  + 8 z + 2
    -------------------------------------------
                         10
                        z
```

Example 3

Figure 7.1(b) shows the plot of a discrete sine function which is given by $f[n] = -10\sin\dfrac{\pi n}{12}$ over $12 \le n \le 18$. Given that the forward transform $F(z) = \dfrac{5\left(2z^5\sin\dfrac{\pi}{12} + z^3\sqrt{2} + z^2\sqrt{3} + 2z\sin\dfrac{5\pi}{12} + z^4 + 2\right)}{z^{18}}$ is for the $f[n]$ which we wish to find.

This example is similar to the example 1. In terms of the unit step sequence we write the expression for the discrete signal as $f[n] =$

$-10\sin\frac{\pi n}{12}\{u[n-12]-u[n-19]\}$ (for inclusion of the last sample we use 19 not 18) so we execute the following applying ongoing symbology and function:

```
>>syms n ↵                    ← Declaring related n as symbolic, n⇔n
>>f=-10*sin(pi*n/12)*(heaviside(n-12)-heaviside(n-19)); ↵   ← f⇔f[n]
>>F=ztrans(f); ↵                                            ← F⇔F(z)
>>pretty(simple(F)) ↵         ← For readable form of F(z) after simplification

                        5    1/2  3    1/2  2                        4
      2 sin(1/12 pi) z  + 2     z  + 3    z  + 2 sin(5/12 pi) z + z  + 2
   5 ---------------------------------------------------------------------
                                    18
                                   z
```

7.4 Inverse Z transforms

Given a Z transform function $F(z)$, its inverse Z transform (by operator Z^{-1}) is obtained in terms of the contour integral as $f[n]=\frac{1}{2\pi j}\oint_C F(z)z^{n-1}dz$ where C is a counterclockwise closed contour in the region of the convergence of $F(z)$ and the contour encircles the origin of the z-plane.

MATLAB counterpart for the inverse Z transform is iztrans (abbreviation for the inverse z transform). To perform inverse Z transform on $F(z)$, we use the command iztrans(vector string of the transform $F(z)$ – appendix B, transform variable, inverse transform variable). We declare the related variables of $F(z)$ by using the syms before applying the iztrans. Command pretty (appendix C.10) shows easy readable output. It is understood that the return function is discrete.

From the table 7.A, we have $Z\{[a^n]\}=F(z)=\frac{z}{z-a}$ so the inverse Z transform is $Z^{-1}[\frac{z}{z-a}]=[a^n]=f[n]$ which happens through the following:

```
>>syms z a n ↵     ← Defining variables of F(z) and return variable as symbolic,
                                     z⇔z, a⇔a, and n⇔n
>>F=z/(z-a); ↵     ← Assigning vector code of F(z) to F, F⇔F(z), F is user-
                                     chosen name
>>f=iztrans(F,z,n); ↵ ← Assigning inverse transform on F to f, f⇔f[n], f is user-
                                     chosen name
>>pretty(f) ↵      ← Displaying the readable form of the string stored in f
                  n
                 a
```

The command iztrans(F) also executes above result because of the default definition. If the F were in w and the return were required in terms of x, the command would be iztrans(F,w,x). Following the same functions and symbology, few more examples are illustrated in the following.

Example A

$X(z)=\frac{z^2}{(z-a)(z-c)}$ has the inverse transform $x[n]=\frac{a^{n+1}-c^{n+1}}{a-c}$ and is implemented as follows:

```
>>syms z a c n ↵
```
←Defining variables of $X(z)$ and return variable as symbolic, z⇔ z ,a⇔ a ,n⇔ n ,c⇔ c

```
>>X=z^2/(z-a)/(z-c); ↵
```
← Assigning vector code of $X(z)$ to X, X⇔ $X(z)$

```
>>x=iztrans(X,z,n); ↵
```
← Assigning inverse transform on X to x, x⇔ $x[n]$

```
>>pretty(x) ↵
```
← Displaying the readable form of the string stored in x

```
        n    n
     a a  - c c
     ------------
      - c + a
```

Example B

Given that the inverse transform of $F(z)=\dfrac{z^{-1}+z^{-2}}{\left(1-\dfrac{1}{2}z^{-1}\right)\left(1+\dfrac{1}{3}z^{-1}\right)}$ is $f[n]=$ $-6\delta[n]+\dfrac{18}{5}\left(\dfrac{1}{2}\right)^n+\dfrac{12}{5}\left(-\dfrac{1}{3}\right)^n$ which we intend to obtain.

It is permissible that we assign the numerator and denominator of $F(z)$ separately and combine them afterwards so that less mistakes happen in typing or coding due to long expression. Let us carry out the following:

```
>>syms z ↵
```
← Defining variables of $F(z)$ as symbolic, z⇔ z

```
>>N=1/z+1/z^2; ↵
```
← Assigning only the numerator code of $F(z)$ to workspace N, N is user-chosen

```
>>D=(1-1/2/z)*(1+1/3/z); ↵
```
← Assigning only the denominator code of $F(z)$ to workspace D, D is user-chosen

```
>>f=iztrans(N/D); ↵
```
← $F(z)$ is formed by N/D, f holds the inverse transform, f⇔ $f[n]$

```
>>pretty(f) ↵
```
← Displaying the readable form of the string stored in f

```
                                    n              n
   -6 charfcn[0](n) + 18/5 (1/2)  + 12/5 (-1/3)
```
← charfcn[0](n)⇔ $\delta[n]$

Example C

Transform look-up table of MATLAB does not contain the inverse of the function like $F(z)=\ln(1-4z)$ therefore the response would be iztrans(log(1-4*z),z, n) while applying the function.

Example D

The transform function $F(z)=\dfrac{3}{(2-\frac{2}{3}z^{-1})^2(2-3z^{-1})(1-4z^{-1})}$ has the inverse counterpart $f[n]=\dfrac{195}{5929}3^{-n}+\dfrac{3}{308}n3^{-n}-\dfrac{729}{1960}\left(\dfrac{3}{2}\right)^n+\dfrac{432\times 4^n}{605}$ which we wish to find.

It is better if we assign the numerator and denominator strings of the given $F(z)$ separately and combine them latter as follows:

```
>>syms z ↵
```
← Defining variables of $F(z)$ as symbolic, z⇔ z

```
>>N=3; ↵
```
← Assigning only the numerator string of $F(z)$ to workspace N

```
>>D=(2-2/3/z)^2*(2-3/z)*(1-4/z); ↵
```
← Assigning only the denominator string of $F(z)$ to workspace D

```
>>f=iztrans(N/D); ↵     ← Z^-1 on F(z) formed by N/D, f holds the string of f[n]
>>pretty(f) ↵           ← Displaying the readable form of the string stored in f

      195       n                  n       729        n    432   n
     ------- (1/3)  + 3/308 (1/3)  n - ------- (3/2)  + ------ 4
      5929                                1960             605
```

7.5 Z transform on difference equations

Differential equations are applicable for continuous time systems but difference equations are for the discrete time systems. The equation $y_k = y_{k-1} + y_{k-2}$ is an example of the difference equation. The equation says that for any k, the output y is the sum of the preceding two terms. Another style of writing the difference equation $y_k = y_{k-1} + y_{k-2}$ is $y[k] = y[k-1] + y[k-2]$.

MATLAB style of representing the discrete dependent variable y_k or $y[k]$ is provided as follows:

```
>>yk=sym('y(k)'); ↵       ← Describing y_k as symbolic and assigning it to yk
>>yk1=sym('y(k+1)'); ↵ ← Describing y_{k+1} as symbolic and assigning it to yk1
>>yk2=sym('y(k+2)'); ↵ ← Describing y_{k+2} as symbolic and assigning it to yk2
                                    ………., and so on.
```

In above implementations the yk, yk1, and yk2 are user-chosen names. Order of a difference equation is the difference between the largest and smallest indices appearing in the equation. Thus the order of difference equation $y_k = y_{k-1} + y_{k-2}$ is $k-(k-2)=2$.

In signal and system course largely we seek for the Z transform system function from a difference equation or a system of difference equations. The procedure is very similar to the one we encountered in determining system functions from differential equations by using the Laplace transform (section 6.4).

Now we concentrate on how the Z transform (section 7.2) applies on using ztrans to various dependent variables of a difference equation.

for $Z[y_k] = Y(z)$,

```
>>yk=sym('y(k)'); ↵
>>ztrans(yk) ↵

ans =

ztrans(y(k),k,z)
```

for $Z[y_{k+1}] = zY(z) - y[0]z$,

```
>>yk1=sym('y(k+1)'); ↵
>>ztrans(yk1) ↵

ans =

z*ztrans(y(k),k,z)-y(0)*z
```

for $Z[y_{k+2}] = z^2Y(z) - y[0]z^2 - y[1]z$,

```
>>yk2=sym('y(k+2)'); ↵
>>ztrans(yk2) ↵

ans =

z^2*ztrans(y(k),k,z)-y(0)*z^2-y(1)*z
```

for $Z[y_{k+3}] = z^3Y(z) - y[0]z^3 - y[1]z^2 - y[2]z$,

```
>>yk3=sym('y(k+3)'); ↵
>>ztrans(yk3) ↵
```

```
ans =

z^3*ztrans(y(k),k,z)-y(0)*z^3-y(1)*z^2-y(2)*z
```
and, so on.

For example we have $Z[\,y_{k+2}\,] = z^2Y(z) - z^2y[0] - y[1]z$ where $Y(z)$ is the Z transform of y_k which is equivalent to the return string ztrans(y(k),k,z). The yk, yk1, yk2, and yk3 are user-chosen variables. You could have executed ztrans(sym('y(k+2)')) instead of assigning to some intermediate variable.

Regarding the initial conditions, the equivalence between the symbolic notation and MATLAB counterparts are $y[0] \Leftrightarrow$ y(0), $y[1] \Leftrightarrow$ y(1), $y[2] \Leftrightarrow$ y(2), and so on.

So far our representations have been focused on positive indices. Situation may come when we have negative indices, few examples of which are the following maintaining ongoing function and symbology:

for $Z[\,y_{k-1}\,] = \frac{1}{z}Y(z)$,

```
>>yk_1=sym('y(k-1)'); ↵
>>ztrans(yk_1) ↵

ans =

1/z*ztrans(y(k),k,z)
```

for $Z[\,y_{k-2}\,] = \frac{1}{z^2}Y(z)$,

```
>>yk_2=sym('y(k-2)'); ↵
>>ztrans(yk_2) ↵

ans =

1/z^2*ztrans(y(k),k,z)
```

for $Z[\,y_{k-3}\,] = \frac{1}{z^3}Y(z)$,

```
>>yk_3=sym('y(k-3)'); ↵
>>ztrans(yk_3) ↵

ans =

1/z^3*ztrans(y(k),k,z)
```

for $Z[\,y_{k-4}\,] = \frac{1}{z^4}Y(z)$,

```
>>yk_4=sym('y(k-4)'); ↵
>>ztrans(yk_4) ↵

ans =

1/z^4*ztrans(y(k),k,z)
```

The yk_1, yk_2, etc are user-chosen variables. Note that unilateral Z transform on the negative indices does not return initial value related strings. Let us see following examples on the Z transform system function (most terminologies of section 6.4 are applicable here) from difference equations.

Example A

The difference equation $y_{k+2} = -4y_{k+1} - 4y_k$ with $y_0 = 1$ and $y_1 = 2$ has the system function $Y(z) = \frac{z^2 + 6z}{z^2 + 4z + 4}$ which we intend to obtain.

As a procedural step we rearrange the given equation so that the right side of the given equation is 0. We put the whole equation (instead of individual dependent variable) as a symbolic object by using the command sym to some user-chosen variable E as follows:

```
>>E=sym('y(k+2)+4*y(k+1)+4*y(k)'); ↵
```
← E holds the equation

Then forward unilateral Z transform is taken by the ztrans as follows:

```
>>T=ztrans(E); ↵
```
← Z transform on E and assigning that to T where T is user-chosen

The string held in T has ztrans(y(k),k,z) for the $Y(z)$ and initial conditions y(0) and y(1) for $y_0 = 1$ and $y_1 = 2$ respectively. For convenience we replace ztrans(y(k),k,z) by Y as follows:

```
>>syms Y ↵          ← Defining Y as symbolic where Y⇔Y(z)
>>T=subs(T,{'ztrans(y(k),k,z)'},Y); ↵ ← Assigning return to T again
>>T=subs(T,{'y(0)','y(1)'},{'1','2'}); ↵ ← Substituting y0 =1 and y1 = 2
>>Y=solve(T,Y); ↵          ← Finding Y(z) by forming an equation T=0
>>pretty(Y) ↵              ← Displaying Y(z)
          z (z + 6)
        ----------------
          2
        z  + 4 z + 4
```

Example B

It is given that $Y(z) = \dfrac{z^2(2z^2 - 3z - 1)}{(z-1)^3(z-2)(z-3)}$ is the system function for the $y[n+2] - 5y[n+1] + 6y[n] = 3 - n^2$ subject to $y[0]$=0 and $y[1]$=2. We wish to verify this.

Turning the right side of the equation to 0, we assign that to E as follows:

```
>>E=sym('y(n+2)-5*y(n+1)+6*y(n)-3+n^2'); ↵
>>T=ztrans(E); ↵       ←T holds the Z transform on E
>>E=subs(T,{'ztrans(y(n),n,z)','y(0)','y(1)'},{'Y','0','2'}); ↵ ← Changing
                                   the transform string and applying the initial data
>>Y=solve(E,'Y'); ↵               ← Find Y(z) from E=0
>>pretty(Y) ↵                     ← Readable form of Y(z)
      2          2
   z (-1 + 2 z  - 3 z)
 ----------------------------
        3   2
 (z - 1)  (z  - 5 z + 6)
```

Example C

The system of difference equations $\left\{\begin{array}{l} y[n+1] - x[n-1] = 2^n \\ x[n] - 9y[n] = 3\delta[n] \end{array} \quad y[0] = 3\right\}$ has the system functions $X(z) = \dfrac{3z^2(10z-17)}{(z-2)(z+3)(z-3)}$ and $Y(z) = \dfrac{3z^3 - 5z^2 + 3z - 6}{(z-2)(z+3)(z-3)}$ which we intend to obtain.

The reader is suggested to go through the sections 6.4 and 6.7 before solving this problem. Solution procedure is similar to those in the two sections. First we enter given equations after rearranging to E1 and E2 respectively as follows:

```
>>E1=sym('y(n+1)-x(n-1)-2^n'); ↵              ← Enter equation 1 to E1
>>E2=sym('x(n)-9*y(n)-3*charfcn[0](n)'); ↵    ← Enter equation 2 to E2
```

Then we take the transform to each equation and again assign it to like name variable as follows:

```
>>E1=ztrans(E1); ↵          ← Z transform on the equation 1
>>E2=ztrans(E2); ↵          ← Z transform on the equation 2
```

Long strings of the transform i.e. ztrans(x(n),n,z) and ztrans(y(n),n,z) are to be replaced by X and Y respectively so we declare them as symbolic as follows:

```
>>syms X Y ↵
```

Substituting the long strings and assigning again to the like name variables take place as follows:

```
>>E1=subs(E1,{'ztrans(y(n),n,z)','ztrans(x(n),n,z)'},{Y,X}); ↵
>>E2=subs(E2,{'ztrans(y(n),n,z)','ztrans(x(n),n,z)'},{Y,X}); ↵
>>E1=subs(E1,{'y(0)'},'3'); ↵        ← Initial condition only on the equation 1
>>R=solve(E1,E2,X,Y) ↵               ← Calling solve for solution

R =                                  ← X(z) and Y(z) are returned as a structured array
        X: [1x1 sym]
        Y: [1x1 sym]
>>X=R.X; Y=R.Y; ↵                    ← Picking up individual component and
                                        assigning to like name variable again
>>pretty(X) ↵                        ← Displaying the readable form of X, X⇔ X(z)
         2
        z  (-17 + 10 z)
    3 -------------------------
             2    3
      -2 z  + z  - 9 z + 18
>>pretty(Y) ↵                        ← Displaying the readable form of Y, Y⇔ Y(z)
               3      2
       3 z + 3 z  - 5 z  - 6
      -------------------------
             2    3
      -2 z  + z  - 9 z + 18
```

7.6 Solving difference equations for discrete systems

Built-in function rsolve finds the sequence solution from a difference equation but in maple (appendix C.5). The function is applied with the syntax rsolve({equation, initial condition 1, initial condition 2, so on}, dependent variable). Let us see the following three examples in this regard.

Example A

The difference equation $y_{k+2} = -4y_{k+1} - 4y_k$ subject to $y_0 = 1$ and $y_1 = 2$ has the solution $y_k = 3(-2)^k + (-2k-2)(-2)^k$ which we intend to obtain.

As a procedural step, first we assign given difference equation in maple as follows:

```
>>maple('E:=y(k+2)=-4*y(k+1)-4*y(k)'); ↵      ← k⇔ k
```

In just conducted execution the E is any user-chosen variable and := is the maple assignment operator. Coding of the dependent variable follows the section 7.5 cited technique for example y_{k+2} ⇔y(k+2). Then we call the rsolve within maple with the said syntax as follows:

```
>>S=maple('rsolve({E,y(0)=1,y(1)=2},y(k))'); ↵   ← Section 7.5 for initial
                                                    condition entering
```

We assigned the return from maple to some user-chosen variable S but in MATLAB which holds the code of the solution. The command sym on the S provides the character readable form in conjunction with the pretty as follows:

```
>>pretty(sym(S)) ↵
              k                k
   3 (-2)  + (-2 k - 2) (-2)
```

Example B

The difference equation $y[n+2]-5y[n+1]+6y[n]=3-n^2$ subject to initial conditions $y[0]=0$ and $y[1]=2$ has the solution $y[n]=-2^{n+1}+3^{1+n}-(n+1)\left(\frac{n}{2}+1\right)$ which we intend to obtain.

We assign the given difference equation to E in maple as follows:

```
>>maple('E:=y(n+2)-5*y(n+1)+6*y(n)=3-n^2'); ↵        ← n⇔n ,y(n) ⇔ y[n]
```

Then we call the rsolve with earlier cited syntax as follows:

```
>>S=maple('rsolve({E,y(0)=0,y(1)=2},y(n))'); ↵
>>pretty(sym(S)) ↵
           n      n
   -2 2  + 3 3  - (n + 1) (1/2 n + 1)
```

Example C

A negative index difference equation is also solvable by the rsolve. The equation $y[n-2]-2y[n-1]+y[n]=\sin n$ subject to the initial conditions $y[0]=0$ and $y[1]=0$ has the solution $y[n]=\dfrac{\sin(n+1)-n\sin 2+n\sin 1-\sin 1}{2\cos 1-2}$ which we intend to obtain.

Maintaining ongoing symbology and function we exercise the following:

```
>>maple('E:=y(n-2)-2*y(n-1)+y(n)=sin(n)'); ↵  ← n⇔n , y(n)⇔ y[n]
>>S=maple('rsolve({E,y(0)=0,y(1)=0},y(n))'); ↵
>>pretty(simple(sym(S))) ↵                    ← simple is used for simplification

   sin(n + 1) - sin(2) n + sin(1) n - sin(1)
   -----------------------------------------
                2 cos(1) - 2
```

With this example we bring an end to the chapter.

Chapter 8
SYSTEM IMPLEMENTATION

Introduction

The definition of a system is vital to the context of the text. What constitutes a system obviously an electrical system? Broadly speaking, any black box which contains mainly resistor, inductor, and capacitor in conjunction with other electrical elements is a system. In order to study such system we apply computer means to define it that is where our focus is. Widely exercised system terms in the course have been given computer codes or proper black box representations in this chapter by the following but not limited to:

- Electrical systems and their defining codes/blocks
- Different systems evolving from space-space, feedback, transfer functions, or other forms
- Way to develop circuit system either from expression or block diagram

8.1 What is a system?

In engineering study the concept of a system is extremely important. Almost all branches of science and engineering share a common terminology source-system-sink. The source usually indicates what we apply to the system and the sink indicates what quantity we are getting form the system. Of coarse the source-system-sink terminology is problem oriented. Think about the stock market, the money we put is the source, stock market is the system, and the investment return is the sink.

Maybe the example is too silly. Let us see the passband filter circuit of the figurer 8.1(a) in which we apply the voltage V_i and interested to see the voltage V_0 across the capacitor therefore the V_i, the whole passband filter circuit, and V_0 are source, system, and sink respectively.

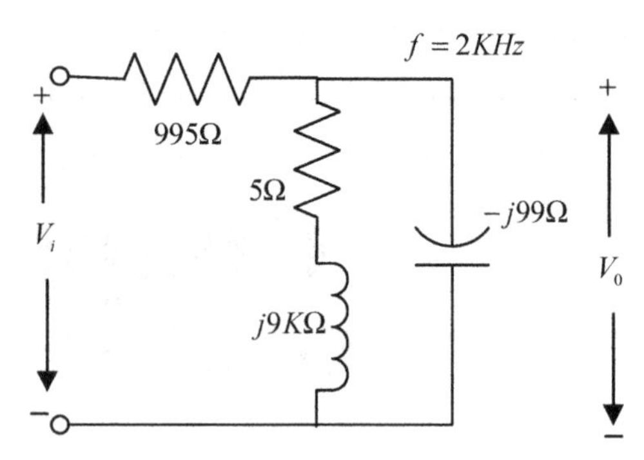

Figure 8.1(a) A passband filter circuit

The V_i and V_0 do not have to be a voltage, they can be other quantities too for example current. Here in the passband filter circuit we have one input and one output but multiple inputs and outputs can also be associated with the circuit. Figure 8.1(b) shows the same filter circuit with one input V_i and two outputs V_0 and I_0. Also the number of inputs can be more than one.

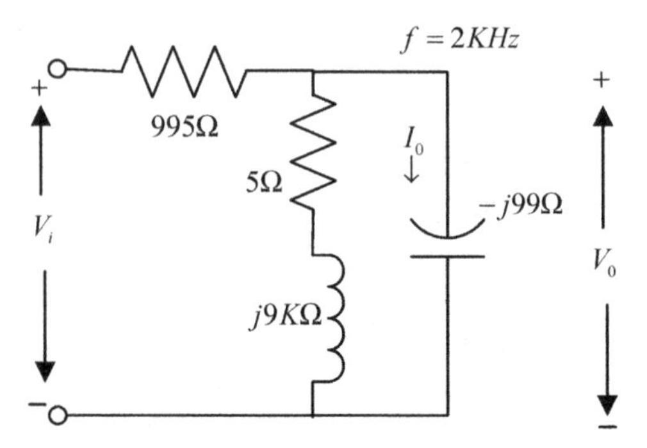

Figure 8.1(b) The passband filter circuit of figure 8.1(a) with two outputs

The concept of system is shown in figure 8.1(c). The whole circuit information is hidden in the system. Once we know our system, we can apply different V_i and investigate its corresponding response of V_0 and I_0 from the system.

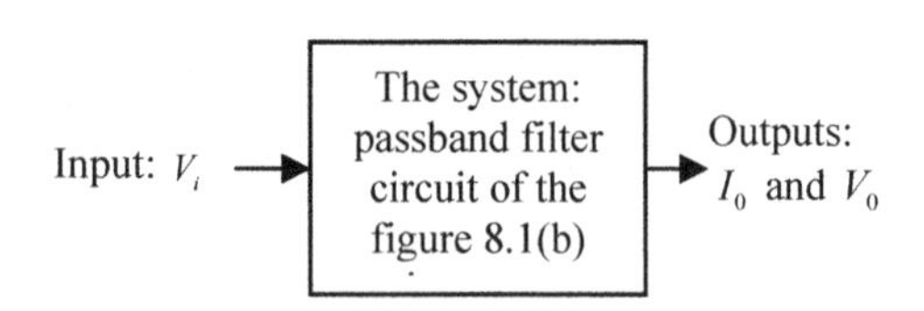

Figure 8.1(c) The passband filter circuit of the figure 8.1(b) forms a system

When do we construct a system? A similar engineering problem often follows a definitive mathematical model. We form a system from the mathematical model and study it for different inputs rather than perform physical experiments, which is very convenient in all regards – cost, factors due to physical implementation, etc.

8.2 How to define a continuous system?

To the context of signal and system course, a system is defined primarily from the transfer function $H(s)$ expression or from its state space model $\{A, B, C, D\}$. It is understood that the $H(s)$ is expressible in numerator and denominator polynomial forms. The MATLAB built-in function tf (abbreviation for the transfer function) is applied to define the $H(s)$ with the syntax tf(numerator polynomial coefficients as a row matrix, denominator polynomial coefficients as a row matrix) where the polynomial coefficients must be in descending order and any missing coefficient is set to 0.

Suppose a continuous system is represented by the transfer function $H(s)=\frac{7s^3-7s+42}{s^4-118s^2-240s}$. The example $H(s)$ has the polynomial coefficient representation as [7 0 -7 42] and [1 0 -118 -240 0] for the numerator and the denominator respectively. Having known so, we define the system as follows:

```
>>H=tf([7 0 -7 42],[1 0 -118 -240 0]) ↵

Transfer function:
   7 s^3 - 7 s + 42
----------------------------
s^4 - 118 s^2 - 240 s
```

In above execution we assigned the return from the tf to the workspace H (it can be any user-supplied variable). If we append a semicolon at the end of the statement i.e. H=tf([7 0 -7 42],[1 0 -118 -240 0]);, the functional popup is not displayed. However the variable H holds the $H(s)$ information as a system.

Not always is the transfer function in numerator-denominator form. When the transfer function is given in pole-zero-gain form, we employ the MATLAB built-in function zpk with the syntax zpk(zeroes as a row matrix, poles as a row matrix, gain as a single number). For example the $\left\{\begin{matrix}\text{zeroes}: 2, 3, -1\\ \text{poles}: 5, 0, 8, 6\\ \text{gain}: 7\end{matrix}\right\}$ forms the system function $H(s)$ as $\frac{7(s-2)(s-3)(s+1)}{(s-5)s(s-8)(s-6)}$ which we enter as follows:

```
>>z=[2 3 -1]; ↵     ← Assigning the zeroes as a row matrix to z, z is user-chosen
>>p=[5 0 8 6]; ↵    ← Assigning the poles as a row matrix to p, p is user-chosen
>>k=7; ↵            ← Assigning the gain as a scalar to k, k is user-chosen
>>H=zpk(z,p,k) ↵ ← Calling the zpk with the mentioned syntax and assigned the
                     return from zpk to H where H is user-chosen variable
Zero/pole/gain:
7 (s-2) (s-3) (s+1)
-----------------------
s (s-5) (s-6) (s-8)
```

We could have executed all commands in one line like H=zpk([2 3 -1],[5 0 8 6],7) instead of intermediate assigning. Appending a semicolon at the end of the statement i.e. H=zpk(z,p,k); does not show the functional popup. The last variable H holds the $H(s)$ information as a system from the pole-zero-gain function.

Mixed form like $\frac{s+42}{(s^4+3s^2-240)(s+2)(2s-1)}$ is manipulated until it takes either form i.e. the numerator-denominator or the pole-zero-gain. Polynomial multiplication is required in such system function which can be conducted by using the built-in function conv with the syntax conv(first polynomial as a row matrix, second polynomial as a row matrix) but maintaining descending power of s and setting missing coefficient to 0 in the representation.

The mixed function had better be in numerator-denominator form. The multiplication of the denominator factor $(s+2)(2s-1)$ then takes place by conv([1

2],[2 -1]) with that the whole $(s^4+3s^2-240)(s+2)(2s-1)$ is obtained by conv([1 0 3 0 -240],conv([1 2],[2 -1])). Knowing so, the complete system H we form by H=tf([1 42],conv([1 0 3 0 -240],conv([1 2],[2 -1]))). In order to avoid typing mistakes, intermediate assignee variables can be used.

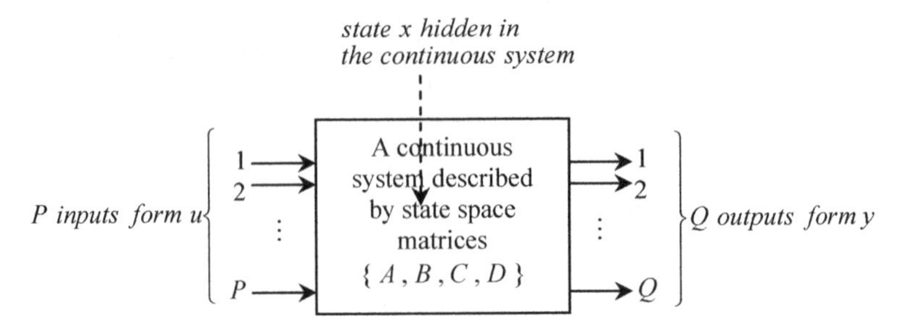

Figure 8.1(d) State space model of a continuous system

As quoted earlier a continuous system is also modeled by the state space form, governing equation of which is $\begin{Bmatrix} \dot{x}=Ax+Bu \\ y=Cx+Du \end{Bmatrix}$ where $\{A, B, C, D\}$ are called the state space matrices. The state x (in general is a column vector) is hidden in the continuous system and is associated with the internal parameters of the system.

For instance if the system is an electric circuit, the states are the inductor current and capacitor voltage. If the system is a mechanical one, the velocity of the damping and the displacement due to the compliance constitute the states. The matrix A carries the system parameter information. For an electric circuit, the A is purely a function of resistance, inductance, and capacitance. For a mechanical object, the A is composed of inertia, damping, and compliance. In general orders of A, B, C, and D are $N\times N$, $N\times P$, $Q\times N$, and $Q\times P$ respectively, all matrices have real constant elements, P and Q are the numbers of inputs and outputs respectively, and P, Q, and N are integers. Figure 8.1(d) shows the schematics of the state space representation.

MATLAB built-in function ss (abbreviation for state space) defines a continuous system from its state space matrix representation with the syntax ss(A, B, C, D). Matrix entering discussion is seen in subsection 1.1.2.

A continuous system described by $\{A, B, C, D\}$ — $u(t)$ → $y(t)$

Figure 8.1(e) A single input - single output continuous system

A single input and single output system like the figure 8.1(e) is defined by $A=\begin{bmatrix} -2 & -1 \\ 2 & 3 \end{bmatrix}$, $B=\begin{bmatrix} -1 \\ -2 \end{bmatrix}$, $C=[-2 \quad 1]$, and $D=[-1]$ which we wish to define.

Let us go through the following in this regard:

```
>>A=[-2 -1;2 3]; B=[-1;-2]; C=[-2 1]; D=-1; ↵
```

```
>>H=ss(A,B,C,D) ↵   ← Calling the ss with the mentioned syntax and the return is
                      assigned to H where H is user-chosen variable
a =                 ← Meaning A
         x1  x2
      x1 -2  -1
      x2  2   3
b =                 ← Meaning B
          u1
      x1  -1
      x2  -2
c =                 ← Meaning C
       x1   x2
   y1 -2     1
d =                 ← Meaning D
         u1
    y1   -1
Continuous-time model.
```

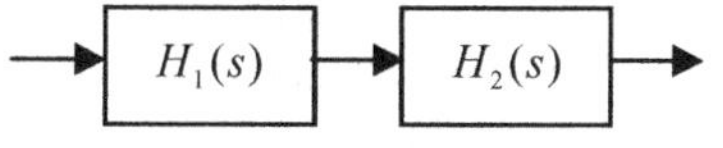

Figure 8.2(a) Two system functions connected in series

In last implementation the first command line is basically the entering of given matrices to like name variables for example A to A. Usage of H=ss(A,B, C,D); does not show the matrix popup. The x1 and x2 mean two states. The u1 and y1 mean one input and one output respectively. The workspace H holds the continuous system information of the figure 8.1(e) which we wanted.

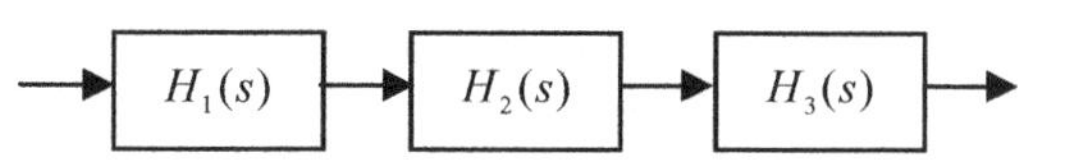

Figure 8.2(b) Three system functions connected in series

8.3 Series/parallel systems

System functions can be interconnected in series or parallel form. In either case built-in function is embedded in MATLAB to find the equivalent system function. Equivalent means we can replace all constituent system functions by a single one.

♦♦ Seriesly connected system functions

As an example figure 8.2(a) shows two system functions $H_1(s)=\frac{5s^2-s+1}{s^3-1}$ and $H_2(s)=\frac{5.43(s-3)}{(s-1)(s+4)}$ connected in series. Their s domain equivalent system function is given by $H_{eq}(s)=H_1(s)\ H_2(s)=\frac{27.15s^3-86.88s^2+21.72s-16.29}{s^5+3s^4-4s^3-s^2-3s+4}$ which we wish to implement in MATLAB.

MATLAB built-in function series helps us compute the series equivalent system function with the syntax series(system 1 representing variable name, system

2 representing variable name). Section 8.2 explained tf and zpk define the $H_1(s)$ and $H_2(s)$ as follows respectively:

>>H1=tf([5 -1 1],[1 0 0 -1]); ↵ ← $H_1(s)$ is defined from numerator and denominator coefficients and assigned to workspace H1 where H1 is user-chosen name and H1⇔ $H_1(s)$

>>H2=zpk(3,[1 -4],5.43); ↵ ← $H_2(s)$ is defined from pole-zero-gain and assigned to workspace H2 where H2 is user-chosen name and H2⇔ $H_2(s)$

>>Heq=series(H1,H2) ↵ ← $H_{eq}(s)$ is computed by the series and assigned to Heq where Heq is user-chosen name and Heq⇔ $H_{eq}(s)$

```
Zero/pole/gain:
27.15 (s-3) (s^2  - 0.2s + 0.2)
--------------------------------------
 (s-1)^2 (s+4) (s^2  + s + 1)
```

As the execution says the Heq contents are in the pole-zero-gain form. If we exercise the tf on the Heq, we find the equivalent system function as follows:

>>tf(Heq) ↵

```
Transfer function:
27.15 s^3 - 86.88 s^2 + 21.72 s - 16.29
---------------------------------------------------     ← the H_eq(s) we expected
  s^5 + 3 s^4 - 4 s^3 - s^2 - 3 s + 4
```

Suppose we have three system functions in series like the figure 8.2(b) where $H_3(s) = \dfrac{3.2(s-2.2)}{(s+1)(s+1.5)}$ and $H_1(s)$ and $H_2(s)$ are the foregoing ones. With these three system functions, we have the equivalent $H_{eq}(s) = H_1(s)\ H_2(s)\ H_3(s)$ by using the following:

>>H3=zpk(2.2,[-1 -1.5],3.2); ↵ ← $H_3(s)$ is defined from pole-zero-gain and assigned to workspace H3 where H3 is user-chosen name and H3⇔ $H_3(s)$

>>Heq=series(series(H1,H2),H3); ↵

In the last command the inner series(H1,H2) finds the series equivalent of $H_1(s)$ and $H_2(s)$ and the outer series does the resulting one and $H_3(s)$. The last Heq

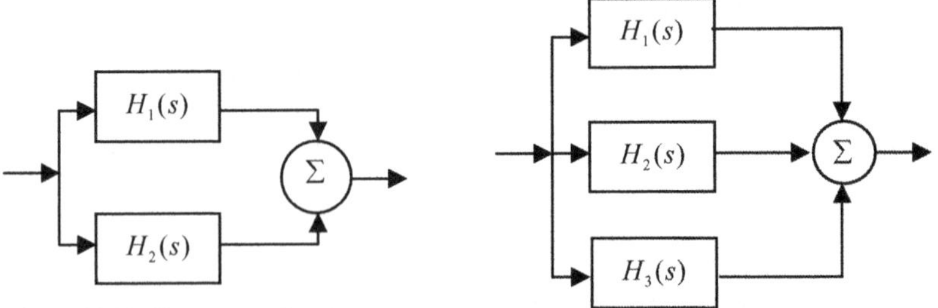

Figure 8.2(c) Two system functions connected in parallel

Figure 8.2(d) Three system functions connected in parallel

keeps the equivalent of the three system functions. If you intend to view it, execute tf(Heq) at the command prompt.

♦ ♦ Parallelly connected system functions

Figure 8.2(c) shows two system functions connected in parallel. Taking the example system functions $H_1(s)$ and $H_2(s)$ of the series connection into account, we should have the equivalent as $H_{eq}(s) = H_1(s) + H_2(s) = \dfrac{10.43s^4 - 2.29s^3 - 22s^2 + 1.57s + 12.29}{s^5 + 3s^4 - 4s^3 - s^2 - 3s + 4}$. MATLAB built-in function parallel determines the equivalent system function of two systems connected in parallel with the syntax parallel(system 1 representing variable name, system 2 representing variable name). We know that the $H_1(s)$ and $H_2(s)$ are stored in the workspace variables H1 and H2 respectively therefore we execute the following:

>>Heq=parallel(H1,H2); ↵ ← $H_{eq}(s)$ is computed by the parallel and assigned to Heq where Heq is user-chosen name and Heq⇔ $H_{eq}(s)$

Just to view the computation in coefficient form, let us carry out the following:

```
>>tf(Heq) ↵

Transfer function:
10.43 s^4 - 2.29 s^3 - 22 s^2 + 1.57 s + 12.29
------------------------------------------------------------
        s^5 + 3 s^4 - 4 s^3 - s^2 - 3 s + 4
```

Figure 8.2(d) presents three system functions connected in parallel whose equivalent is easily found by exercising Heq=parallel(parallel (H1,H2),H3); at the command prompt based on the foregoing $H_1(s)$, $H_2(s)$, and $H_3(s)$ where the symbols have the previously mentioned meanings and mathematically Heq returns $H_1(s) + H_2(s) + H_3(s)$.

Figure 8.2(e) A negative feedback system

8.4 Feedback systems

A feedback system is composed of forward path system function $G(s)$ and feedback path system function $H(s)$. Figures 8.2(e) and 8.2(f) show the feedback system diagram for negative and positive feedbacks respectively. In each configuration the $R(s)$ and $Y(s)$ represent the system functions of the input and output respectively.

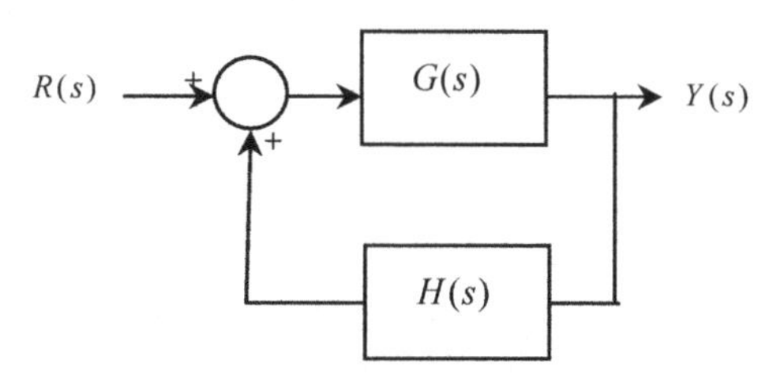

Figure 8.2(f) A positive feedback system

MATLAB built-in function feedback implements both feedback systems of the figures 8.2(e) and 8.2(f). The syntax we apply is feedback($G(s)$, $H(s)$,type of feedback). The $G(s)$ or $H(s)$ is entered by employing the section 8.2 mentioned built-in functions regardless of the numerator-denominator, pole-zero-gain, or state-space form. The type of feedback is +1 for positive feedback and -1 for negative feedback. Let us see the following examples in this regard.

⬧⬧ Negative feedback

In figure 8.2(e) when $G(s)=\frac{3s}{2s^2+3s-4}$ and $H(s)=\frac{2}{s+1}$, it is given that the overall system function is $\frac{Y(s)}{R(s)}=\frac{G(s)}{1+G(s)H(s)}=\frac{3s^2+3s}{2s^3+5s^2+5s-4}$ which we wish to obtain.

Applying section 8.2 cited tf we enter the $G(s)$ and $H(s)$ as follows:

```
>>G=tf([3 0],[2 3 -4]); ↵     ← G holds G(s), G is user-chosen
>>H=tf(2,[1 1]); ↵            ← H holds H(s), H is user-chosen
```

After that we call the feedback with just cited syntax and assign the return from the function to E (some user-chosen variable) as follows:

```
>>E=feedback(G,H,-1) ↵   ← E holds the Y(s)/R(s)

Transfer function:
        3 s^2 + 3 s
------------------------------
2 s^3 + 5 s^2 + 5 s – 4
```

Usage of command E=feedback(G,H,-1); does not display the functional popup.

⬧⬧ Positive feedback

It is given that the overall system function is $\frac{Y(s)}{R(s)}=\frac{G(s)}{1-G(s)H(s)}=\frac{3s^2+3s}{2s^3+5s^2-7s-4}$ for the figure 8.2(f) with negative feedback mentioned system functions which we wish to obtain. All we need is change the third input argument of the function as follows:

```
>>E=feedback(G,H,1) ↵   ← E holds the Y(s)/R(s)

Transfer function:
        3 s^2 + 3 s
-----------------------------
2 s^3 + 5 s^2 - 7 s - 4
```

8.5 Modeling connected continuous systems

In previous sections we mainly employed built-in MATLAB functions to define various forms of systems. SIMULINK (chapter 1) keeps pre-designed blocks for implementing interconnected continuous systems. The pre-designed blocks are for specific type of system functions. Also there are dedicated blocks for the input to and output from the system as explained in the following.

♦♦ System function from numerator-denominator form

If a system function is in numerator-denominator form, the block Transfer Fcn models the system. Open a new SIMULINK model file (subsection 1.2.2) and get the Transfer Fcn block in the model file following the link mentioned in appendix A. The block's icon appearance is seen in figure 8.3(a). Doubleclick the block to see its parameter window like the figure 8.3(d) and here in this window we enter the system function data.

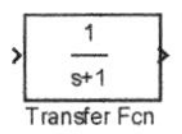

Figure 8.3(a) Icon appearance of Transfer Fcn block

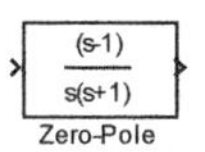

Figure 8.3(b) Icon appearance of Zero-Pole block

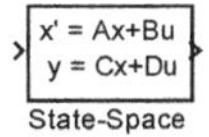

Figure 8.3(c) Icon appearance of State-Space block

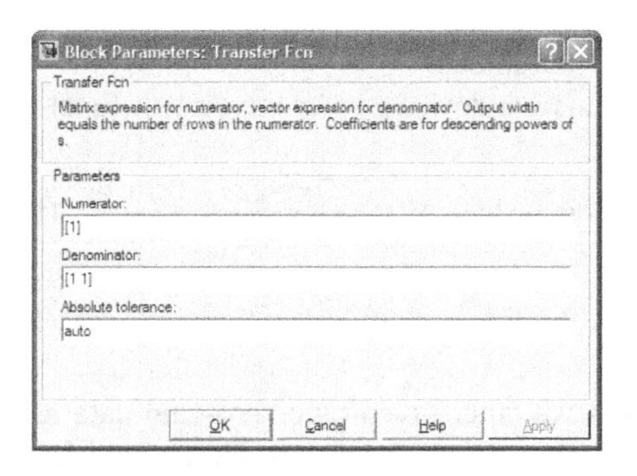

Figure 8.3(d) Block parameter window of the Transfer Fcn

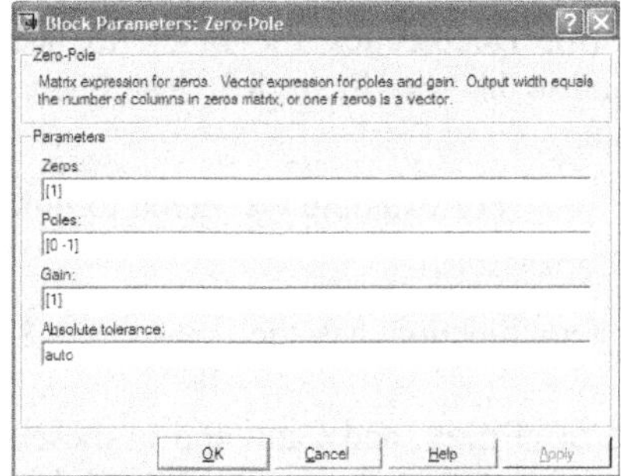

Figure 8.3(e) Block parameter window of the Zero-Pole

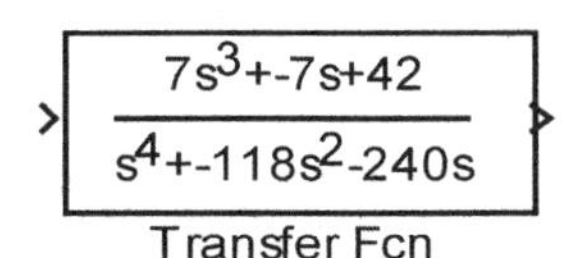

Figure 8.3(f) Transfer Fcn models $H(s)$

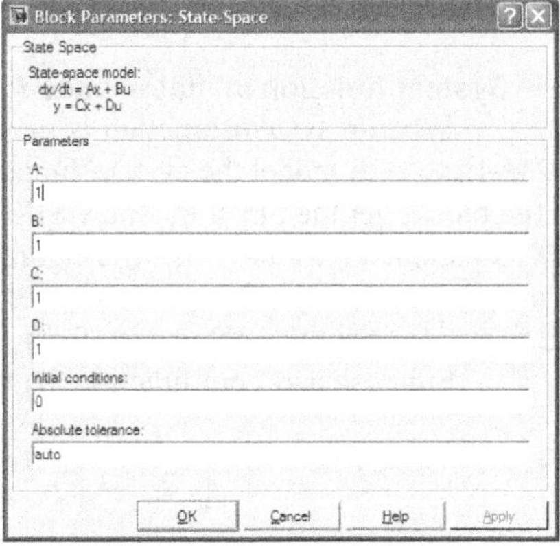

Figure 8.3(g) Block parameter window of the State-Space

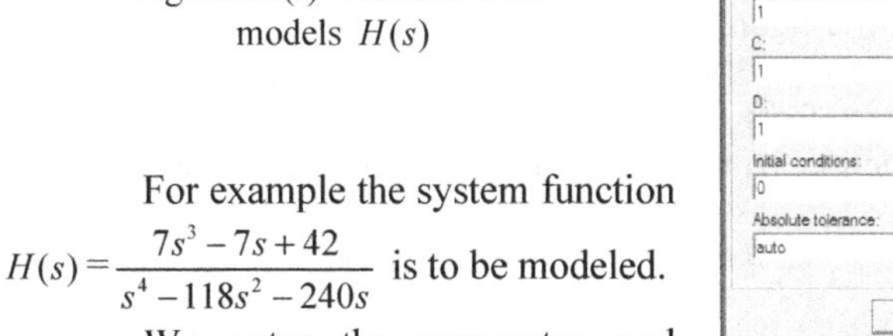

For example the system function $H(s)=\dfrac{7s^3-7s+42}{s^4-118s^2-240s}$ is to be modeled.

We enter the numerator and denominator polynomial coefficients as a row matrix in the slots of Numerator and Denominator of the figure 8.3(d) respectively where polynomial coefficients must be in descending order and any missing coefficient is set to 0. The example $H(s)$ has the polynomial coefficient representation as [7 0 -7 42] and [1 0 -118 -240 0] for numerator and denominator respectively. Having known so, we enter these two row matrices in the respective slot of the figure 8.3(d) deleting the default values. One important point is to be kept in mind that the degree of the numerator polynomial must not be more than the degree of denominator. However upon entering the row matrices you find the block appearance changed. You see den(s) in the denominator because the expression overfits the default block size. Enlarge the block (subsection 1.2.3) to see its contents like the figure 8.3(f) i.e. the Transfer Fcn of the figure 8.3(f) models the $H(s)$.

♦♦ System function from pole-zero-gain form

If a system function is in pole-zero-gain form, the Zero-Pole block models the system. Open a new SIMULINK model file and get the block in the model file following the link mentioned in appendix A. The block's icon appearance is seen in figure 8.3(b). Doubleclick the block to see its parameter window like the figure 8.3(e) and here in this window we enter the system function data.

$$\frac{7(s-2)(s-3)(s+1)}{(s-5)s(s-8)(s-6)}$$

Zero-Pole

Figure 8.3(h) Zero-Pole models $H(s)$

For example we intend to model the system $H(s)=\dfrac{7(s-2)(s-3)(s+1)}{(s-5)s(s-8)(s-6)}$. This system function has the $\begin{Bmatrix}\text{zeroes}: 2,\ 3,\ -1\\ \text{poles}: 5,\ 0,\ 8,\ 6\\ \text{gain}: 7\end{Bmatrix}$. In figure 8.3(e) you find the slots for Zeros, Poles, and Gain and in these slots we enter given pole-zero-gain data as a row matrix therefore enter [2 3 -1], [5 0 8 6], and 7 as the Zeros, Poles, and Gain in the parameter window of the figure 8.3(e) deleting the default values respectively. In doing so you see the block contents as zeros(s) or poles(s) due to overfitting of the entered data to the default block size. Enlarge the block by the mouse to see its appearance as shown in figure 8.3(h) which models the given system $H(s)$.

♦♦ System function in state-space form

When a system function is given in state-space form, we employ the block State-Space to model the system function. Figure 8.3(c) shows the icon appearance of the block, get the block in a new SIMULINK model file. Doubleclick the block to see its parameter window like the figure 8.3(g). Here in this window we enter given matrix (subsection 1.1.2) information of the system. The reader is referred to section 8.2 for more about the state-space form.

Suppose a system function $H(s)$ is characterized by the state-space matrices $\{A, B, C, D\}$ where $A=\begin{bmatrix}-2 & -1\\ 2 & 3\end{bmatrix}$, $B=\begin{bmatrix}-1\\ -2\end{bmatrix}$, $C=[-2 \quad 1]$, and $D=[-1]$ and which we intend to model.

Deleting default values by using the keyboard from the parameter window of the figure 8.3(g), we enter given matrix information as [-2 -1;2 3], [-1;-2], [-2 1], and -1 in the slots of A, B, C, and D for A, B, C, and D respectively which still keeps the block appearance like the figure 8.3(c) and thereby modeling the $H(s)$.

♦♦ System function in mixed form

Sometimes given system functions are in mixed form i.e. combination of pole-zero-gain and numerator-denominator. In section 8.2 we illustrated how the conv function is used for polynomial multiplication in order to define the numerator-denominator form. We apply the same technique with the Transfer Fcn block.

Mixed form system like $H(s) = \frac{s+42}{(s^4+3s^2-240)(s+2)(2s-1)}$ has the numerator and denominator polynomial coefficients as [1 42] and conv([1 0 3 0 -240],conv([1 2],[2 -1])) and which are entered in the slots of the Numerator and Denominator of the parameter window in figure 8.3(d) respectively. Having entered the coefficients and enlarged the block, we see the system $H(s)$ modeled like the figure 8.3(i).

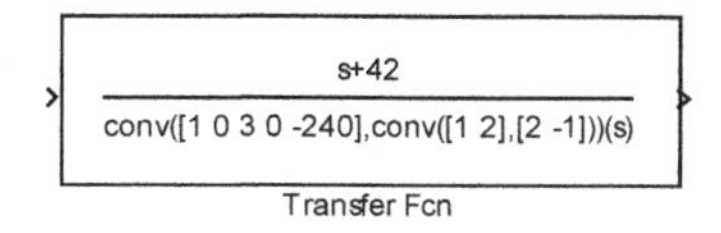

Figure 8.3(i) Modeling mixed form system by Transfer Fcn

Since SIMULINK interacts with MATLAB, we can put the numerator and denominator coefficients in the workspace first if they are long and call the coefficients afterwards in the respective slot in the parameter window. For this example let us carry out the following:

```
>>n=[1 42]; ↵                  ← Assigning numerator coefficients of s+42 to n
>>d1=[1 0 3 0 -240]; ↵         ← Assigning coefficients of s^4+3s^2-240 to d1
>>d2=conv([1 2],[2 -1]); ↵     ← Assigning multiplication of (s+2)(2s-1) to d2
>>d=conv(d1,d2); ↵             ← Forming the complete denominator and assigned to d
```

In above executions, the n, d1, d2, and d are user-chosen names. After that you can enter just n and d in the slots of the Numerator and Denominator of the parameter window in figure 8.3(d) to form the system respectively.

♦♦ Feedback system

In section 8.4 we addressed the computation of system function pertaining to feedback systems. Earlier mentioned Transfer Fcn, Zero-Pole, or State-Space in conjunction with the Sum block (link and appearance in the appendix A) models a feedback system in SIMULINK. Figure 8.2(e) or 8.2(f) shown system function $G(s)$ or $H(s)$ is modeled by any of the three SIMULINK blocks.

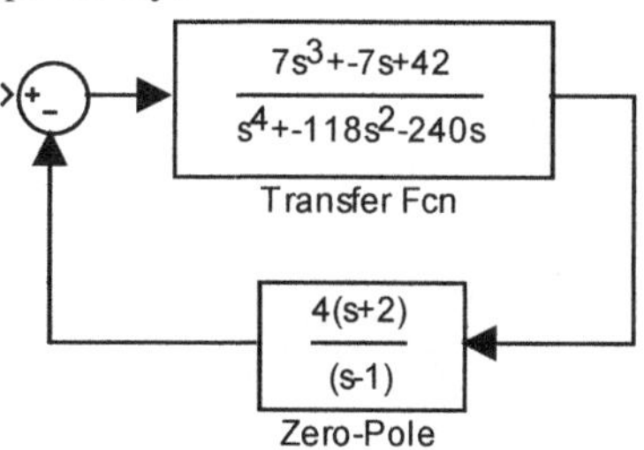

Figure 8.4(a) Modeling the negative feedback system of figure 8.2(e)

For example the negative feedback system of figure 8.2(e) is to be modeled with $G(s) = \frac{7s^3-7s+42}{s^4-118s^2-240s}$ and $H(s) = \frac{4(s+2)}{s-1}$.

The $G(s)$ and $H(s)$ are modeled by the Transfer Fcn and Zero-Pole due to given form respectively. Open a new SIMULINK model file, get one Transfer Fcn, one Zero-Pole, and one Sum blocks in the model file. Doubleclick the Transfer Fcn and enter its Numerator and Denominator as [7 0 -7 42] and [1 0 -118 -240 0] in the parameter window respectively for modeling the $G(s)$. Then doubleclick the Zero-Pole and enter its Zeros, Poles, and Gain as -2, 1, and 4 in the parameter window for modeling the $H(s)$ respectively. Enlarge each of the Transfer Fcn and Zero-Pole to see its contents.

The Sum by default has two inputs and which are related with the List of signs in the parameter window of the block (indicated by ++). On doubleclicking the Sum if we modify that to +-, then the negative port receives the feedback signal. Nevertheless the feedback signal flow in figure 8.2(e) is from right to left so rightclick on the Zero-Pole and click the Flip Block under the Format popup to have the proper orientation of the block in the model. Place so designed blocks relatively and connect them like the figure 8.4(a) which essentially models the negative feedback system.

For positive feedback system of figure 8.2(f) we do not have to change the List of signs of the Sum block and leave it as default ++.

❖ Input to and output from the system

One important point has to be stressed here. Even though we model the system function in Laplace transform domain, system analysis takes place in the time domain in SIMULINK. Accordingly input supplied to and output returned from the system are also in the time domain.

In chapter 3 we have illustrated various types of signal modeling. Any signal of the chapter 3 can be the input to a continuous system. The output from the system can be sent to a Scope (subsection 1.2.5) to view the system return as a function of time.

System functions: $G_1(s)=20$, $G_2(s)=\dfrac{0.3}{s^2+5s+12}$, $G_3(s)=\dfrac{s+0.01}{(s+0.1)(s+0.02)}$, $G_4(s)=s$, and $G_5(s)=\{A, B, C, D\}$ where $A=1$, $B=1.1$, $C=10^{-5}$, and $D=0.5$

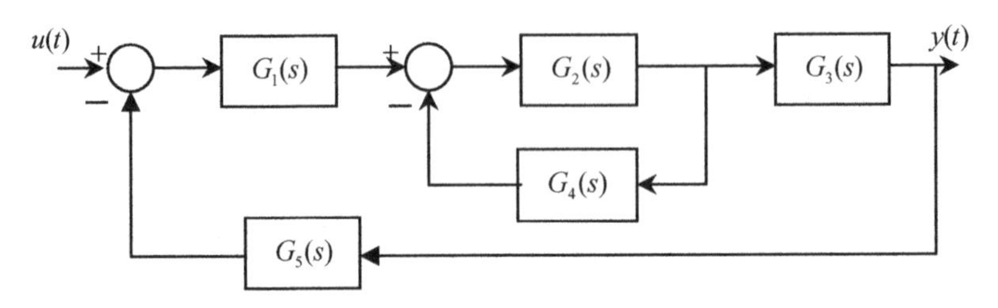

Figure 8.4(b) An interconnected system

❖ Interconnected system

Now we wish to introduce the modeling of an interconnected system.

Figure 8.4(b) shows an interconnected system which is composed of five modular system functions as placed on the top of the figure. Figure 8.4(c) shows the SIMULINK model of the interconnected system in figure 8.4(b). In order to construct the model, we carry out the following:

⇒ open a new simulink model file, bring a Gain block (appendix A) in the model file to model the system $G_1(s)$, doubleclick the Gain, and enter its Gain as 20 in the parameter window of the block

⇒ get a Step block in the model file to model the input $u(t)$, doubleclick the block (figure 3.1(g)) to change its Step time from default 1 to 0 for $u(t)$, and rename the block as u(t)

⇒ get one Sum block in the model file, doubleclick the block to change its List of signs from default ++ to + –, and copy and paste the block to get another one in the model file (because there are two summing points in figure 8.4(b))

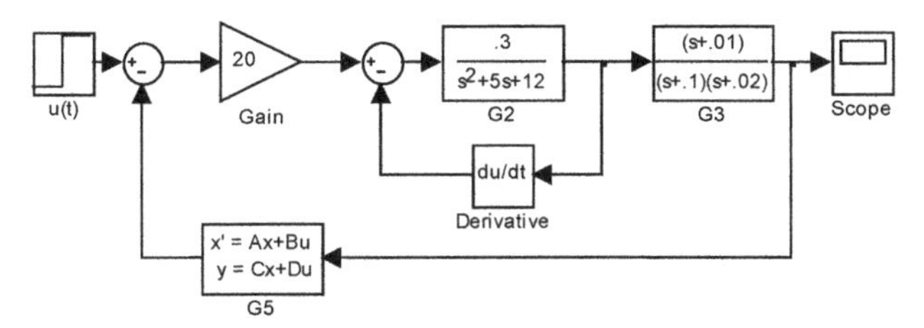

Figure 8.4(c) Simulink model for the continuous system of figure 8.4(b)

⇒ get one Transfer Fcn block in the model file for $G_2(s)$, doubleclick the block to set its Numerator and Denominator as 0.3 and [1 5 12] respectively, and rename (subsection 1.2.3) the block as G2 just to make sense with given name

⇒ get one Zero-Pole block in the model file for $G_3(s)$, doubleclick the block to enter its Zeros, Poles, and Gain as -0.01, -0.1 and -0.02, and 1 respectively, and rename the block as G3

⇒ a derivative in time domain is equivalent to s in the transform domain so we get one Derivative block in the model file for $G_4(s)$, rightclick on the Derivative, and click the Flip Block in the Format popup because of the signal flow of $G_4(s)$ in the system

⇒ get one State-Space block in the model file for $G_5(s)$, doubleclick the block to enter its A, B, C, and D as 1, 1.1, 1e-5, and 0.5 in the parameter window respectively, rightclick on the State-Space, click the Flip Block under the Format popup because of the signal flow of $G_5(s)$ in the system, and rename the block as G5

⇒ get one Scope block in the model file for the $y(t)$

Finally place various blocks and connect them relatively according to the figure 8.4(c) which represents the model of the interconnected system in figure 8.4(b).

8.6 Defining a circuit system

Electric circuits form a continuous system. Starting from a given circuit we write its governing equations by using Kirchhoff's voltage law (KVL), Kirchhoff's current law (KCL), and Ohm's law. While writing these equations different circuit analysis techniques are employed like nodal voltage method, loop current method, or other. We assume that the reader is familiar with these laws and methods due to the nature and scope of the course. The input to and output from a circuit are problem dependent. We must seek for number of governing equations equal to required number of outputs at least. Governing equation number might be greater than the number of outputs required. Unknown variables associated with the governing equations also play a role in the system definition. To define such continuous system, we commence with the time domain. Let us see following examples in defining a circuit system.

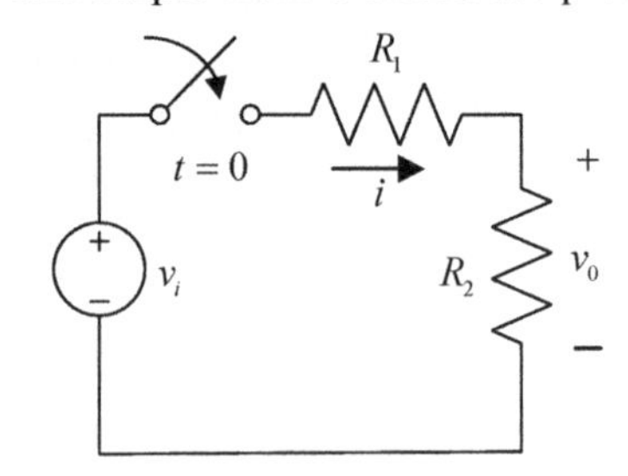

Figure 8.5(a) A series circuit with input and output

Example 1:

Figure 8.5(a) shows a series circuit from which two outputs v_0 and i are required when an input voltage v_i is applied at $t=0$.

v_i → Continuous system with governing equations:
(1) $v_i = (R_1 + R_2)i$
(2) $v_0 = R_2\, i$
→ v_0, → i

Figure 8.5(b) Continuous system for the circuit in figure 8.5(a)

Since we need two outputs, there must be at least two equations to define the circuit system. KVL around the circuit provides $v_i = (R_1 + R_2)i$ and Ohm's law provides $v_0 = R_2\, i$. Figure 8.5(b) depicts the system formation on the circuit in figure 8.5(a). If we have the system of the figure 8.5(b), we can forget about the circuit in figure 8.5(a).

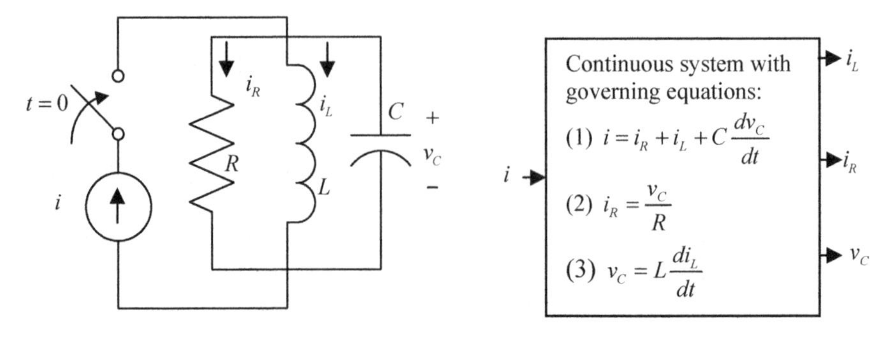

Figure 8.5(c) A current source is to be connected at t =0 to a parallel circuit

Figure 8.5(d) Continuous system for the circuit in figure 8.5(c)

Example 2:

Figure 8.5(c) shows a circuit in which a current source is to be connected at $t=0$ to a parallel R - L - C circuit. We wish to develop the circuit system.

From the circuit we need three quantities – i_L, v_C, and i_R so there should be at least three equations and there is only one input which is i.

Writing KCL at the parallel R - L - C circuit we have $i = i_R + i_L + C\frac{dv_C}{dt}$. The voltage across the capacitor is also the voltage across the resistor and inductor so Ohm's law provides us $i_R = \frac{v_C}{R}$ and $v_C = L\frac{di_L}{dt}$. With these three equations we have the system for the circuit in figure 8.5(c) depicted in figure 8.5(d).

Example 3:

Last two examples illustrated only time domain form of a continuous system. We can define the governing equations through the transform or s domain. In example 2 we wish to define the governing equations in transform domain.

To do so, we apply table 8.A quoted relationships. That is an inductor L is replaced by sL and a capacitor C is replaced by $\frac{1}{sC}$ as if each one is an impedance. After making such change, we obtain the governing equations as shown in figure 8.5(e).

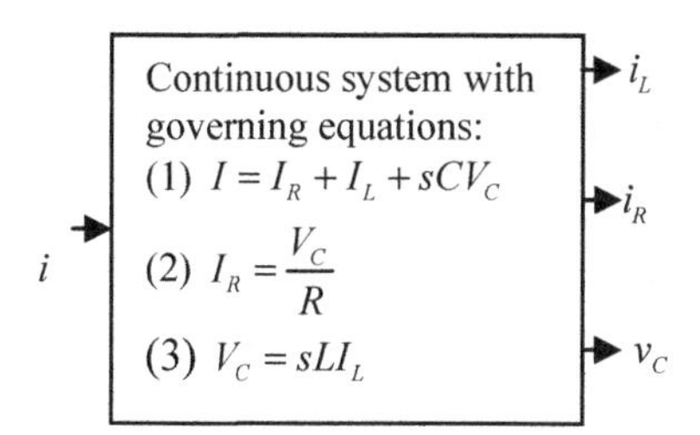

Figure 8.5(e) Continuous system for the circuit in figure 8.5(c) in transform domain

Although we define the governing equations in transform domain but we are often interested to apply the input and to see the output in time domain that is why we did not change the time domain symbology in figure 8.5(e). If we have i (appropriately $i(t)$) in time domain, the I (appropriately $I(s)$) represents the same in the transform domain. Similar interpretation follows for the other variables in the figure 8.5(e).

Example 4:

In example 3 suppose we need only i_L and v_C, still we need to write all three equations. The reader might say why do not we take equations 1 and 3? Well the i_R is present in the equation 1 which is an alien variable when only i_L and v_C are considered. Linear equation theory says – number of equations must be equal to the number of unknown variables in order to have a solution. This is the reason why we mentioned at least in earlier discussions.

8.7 System formation by expression from electric circuits

In section 8.6 (prerequisite of this section) we have prepared the foundation of forming a continuous system from an electric circuit. In this section we illustrate the system formation by expression on electric circuits in the transform domain. The reader might ask when this sort of analysis becomes useful. If we have a circuit system and if we wish to investigate the required output quantities (frequently which are voltage and current) subject to varying input (s), this input-output or block diagram based analysis is the perfect tool for that. Or in other words we conduct virtual circuit experiment by using the system.

Our objective is to find the circuit system function $H(s)$ starting from the circuit's time or transform domain (i.e. s) equations. Repeatedly appendix D cited solve will be used in forming such system in the following. Usage of the solve takes place here in a slightly different manner. We declare all related variables by the built-in command **syms** beforehand and organize the equations in such a way that the right side of any equation is zero. When we apply the **syms**, we do not put the equations under the quote. Suppose $H(s)$ is to be formed from $\frac{V_0(s)}{V_i(s)}$ then $V_0(s)$ is unknown or variable of interest and we treat the $V_i(s)$ as known. The $H(s)$ can be voltage to voltage, current to current, voltage to current, or other. Let us go through the following examples in this regard.

Example 1:

With R_1=5Ω and R_2=7Ω in the series circuit of figure 8.5(a), the v_0 to v_i and i to v_i system functions are $\frac{V_0(s)}{V_i(s)}=\frac{7}{12}$ and $\frac{I(s)}{V_i(s)}=\frac{1}{12}$ respectively which we intend to obtain.

Since there is no inductor and capacitor in the circuit, the equation will be linear algebraic (no derivative or integral term). Due to linearity time domain equation is also the s domain equation. Writing KVL on the circuit in figure 8.5(a) we have v_i=12i and v_0=7i i.e. two equations and three variables. Our variables of interest are v_0 and i so we go through the following:

```
>>syms vi vo c ↵ ← Declaring the related variables by syms, vi⇔vi, vo⇔v0,
                       and c⇔i, vi, vo, and c are user-chosen
>>e1=vi-12*c; ↵ ← Assigning first rearranged equation to e1, e1 is user-chosen
>>e2=vo-7*c; ↵ ← Assigning second rearranged equation to e2, e2 is user-chosen
>>O=solve(e1,e2,vo,c); ↵ ← Calling the solve with variable of interest and
                       assigning the return to O, O is user-chosen
>>O.vo ↵              ← Calling the O.vo or V0(s)
ans =
7/12*vi          ← Meaning V0(s) = 7/12 Vi(s)
>>O.c ↵          ← Calling the O.c or I(s)
ans =
1/12*vi          ← Meaning I(s) = 1/12 Vi(s)
```

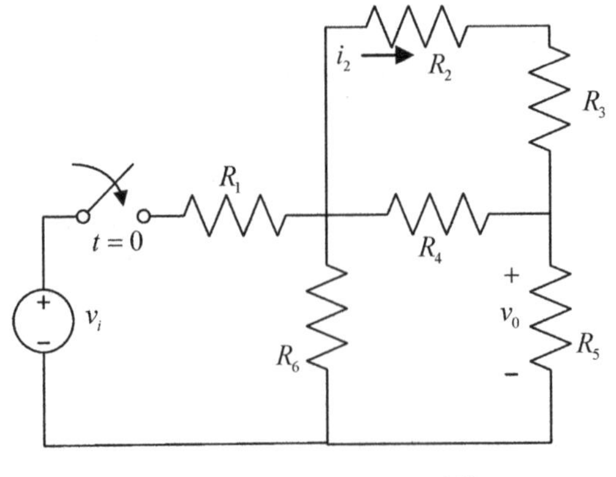

Figure 8.6(a) A series parallel resistive circuit forms a system (left side figure)

Example 2:

With R_1=3Ω, R_2=4Ω, R_3=5Ω, R_4=6Ω, R_5=7Ω, and R_6=8Ω in the series-parallel circuit of figure 8.6(a), the v_0 to v_i and i_2 to v_i system functions are $\frac{V_0(s)}{V_i(s)} = \frac{280}{703}$ and $\frac{I_2(s)}{V_i(s)} = \frac{16}{703}$ respectively which we intend to obtain.

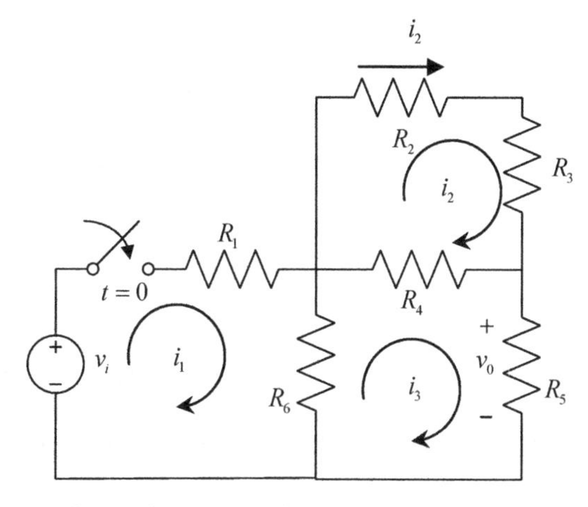

Figure 8.6(b) The circuit of the figure 8.6(a) with loop current labeling (left side figure)

The reader can employ any method for the circuit solution. We applied loop current method whose labeling is shown in figure 8.6(b). Based on that the three loop equations along with the output are as follows:

Loop 1: $v_i = R_1 i_1 + R_6(i_1 - i_3)$,

Loop 2: $(R_2 + R_3 + R_4)i_2 - i_3 R_4 = 0$,

Loop 3: $(R_4 + R_5 + R_6)i_3 - i_2 R_4 - i_1 R_6 = 0$, and

Output equation: $v_0 = R_5 i_3$.

We have four equations and five variables – v_i, i_1, i_2, i_3, and v_0. We can express the i_1, i_2, i_3, and v_0 in terms of v_i from which we ignore the i_1 and i_3 which are not required. Let us go through the following:

```
>>syms vi vo c1 c2 c3 ↵
```
← Declaring the related variables by the **syms**, vi⇔ v_i, vo⇔ v_0, c1⇔ i_1, c2⇔ i_2, c3⇔ i_3, vo, vi, c1, c2, and c3 are user-chosen

```
>>R1=3;R2=4;R3=5;R4=6;R5=7;R6=8; ↵
```
← Assigning the resistance value to like name variable like R_1 to R1

```
>>e1=vi-R1*c1-R6*(c1-c3); ↵
```
← Assigning rearranged loop equation 1 to e1

```
>>e2=c2*(R2+R3+R4)-c3*R4; ↵
```
← Assigning loop equation 2 to e2

```
>>e3=c3*(R4+R5+R6)-c2*R4-c1*R6; ↵
```
← Assigning loop equation 3 to e3

```
>>e4=vo-c3*R5; ↵
```
← Assigning rearranged output equation to e4

```
>>O=solve(e1,e2,e3,e4,vo,c1,c2,c3); ↵
```
← Calling the **solve** with variables of interest and assigning the return to O

```
>>O.vo ↵
```
← Calling the O.vo or $V_0(s)$

```
ans =

280/703*vi
```
← Meaning $V_0(s) = \frac{280}{703} V_i(s)$

```
>>O.c2 ↵
ans =
16/703*vi
```

← Calling the O.c2 or $I_2(s)$

← Meaning $I_2(s) = \dfrac{16}{703} V_i(s)$

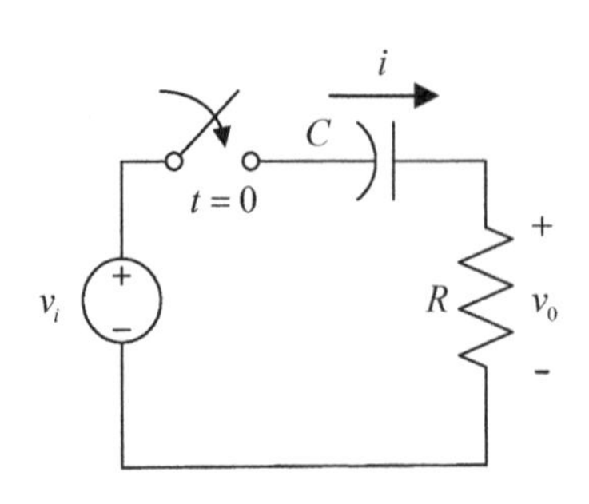

Figure 8.6(c) An R - C circuit

Figure 8.6(d) The circuit of the figure 8.6(c) in s domain

Example 3:

With R=4 $K\Omega$ and $C = 0.5mF$ in circuit of figure 8.6(c), the v_0 to v_i and i to v_i system functions are $\dfrac{V_0(s)}{V_i(s)} = \dfrac{2s}{1+2s}$ and $\dfrac{I(s)}{V_i(s)} = \dfrac{s}{2000(1+2s)}$ respectively which we intend to obtain.

When we have an inductor or a capacitor present in a circuit, governing equations are not linear any more because of integral (for capacitor) and derivative (for inductor) operators. To solve such system, we first draw the s domain equivalent of the given circuit and then apply the solve to find the solution.

We know that a capacitor C in time domain turns to $\dfrac{1}{sC}$ in s domain when the input is voltage and there is no change in the resistance that is how figure 8.6(d) represents the s domain equivalent of the circuit in figure 8.6(c). Based on the figure 8.6(c), we write the KVL as follows:

$$V_i(s) = \left(R + \frac{1}{sC} \right) I(s) \text{ and } V_0(s) = RI(s).$$

There are two equations and three unknowns – $V_i(s)$, $V_0(s)$, and $I(s)$. The variables of interest are $V_0(s)$ and $I(s)$ – knowing so we go through the following:

```
>>syms s vi vo c ↵
```

← Declaring the related variables by the syms, vi⇔ $V_i(s)$, vo⇔ $V_0(s)$, c⇔ $I(s)$, s⇔ s, all variables are user-chosen

Note that in symbology the c and C are different in MATLAB and they refer to $I(s)$ and C respectively. As usual we proceed with the following:

```
>>R=4e3; C=0.5e-3; ↵
>>e1=vi-(R+1/s/C)*c; ↵
>>e2=vo-c*R; ↵
>>O=solve(e1,e2,vo,c); ↵
```

← Entering values of the circuit parameters to like name variables like R to R, appendix B for coding

← Entering the rearranged first equation to e1

← Entering the rearranged second equation to e2

← Calling the solve with variable of interest and assigning the return to O

```
>>O.c ↵                        ← Calling the O.c or I(s)

ans =

1/2000*vi*s/(1+2*s)            ← The code of I(s)
>>pretty(O.c) ↵                ← Appendix C.10 cited pretty shows readable form

              vi s
     1/2000 ---------
              1 + 2 s
>>pretty(O.vo) ↵
              vi s
          2 ---------
             1 + 2 s
```

← Meaning $I(s)=\dfrac{s}{2000(1+2s)}V_i(s)$

← For $V_0(s)$

← Meaning $V_0(s)=\dfrac{2s}{1+2s}V_i(s)$

Example 4:

With R_1=7Ω, R_2=10Ω, $C=1F$, and L=2.5 H in circuit of figure 8.6(e), the v_C to v_i and i_L to v_i system functions are $\dfrac{V_C(s)}{V_i(s)}=\dfrac{7(4+s)}{28+17s+70s^2}$ and $\dfrac{I_L(s)}{V_i(s)}=\dfrac{4(1+7s)}{28+17s+70s^2}$ respectively which we wish to obtain.

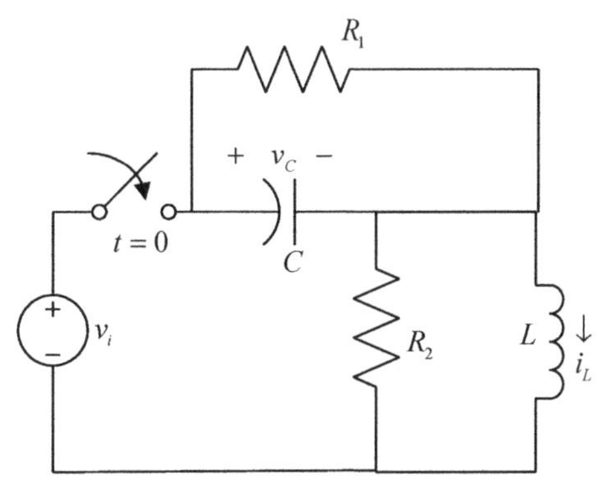

Figure 8.6(e) A series parallel circuit

The circuit can be solved by many techniques. We are planning to apply the node voltage method for the solution. For the node voltage method we labeled the circuit as depicted in figure 8.6(f) in transform or s domain, one extra variable V_L (whose time domain counterpart is v_L and indicated by a bold dot in figure 8.6(f)) is included for easy equation writing.

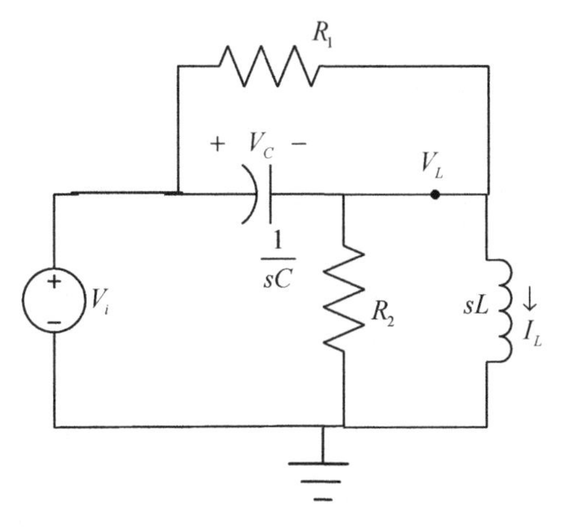

Figure 8.6(f) The circuit of the figure 8.6(e) in transform domain

Let us write the governing equations of the circuit in the transform domain as follows:

KVL at input (equation 1):

$$V_i - V_L = V_C$$

Voltage across the inductor (equation 2):

$$V_L = sLI_L$$

KCL at the node V_L

(equation 3):

$$\frac{V_L - V_i}{R_1} + I_L + \frac{V_L}{R_2} = V_C sC$$

Although we do not need V_L, we treat the V_L, I_L, and V_C as variables of interest and express each of which in terms of the others in the three equations. We carry out the following by employing ongoing symbology and function:

>>R1=7; C=1; R2=10; L=2.5; ↵ ← Entering values of the circuit parameters to like name variables like R_1 to R1

>>syms s Vi Vc IL VL ↵ ← Declaring the related variables by the syms, Vi⇔ $V_i(s)$, Vc⇔ $V_C(s)$, IL⇔ $I_L(s)$, s⇔ s, and VL⇔ $V_L(s)$

In just conducted executions all variables like Vi, Vc, etc are user-chosen. As a continuation the next is to enter the governing equations as follows:

>>e1=Vi-VL-Vc; ↵ ← Entering the rearranged first equation to e1

>>e2=VL-s*L*IL; ↵ ← Entering the rearranged second equation to e2

>>e3=(VL-Vi)/R1+IL+VL/R2-Vc*s*C; ↵ ← Entering the rearranged third equation to e3

>>O=solve(e1,e2,e3,VL,Vc,IL); ↵ ← Calling the solve with variable of interest and assigning the return to O

>>pretty(O.Vc) ↵ ← Display the O.Vc or $V_C(s)$

```
            Vi (4 + s)
      7 -----------------------
                            2
          28 + 17 s + 70 s
```

>>pretty(O.IL) ↵ ← Display the O.IL or $I_L(s)$

```
           Vi (1 + 7 s)
      4 -----------------------
                            2
          28 + 17 s + 70 s
```

In a similar fashion the reader can find the expression based system function for any other electric circuit.

Table 8.A Time and transform domain equations of voltage and current for three basic elements – R, L, and C in an electric circuit

Element	Time domain relationship	Transform domain relationship
$i_R(t)$ (resistor) + $v_R(t)$ −	$v_R(t) = Ri_R(t)$	$V_R(s) = RI_R(s)$
$i_L(t)$ (inductor) + $v_L(t)$ −	$v_L(t) = L\frac{di_L(t)}{dt}$	$V_L(s) = LsI_L(s)$
$i_C(t)$ (capacitor) + $v_C(t)$ −	$v_C(t) = \frac{1}{C}\int_{p=0}^{p=t} i_C(p)dp$	$V_C(s) = \frac{1}{Cs}I_C(s)$

Table 8.B Elementary system functions of the three basic elements – R, L, and C in an electric circuit

Element	When input is voltage	When input is current
Resistor	$V_R(s) \to \left[\frac{1}{R}\right] \to I_R(s)$	$I_R(s) \to [R] \to V_R(s)$
Inductor	$V_L(s) \to \left[\frac{1}{Ls}\right] \to I_L(s)$	$I_L(s) \to [Ls] \to V_L(s)$
Capacitor	$V_C(s) \to [sC] \to I_C(s)$	$I_C(s) \to \left[\frac{1}{sC}\right] \to V_C(s)$
* Each [ ] in above represents elementary system function of the concerned element		

8.8 Model formation on electric circuits

Main objective of this section is to address the modeling issues of an electric circuit based on block diagram in SIMULINK (chapter 1). Its MATLAB complement is seen in section 8.2. Solely continuous time system modeling is presented in section 8.5 which is the prerequisite for this section. Staring from a given circuit we write its transform domain based equation from input to output and look for appropriate block of SIMULINK to implement such circuit. Before modeling in SIMULINK, following important points must be considered.

Input to the circuit:

Input to a circuit is completely problem dependent. For example, in the circuit of the figure 8.6(a) the input is labeled as v_i or voltage. The v_i can be any time domain wave shape of the chapter 3.

Output from the circuit:

Output from a circuit is also problem dependent. For example the circuit of the figure 8.6(a) has two outputs – v_0 and i_2 which are assumed to be in time domain. We can view the output by using a Scope block (subsection 1.2.5) of SIMULINK.

Elementary system function:

Tables 8.A and 8.B show the necessary voltage and current relationships both in time and in transform domains of the three basic circuit elements – R, L, and C. It should be pointed out that in SIMULINK the input and output are always in time domain despite the modeling of the circuit system takes place in transform domain.

Concerning the table 8.B if the elementary system function is R or $\frac{1}{R}$ we apply the Gain block of SIMULINK. Again if the elementary system function is $\frac{1}{sL}$ or $\frac{1}{sC}$ we use the Transfer Fcn block of SIMULINK. For the third type elementary system function i.e. sL or sC we engage the Derivative block in conjunction with one Gain block where the Gain conceives the L or C value. Appendix A provides the block links and icon appearances of all these blocks.

Summing point in the block diagram:

This is the critical point in modeling a circuit system. Figures 8.7(a) and 8.7(b) present two input (can be more than 2) summing points. Usually they are for the negative and positive feedbacks respectively. The reader might ask how should I know which type of summing point will be used? We can provide some clues for that. The first clue is the summing point is completely circuit dynamics dependent that is how the transform domain equations involving KVL, KCL, and Ohm's law develop. The second clue is only the like quantity is added or subtracted by the summing point. Thirdly it is always desirable to express the output quantity as output =[some function]×input from given governing equation.

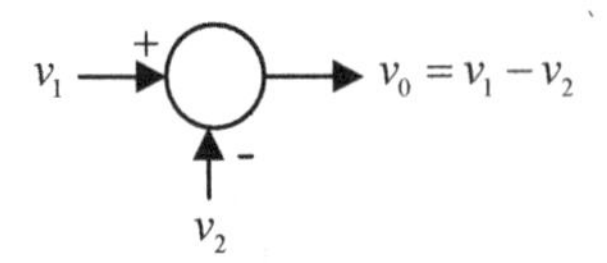

Figure 8.7(a) Summing point for subtraction

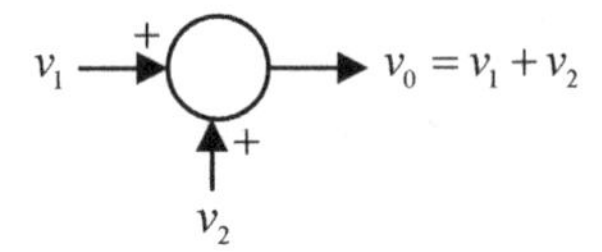

Figure 8.7(b) Summing point for addition

For example $100[V_1(s)+V_2(s)]=E(s)$ or $100[I_1(s)+I_2(s)]=E(s)$ can be modeled by the summing point of the figure 8.7(b) where V and I correspond to voltage and current respectively. Neither the summing point in figure 8.7(a) nor the one in figure 8.7(b) models the equation like $100[V(s)+I(s)]=E(s)$ unless $I(s)$ is changed to voltage by applying some conversion gain.

Example of fitting some equation to the summing point:

Now we provide some examples on transform domain equations which we fit by the summing points.

Drill 1:

$100[V_1(s)+V_2(s)] = E(s)$ can be written as $E(s) = 100\times[V_1(s)+V_2(s)]$ from which we derive the block diagram of figure 8.7(c).

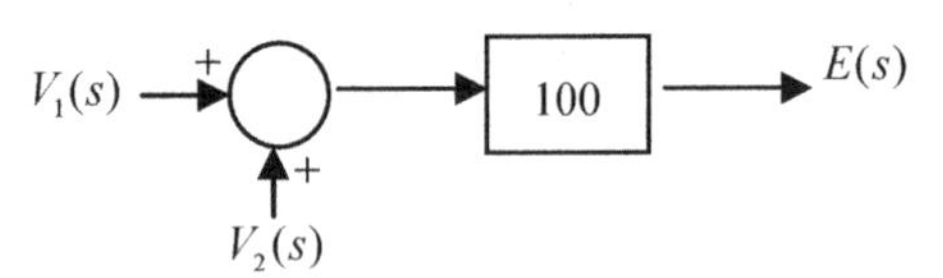

Figure 8.7(c) Fitting the equation $100[V_1(s)+V_2(s)]=E(s)$ by summing point

Drill 2:

$s[I_1(s)+I_2(s)] = 34V(s)$ can be written as $V(s) = \frac{s}{34}\times[I_1(s)+I_2(s)]$ from which we derive the block diagram of figure 8.7(d).

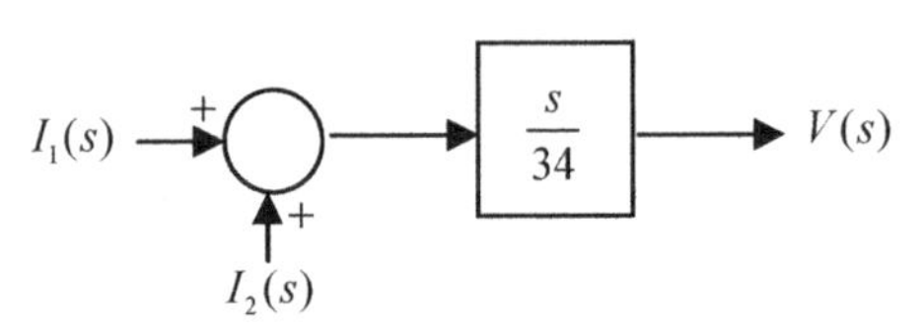

Figure 8.7(d) Fitting the equation $s[I_1(s)+I_2(s)]=34V(s)$ by summing point

Drill 3: $\frac{V(s)-RI_1(s)}{sL} = -67I_2(s)$ can be written as $I_2(s) = \frac{1}{-67sL} \times [V(s) - RI_1(s)]$ from which we derive the block diagram of the figure 8.7(e).

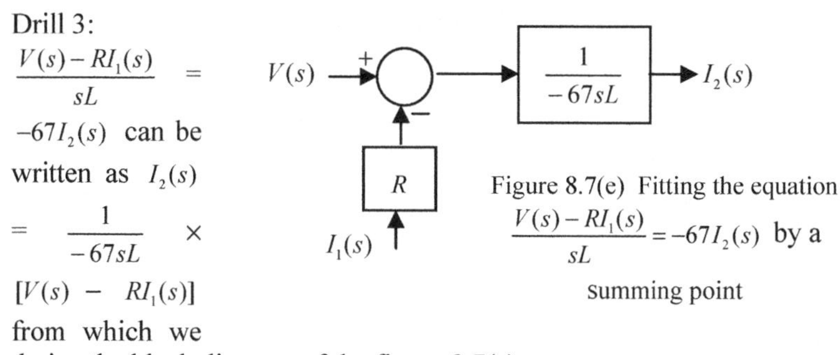

Figure 8.7(e) Fitting the equation $\frac{V(s)-RI_1(s)}{sL} = -67I_2(s)$ by a summing point

Role of an intermediate variable:

Apart from the input and output variables, frequently we include intermediate variables to facilitate our modeling. Usually these variables have expression significance but we do not put them in the model.

Direction of the signal flow:

Whatever signal flows through our circuit system – voltage or current, blocks in a diagram for the circuit modeling are unidirectional.

We paved the way for modeling a circuit in SIMULINK by addressing all relevant issues. We assume that the reader has gone through earlier discussions in this section. As a procedural step, we write the governing equations of a given circuit in transform domain and fit the equations to appropriate blocks of SIMULINK as will be demonstrated by choosing some examples in the following.

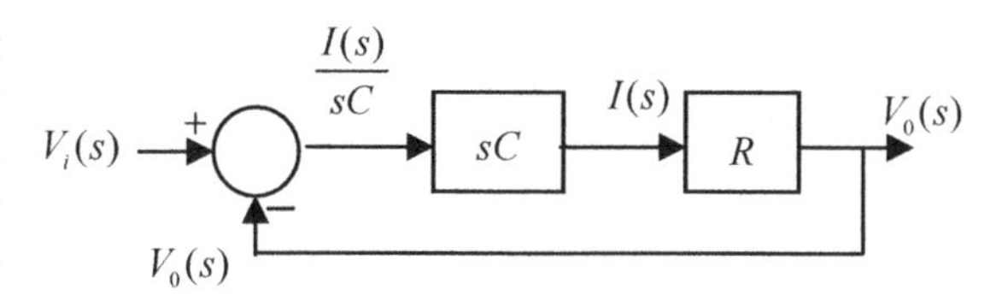

Figure 8.7(f) Modeling the circuit of figure 8.6(c)

❖❖ **Example 1**

With R=4 $K\Omega$ and $C=0.5mF$ we intend to form a system for the circuit in figure 8.6(c) for v_0 and i in SIMULINK.

The governing equations in transform domain are $V_i(s)=\left(R+\frac{1}{sC}\right)I(s)$ and $V_0(s)=RI(s)$. It is always desirable to have like quantity in equation while modeling (like voltage-voltage or current-current). Substituting the second equation to the first we obtain $V_i(s)-V_0(s)=\frac{I(s)}{sC}$. One obtains $I(s)$ from $\frac{I(s)}{sC}$ by a gain of sC. Again $V_0(s)$ is obtained from $I(s)$ by a gain of R. The sC indicates a derivative or s with the gain C. However figure 8.7(f) shows the model of the circuit in figure 8.6(c) that is what we implement in SIMULINK as follows:

⇒ Open a new SIMULINK model file (subsection 1.2.2)

⇒ Get one Sum, one Derivative, and two Gain blocks in the model file following the link mentioned in appendix A

⇒ Rename the Gain blocks as C and R to maintain consistency (subsection 1.2.3)

⇒ Doubleclick the block C, enter its Gain as 0.5e-3 (appendix B for coding) in the parameter window of the block, and enlarge the C by the mouse to see its contents

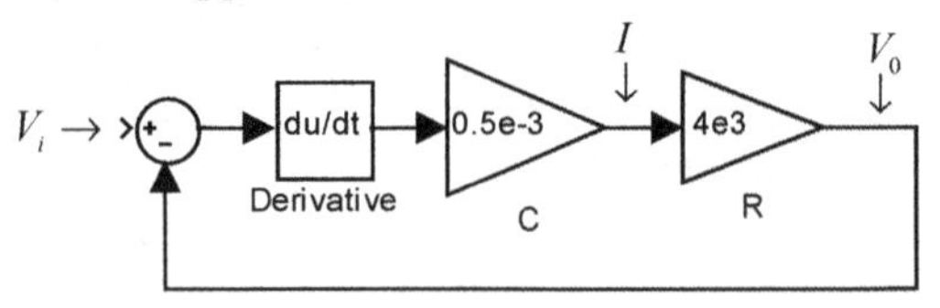

Figure 8.7(g) Simulink model for the circuit in figure 8.6(c)

⇒ Doubleclick the block R, enter its Gain as 4e3 in the parameter window of the block, and enlarge the R to see its contents

⇒ Doubleclick the Sum and change its List of signs from default ++ to +- in the parameter window for entering the feedback signal

⇒ Place the blocks relatively and connect them according to the figure 8.7(g) which is the model for the electric circuit of the figure 8.6(c)

In the figure we have shown the V_i, I, and V_0 points by arrow marks.

Figure 8.8(a) A two stage RC filtering circuit (right side figure)

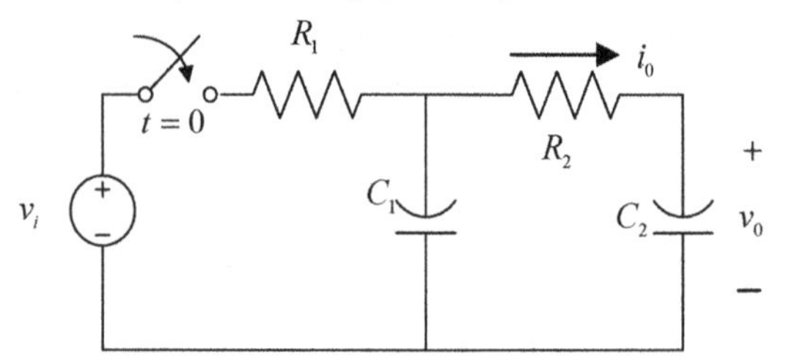

Figure 8.8(b) The circuit of figure 8.8(a) in transform domain (right side figure)

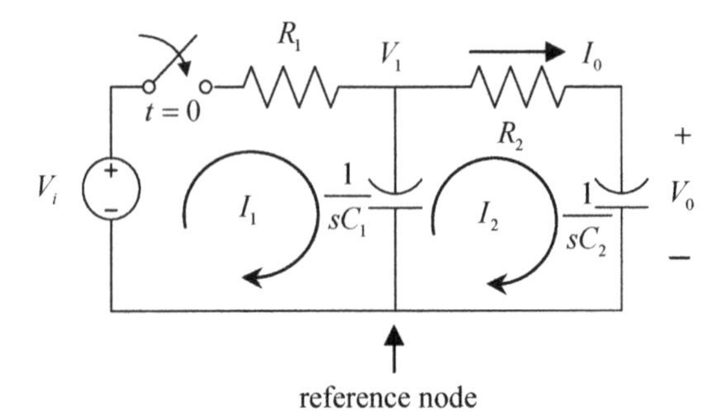

♦♦ Example 2

Our objective is to form a system in SIMULINK for the two stage RC filtering circuit of figure 8.8(a) for the v_0 and i_0 considering $R_1 = 4\ K\Omega$, $R_2 = 5\ K\Omega$, $C_1 = 0.5mF$, and $C_2 = 0.9mF$.

We apply both the loop current and node voltage techniques for easy block diagram implementing on the circuit and the circuit including some intermediate variables V_1, I_1, and I_2 in the transform domain is redrawn in figure 8.8(b) where $I_2 = I_0$. There are four elements in the circuit and their governing equations are as follows:

for R_1: $I_1 = \frac{V_i - V_1}{R_1}$, for C_1: $V_1 = \frac{I_1 - I_2}{sC_1}$,

for R_2: $I_2 = \frac{V_1 - V_0}{R_2}$, and for C_2: $V_0 = \frac{I_2}{sC_2}$.

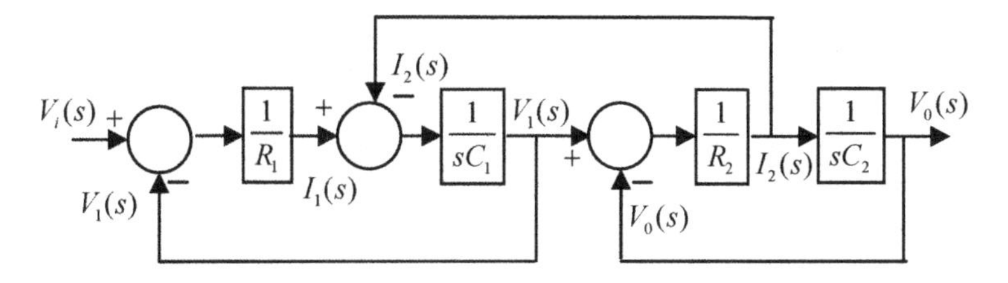

Figure 8.8(c) Modeling the circuit of figure 8.8(a)

Based on the four governing equations in transform domain we put the block diagram into perspective as shown in figure 8.8(c) this is what we need to model in SIMUNLINK and do so as described in the following:

⇒ Open a new SIMULINK model file

⇒ Get three Sum (for the three summing points), two Gain (for $\frac{1}{R_1}$ and $\frac{1}{R_2}$), and two Transfer Fcn (for $\frac{1}{sC_1}$ and $\frac{1}{sC_2}$) blocks in the model file following the link mentioned in appendix A

⇒ Doubleclick the block Gain, enter its Gain as 1/4e3 (for $\frac{1}{R_1}$) in the parameter window, enlarge the Gain to see its contents, doubleclick the block Gain1, enter its Gain as 1/5e3 (for $\frac{1}{R_2}$) in the parameter window, and enlarge the Gain1 to see its contents

⇒ Doubleclick the block Transfer Fcn, enter its Denominator as [0.5e-3

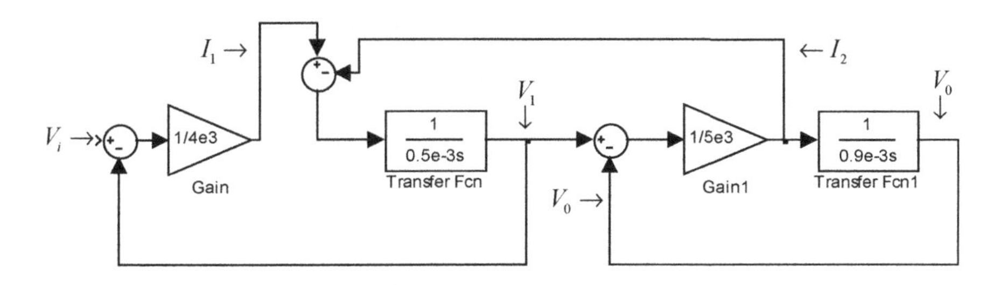

Figure 8.8(d) Modeling circuit system in figure 8.8(a)

0] in the parameter window (section 8.5 for $\frac{1}{sC_1}$), enlarge the Transfer Fcn to see its contents, doubleclick the Transfer Fcn1, enter its Denominator as [0.9e-3 0] in the parameter window (for $\frac{1}{sC_2}$), and enlarge the Transfer Fcn1 to see its contents

⇒ Doubleclick each of the three Sum blocks, change its List of signs from default ++ to +- in the parameter window for entering the feedback signals, rightclick on one Sum block, click the Rotate block under the Format popup, rightclick again on the same Sum block, and click the Flip block under the Format popup (the last two actions for aligning the second summing point of figure 8.8(c) in proper orientation)

⇒ Place the blocks relatively and connect them according to the figure 8.8(d) which is the model for the electric circuit system of the figure 8.8(a)

The required quantities and the intermediate variables (V_i, I_1, I_2, V_1, and V_0 points) on the system are also shown by the arrow marks in the figure 8.8(d).

❖ Drawback of the block diagram approach

In the last two examples we presented the block diagram formation on a circuit system. When we have multiple loops and long circuits, the block diagram approach is somewhat clumsy. The best approach is first apply the expression technique of the section 8.7 on the given circuit and then model the expression in SIMULINK based on the input-output.

However we bring an end to the chapter with this discussion.

Chapter 9
SYSTEM ANALYSIS

Introduction

Once a system is formed, the very next step is to analyze the system for its characteristics. Built-in functions and pre-designed blocks are supplied with MATLAB to facilitate such analysis. The system characteristic requirements can be time related, frequency related, comparison related, or other. As far as the study type is concerned, the characteristic requirement can be continuous or discrete. Comprehensively we aim at introducing the analysis techniques for the following:

- Step response of a continuous system for varying situations
- Impulse response of a continuous time system
- Frequency response of a continuous system
- System stability through pole-zero map

9.1 Step response of a continuous system

What the output should be when a unit step signal is applied to a system (which is represented by Laplace transform system function $H(s)$) is called the step response of the system. Figure 9.1(a) depicts the step response strategy of a continuous system on input $u(t)$ and output $y(t)$. Step response has thorough implication in system analysis. We know that a system (section 8.1) may have its input as voltage, current, force, or other physical quantity. How the output from the system will behave for one unit of the input (for example 1 V voltage or 1 A current) necessitates the study of step response. The step response requirement appears in various forms which we address in the following.

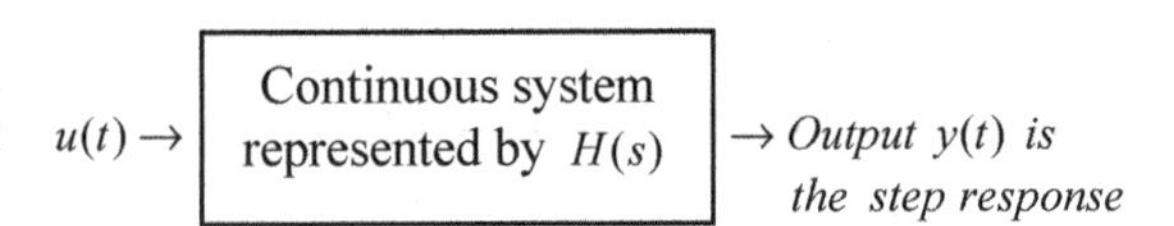

Figure 9.1(a) Concept behind the step response

Expression based step response

In this sort of problem we have the system function $H(s)$ given and look for the inverse Laplace transform of $\frac{H(s)}{s}$ to obtain the t domain expression of $y(t)$. In this regard the reader needs inverse transform finding basics of section 6.8.

Example:

Given that $y(t)=2-2\cos t+4\sin t$ is the step response of the system represented by $H(s)=\frac{2(2s+1)}{s^2+1}$ which we wish to obtain.

We need to have $\frac{H(s)}{s}$ so the inverse transform of $\frac{2(2s+1)}{s(s^2+1)}$ is found as follows:

```
>>syms s ↵
```
← Declaring the related s as symbolic, s⇔ s

```
>>F=2*(2*s+1)/s/(s^2+1); ↵
```
← Code of $\frac{H(s)}{s}$ is assigned to F, F is user-chosen variable

```
>>f=ilaplace(F) ↵
```
← Inverse Laplace transform is taken and assigned to f where f is user-chosen

```
f =

2-2*cos(t)+4*sin(t)
```
← f holds the $2-2\cos t+4\sin t$

Graphical step response

In graphical step response basically we look for the $y(t)$ versus t plot from given system function $H(s)$ (figure 9.1(a)). MATLAB built-in function step provides the graph with the syntax step($H(s)$ representing variable, final bound of the interval assuming that the starting time at t =0) and we do not assign the output of the step to any variable. The $H(s)$ entering happens by applying the section 8.2 cited functions.

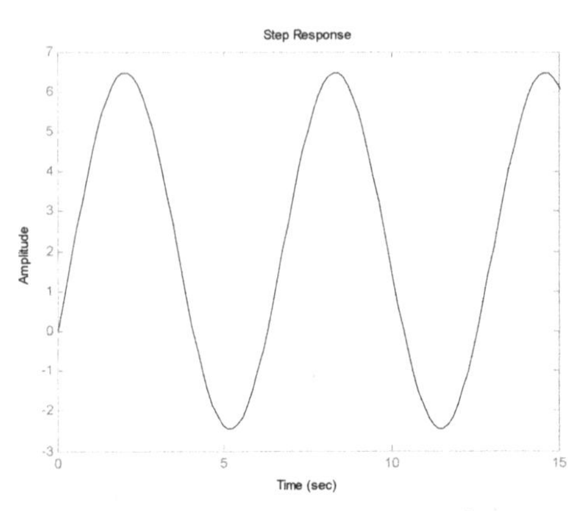

Figure 9.1(b) Step response of a continuous system

Example:

Given that figure 9.1(b) is the step response (i.e. $y(t)$ versus t plot) of the system represented by $H(s)=\frac{2(2s+1)}{s^2+1}$ over $0 \le t \le 15$ sec which we wish to obtain.

We first enter given $\frac{2(2s+1)}{s^2+1}$ to H (user-chosen variable) by the tf of section 8.2 as follows:

```
>>H=tf([4 2],[1 0 1]); ↵
```
← Workspace variable H holds the system $H(s)$, no $u(t)$ information is required

The interval is $0 \le t \le 15$ sec so only the last bound 15 is required as the second input argument of the step so the calling takes place as follows:

```
>>step(H,15) ↵
```
← Returns the figure 9.1(b)

The Amplitude and Time (sec) axes of the figure 9.1(b) refer to the $y(t)$ and t values respectively.

Getting access to step response data

Starting from a system function $H(s)$, we may need $y(t)$ value for some t for the step response of the figure 9.1(b). The function step just illustrated also keeps provision for returning the functional values on some t. Under this circumstance the step has output arguments (subsection 1.1.2) and the syntax is y= step(sytem,time vector) where y is a user-chosen variable name. One limitation of the function is computation must start from t =0 and a single step response value is not returned from the function.

Suppose we wish to find the step response values of $y(t)$ as in figure 9.1(b) for the system $H(s)=\frac{2(2s+1)}{s^2+1}$ over $0 \le t \le 0.5$sec. The t step size has to be provided by the user and let it be 0.1sec. First we form a column vector (subsection 1.1.2) t based on the interval and step size as follows:

```
>>t=[0:0.1:0.5]'; ↵
>>y=step(H,t); ↵
```
← t holds t values from 0 to 0.5 with a step 0.1

← Calling the step for numerical return on earlier H

We can see the return to y and t values (y and t both are a column matrix) side by side by executing the following:

```
>>[t y] ↵

ans =
         0         0
    0.1000    0.4093
    0.2000    0.8345
    0.3000    1.2714
    0.4000    1.7156
    0.5000    2.1625
      ↑         ↑
```
t values $y(t)$ values

← For example at t =0.2, $y(t)$=0.8345

The $y(t)$ values at t =0, 0.1, 0.2, etc is obtained by writing the command y(1), y(2), y(3), so on respectively.

For a single value return we form a two element row matrix considering the first element as 0 and take the second return from y. For example $y(t)$=0.8345 at t=0.2 is needed so exercise the command y=step(H,[0 .2]); first and y(2) afterwards for the functional value.

Step response of multiple systems

You can graph several step responses over common time interval by using the same step function which applies the syntax step(system 1, system 2, so on, last bound of the interval starting from 0).

For example we have the system functions: $H_1(s)=\dfrac{1}{s^2+1}$, $H_2(s)=\dfrac{1}{s^2+0.4s+1}$, and $H_3(s)=\dfrac{1}{s^2+5s+1}$. We intend to plot the step responses of these three systems over $0 \le t \le 20\,\text{sec}$.

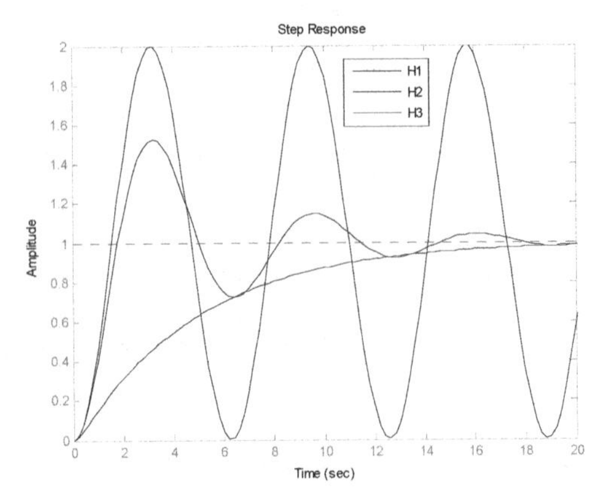

Figure 9.1(c) Step responses of three systems over a common interval

As a first step, we enter the three systems (section 8.2) as follows:

```
>>H1=tf(1,[1 0 1]); ↵
```
← Workspace variable H1 holds the system $H_1(s)$

```
>>H2=tf(1,[1 0.4 1]); ↵
```
← Workspace variable H2 holds the system $H_2(s)$

```
>>H3=tf(1,[1 5 1]); ↵
```
← Workspace variable H3 holds the system $H_3(s)$

The H1, H2, and H3 are all user-chosen names. Then call the step with just mentioned syntax as follows:

```
>>step(H1,H2,H3,20) ↵
```
← Figure 9.1(c) is the outcome

Once the responses are drawn, identification of the curves takes place by using the built-in command legend with the syntax legend(user-supplied text for H1 under quote, user-supplied text for H2 under quote, and so on) and we do so as follows:

```
>>legend('H1','H2','H3') ↵
```

Since the text is written in black and white form for cost reason, you do not see the differentiating mark in figure 9.1(c) but MATLAB figure window does not disappoint you that is certain.

9.2 Impulse response of a continuous system

When a Dirac delta function $\delta(t)$ is applied to a continuous system represented by $H(s)$, the output as a function of time is called the impulse response. Since Laplace transform

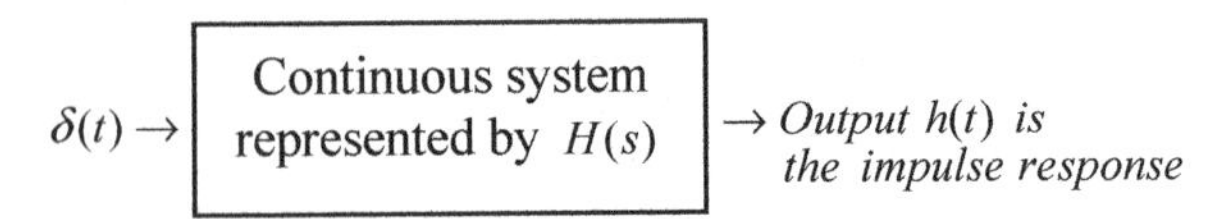

Figure 9.1(d) Concept behind the impulse response

of $\delta(t)$ is 1, impulse response is simply the inverse Laplace transform of $H(s)$. Figure 9.1(d) presents the concept behind the impulse response i.e. our focus is on $h(t)$ versus t.

Section 9.1 is the prerequisite for this section. Whatever execution we did for the step response by using the built-in function step can be conducted for the impulse response by using the built-in function impulse with identical syntax. Just to highlight some problems of the impulse response, we include the following (all symbology and function have the section 9.1 quoted meanings).

Given that $h(t) = 4\cos t + 2\sin t$ is the impulse response of the system represented by $H(s) = \frac{2(2s+1)}{s^2+1}$ which we obtain by exercising the commands syms s, H=2*(2*s+1)/(s^2+1); h=ilaplace(H) where H⇔ $H(s)$ and h⇔ $h(t)$.

The impulse response in graphical form (i.e. $h(t)$ versus t plot) of the system represented by $H(s) = \frac{2(2s+1)}{s^2+1}$ over $0 \le t \le 15\sec$ we obtain by the commands H=tf([4 2],[1 0 1]); impulse(H,15).

Impulse response $h(t)$ values for the system $H(s) = \frac{2(2s+1)}{s^2+1}$ over $0 \le t \le 0.5\sec$ with a step 0.1sec we get by the commands H=tf([4 2],[1 0 1]); t=[0:0.1:0.5]'; h=impulse(H,t);. Latter on we view the data side by side by [t h].

For a single value return e. g. $h(t)$ =4.3176 at t =0.2sec we exercise the command h=impulse(H,[0 .2]); first and h(2) afterwards on the last H.

Plotting the impulse responses of three systems $H_1(s) = \frac{1}{s^2+1}$, $H_2(s) = \frac{1}{s^2+0.4s+1}$, and $H_3(s) = \frac{1}{s^2+5s+1}$ over $0 \le t \le 20\sec$ we conduct by the following: H1=tf(1,[1 0 1]); H2=tf(1,[1 0.4 1]); H3=tf(1,[1 5 1]); impulse(H1,H2,H3,20). Latter on the command legend('H1','H2','H3') identifies each impulse response by color in the figure window.

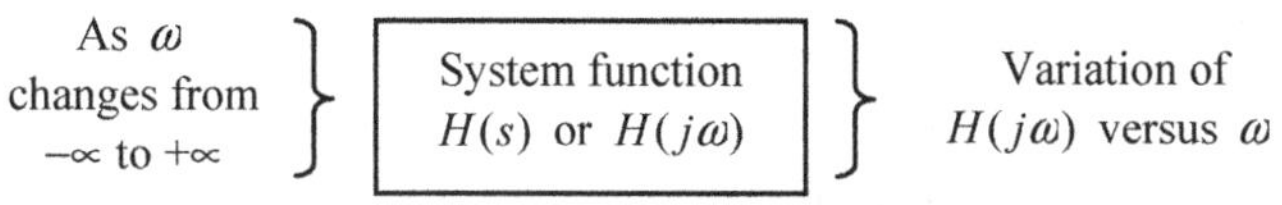

As ω changes from $-\infty$ to $+\infty$	}	System function $H(s)$ or $H(j\omega)$	}	Variation of $H(j\omega)$ versus ω

Figure 9.2(a) Frequency response of a continuous system

9.3 How to calculate the frequency response of a continuous system?

When we have input frequency which varies from $-\infty$ to $+\infty$, what the output response of a continuous system should be is termed as the frequency response and the strategy is schematized in figure 9.2(a). If we have system function $H(s)$, we replace the s by $j\omega$ and work on $H(j\omega)$. In section 8.2, we addressed how a $H(s)$ is defined by the tf, zpk, or ss. The same entering style applies here as well. The built-in function freqresp (abbreviation for the frequency response) helps us compute the $H(j\omega)$ function as will be explained in the following.

Value of the $H(s)$ at a particular angular frequency

Let us consider the system function $H(s)=\frac{s}{s+1}$. For example at ω =4 $rad/\sec$, the $H(j\omega)$ (replacing the s by $j\omega$ in $H(s)$ i.e. $H(j\omega)=\frac{j\omega}{j\omega+1}$) becomes 0.9412+ j 0.2353 which we wish to compute.

The freqresp has a syntax freqresp(system, required angular frequency) which we call as follows:

```
>>H=tf([1 0],[1 1]); ↵        ← Defining H(s) as in section 8.2, H⇔ H(s)
>>R=freqresp(H,4) ↵           ← Workspace R is any user-chosen variable

R =
        0.9412 + 0.2353i               ← R holds the H(j4)
```

Values of the $H(s)$ at a set of angular frequencies

We wish to compute the same $H(j\omega)=\frac{j\omega}{j\omega+1}$ for ω=0, 10, and 100 $rad/\sec$ which should be 0, 0.9901+ j 0.099, and 0.9999+ j 0.01 respectively.

The freqresp also keeps option for returning the system function values for multiple angular frequencies. In the second input argument of the freqresp, now we insert the required frequencies as a row matrix as follows:

```
>>R=freqresp(H,[0 10 100]); ↵ ← Second argument holds frequencies
```

At this point in MATLAB context, the return to R (user-chosen variable name) for the set of frequencies is as a three dimensional array with the first two dimensions empty. The discussion of the three dimensional array is beyond the scope of the text (reference 4). To remove the first two singleton dimensions from the R, we employ the command squeeze on the R as follows:

```
>>V=squeeze(R) ↵              ← The V is any user-chosen variable name

V =
            0
            0.9901 + 0.0990i
            0.9999 + 0.0100i
```

As the return says, computed functional values of the $H(j\omega)$ are available as a column matrix in the workspace variable V for the three frequencies respectively. The V(1), V(2), and V(3) return the three frequencies separately respectively.

Range of $H(s)$ values for a range of angular frequencies

Very often for graphing or analysis reason we need to have $H(j\omega)$ values for a range of frequencies. For instance we intend to obtain the earlier $H(j\omega)$ over $-10 \le \omega \le 10$ $rad/\sec$ with a ω step 0.1 $rad/\sec$. Under this circumstance we generate the ω vector as a row matrix (subsection 1.1.2) using the colon operator by executing first w=-10:0.1:10; and then R=freqresp(H,w); V=squeeze(R); at the command prompt. Though the second input argument of freqresp is a row matrix, the return to V is a column matrix.

If you say I need to see the values side by side, execute [w' V] at the command prompt (w' means transpose of w from row to column) i.e.

```
>>[w' V] ↵

ans =
     -10.0000          0.9901 - 0.0990i       ← e. g. H(−j10)
      -9.9000          0.9899 - 0.1000i
      -9.8000          0.9897 - 0.1010i
      -9.7000          0.9895 - 0.1020i
                  ⋮
         ↑                    ↑
     ω values           H(jω) values
```

♦♦ Separating various components of $H(j\omega)$

Once we have the $H(j\omega)$ calculated, further requirements can be the separation of the real, imaginary, magnitude, and phase angle components of $H(j\omega)$ i.e. $\text{Re}\{H(j\omega)\}$, $\text{Im}\{H(j\omega)\}$, $|H(j\omega)|$, and $\angle H(j\omega)$ which require the uses of the built-in functions real, imag, abs, and angle respectively. For instance values stored in just mentioned V can be separated by using the commands real(V), imag(V), abs(V), and angle(V) respectively – in single or multiple frequency case. The return from the angle is by default in radian. If we wish to see the return in degrees, we use the command 180*angle(V)/pi. We may assign the return from each of these four functions to some chosen name for further numerical processing.

♦♦ Decibel (dB) values of the $H(j\omega)$ magnitudes

Sometimes it is desirable to have the decibel (dB) values of the magnitude $|H(j\omega)|$ which happens through the command 20*log10(abs(V)) on the earlier V where $\log_{10} x$ has the code log10(x) and the decibel is defined as $20\log_{10}|H(j\omega)|$.

♦♦ Horizontal frequencies in terms of Hz instead of rad/sec

We might be interested in Hertz (Hz) frequency instead of angular one ($rad/\sec$) for what reason we employ f=w/2/pi where f is user-chosen and holds the Hertz frequency as column matrix (because $f=\frac{\omega}{2\pi}$).

♦♦ Bypassing $\log_{10} 0=-\infty$ point in dB values

In the decibel plot it is very common that $|H(j\omega)|=0$ for some frequency. Since $\log_{10} 0=-\infty$, the computer prints some error message. Under this type of situation, we add a negligible positive quantity epsilon to the $|H(j\omega)|$ values, which has the MATLAB code eps. For the ongoing system function, we should use 20*log10(eps+abs(V)).

◆◆ Suppressing high dB values of the $H(j\omega)$ magnitudes

Even though $\log_{10} 0 = -\infty$ is overcome by using the eps, the values of the 20*log10(eps) become too much negative like –300 dB or so which has no practical importance. In most system analysis, we restrict the lowest dB as –50 or –60dB. If any dB value is less than –50 or –60, we force that to be –50 or –60. This sort of dB axes manipulation needs some programming technique. Before we do that, let us assign the calculated dB values of ongoing V to some variable D (any user-chosen name) as follows:

```
>>D=20*log10(eps+abs(V)); ↵
```
← D holds the dB values as a column matrix

Let us say any dB value stored in D less than –50 will be set to –50. For this purpose the function find (appendix C.3) becomes useful as follows:

```
>>r=find(D<=-50); D(r)=-50; ↵
```

Concerning above implementation, the r (any user-chosen variable name) holds the integer position index of the column matrix D where $20\log_{10}|H(j\omega)| \leq -50$ (conducted by the command r=find(D<=-50);). Only $20\log_{10}|H(j\omega)| \leq -50$ elements in the D are set to –50 by writing the command D(r)=-50;. Thus the last D holds the expected $20\log_{10}|H(j\omega)|$ values as a column matrix.

◆◆ Normalization of $H(j\omega)$ magnitudes with respect to maximum

Some system function may not have $|H(j\omega)|$ values ranging from 0 to 1. If dB variation is needed from 0 to some value, the $|H(j\omega)|$ must be between 0 and 1. For example the system $H(s) = \frac{3s}{s+1}$ shows non 0-1 variation. Let us form the system as follows:

```
>>H=tf([3 0],[1 1]); ↵
```
← Defining $H(s)$ as in section 8.2, H⇔ $H(s)$

We wish to find $|H(j\omega)|_{max}$ over the interval $-10 \leq \omega \leq 10$ $rad/\sec$ with a ω step 0.1 $rad/\sec$. First we determine the $|H(j\omega)|$ values by applying earlier functions at the indicated ω points as follows:

```
>>w=-10:0.1:10; R=freqresp(H,w); V=squeeze(R); ↵
```

We know that the last V is holding complex $H(j\omega)$ values as a column matrix. Appendix C.4 cited max finds the maximum of $|H(j\omega)|$ as follows:

```
>>M=max(abs(V)) ↵

M =

        2.9851
```
← M holds the $|H(j\omega)|_{max}$ value

Normalization means finding the $\frac{H(j\omega)}{|H(j\omega)|_{max}}$ values so we just divide the V by the M as follows:

```
>>S=V/M; ↵
```
← S holds the normalized complex $H(j\omega)$ values where S is user-chosen variable

Obviously the S is a column matrix as well.

9.4 How to graph the frequency response of a continuous system?

Last section demonstrates frequency response of a continuous system for the most part on calculation context. This section is all about the graphing of the frequency response of a continuous system.

Frequency response of a continuous system $H(j\omega)$ in graphical form has four components namely $\text{Re}\{H(j\omega)\}$, $\text{Im}\{H(j\omega)\}$, $|H(j\omega)|$, and $\angle H(j\omega)$ versus ω over certain interval of ω. To graph the frequency response, there can be two options – either you get the data and plot on your own or use the ready made function like bode, both of which are addressed in the following.

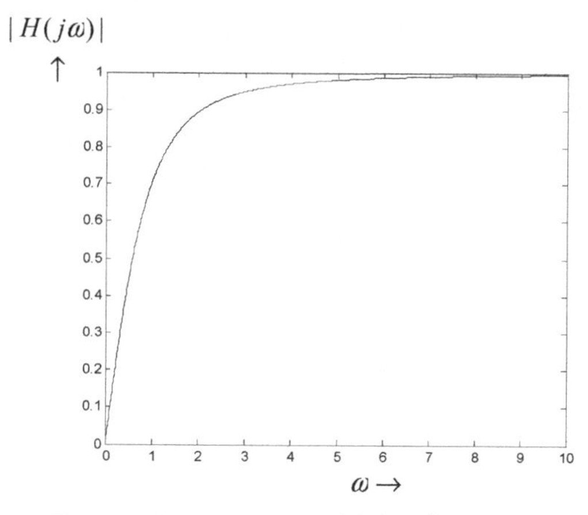

Figure 9.2(b) The plot of the $|H(j\omega)|$ versus ω

9.4.1 Getting data first and plotting the response afterwards

In section 9.3 we have addressed computations all along on $H(s)=\frac{s}{s+1}$ or $H(j\omega)=\frac{j\omega}{j\omega+1}$. We wish to plot the $|H(j\omega)|$ versus ω over $0 \le \omega \le 10 rad/\sec$ by choosing a ω step $0.01\ rad/\sec$.

First we follow the techniques of section 9.3 for obtaining the system function frequency magnitude data as follows:

```
>>H=tf([1 0],[1 1]); ↵      ← Defining the H(s) as in section 8.2, H⇔H(s)
>>w=[0:0.01:10]'; ↵         ← The w holds the ω values as a column matrix
                                          (subsection 1.1.2)
>>R=freqresp(H,w); ↵        ← freqresp calculates H(jω) and assigns to R
>>V=squeeze(R); ↵           ← Removing singleton dimensions from R because of
                  being 3D array, V holds H(jω) complex values as a column matrix
```

After that we employ the appendix F cited plot command as follows:

```
>>plot(w,abs(V)) ↵          ← Graphing the |H(jω)| versus ω like figure 9.2(b)
```

The plot has two input arguments, the first and second of which are the ω and $|H(j\omega)|$ data as a row or column matrix respectively. In a similar fashion the commands plot(w,real(V)), plot(w,imag(V)), and plot(w,angle(V)) graph the spectra $\text{Re}\{H(j\omega)\}$ versus ω, $\text{Im}\{H(j\omega)\}$ versus ω, and $\angle H(j\omega)$ versus ω respectively (graphs are not shown for space reason). Not only that, decibel spectrum of the section 9.3 (whose values are in D in the section) can also be graphed by the command plot by using plot(w,D) regardless of with and without dB suppression.

9.4.2 Employing ready made bode plotter

MATLAB is so resourceful in programming and built-in function sense that we may have different options for the same type of problem. The ready made tool means we use the bode plot to see the system frequency response but mainly the magnitude and phase ones. The function we employ is bode which keeps provision for accepting different input-output arguments as will be explained in the following.

Graphing a system function:

Suppose we wish to plot the $H(j\omega)$ over ω for the continuous system $H(s)=\frac{s}{s+1}$ (or $H(j\omega)=\frac{j\omega}{j\omega+1}$) from 0.1 to 100 $rad/\sec$ by using the command bode and do so at the command prompt as follows:

>>H=tf([1 0],[1 1]); ↵ ← Defining the $H(s)$ as in section 8.2, H⇔ $H(s)$

>>bode(H,{0.1,100}) ↵

The bode has two input arguments, the first and second of which are the system function assignee name and the ω interval description respectively. In the ω interval description we use the second brace (not the first, nor the third brace). We input only the bounds of the interval, ω step size is automatically chosen by the bode. Figure 9.2(c) presents the bode plot for the example $H(j\omega)$. By default the bode returns the decibel spectrum $20\log_{10}|H(j\omega)|$ and phase spectrum $\angle H(j\omega)$ in degrees versus ω together.

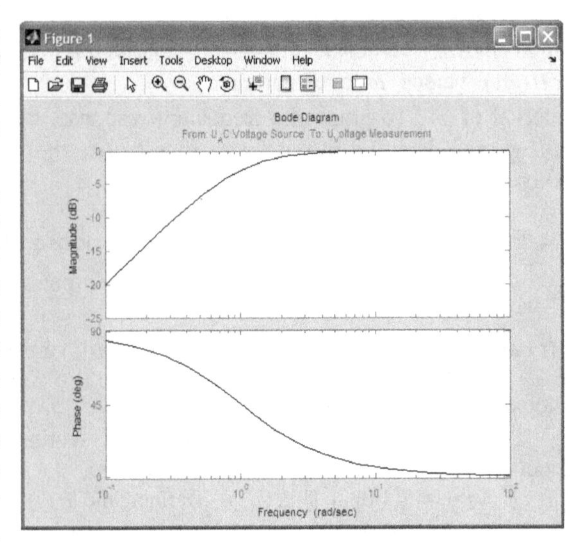

Figure 9.2(c) Bode plot of a continuous system

There are different options hidden in the bode plot of figure window 9.2(c). Rightclick on the mouse in the plot area of the figure 9.2(c), find the Properties in a popup, and click the Properties to see the prompt window like the figure 9.2(d).

To graph $|H(j\omega)|$ versus ω over the same interval, we click the Units menu of the figure 9.2(d) and select Absolute in the Magnitude in popup.

To graph $|H(j\omega)|$ versus frequency in Hertz over the same interval, we click the Units menu of the figure 9.2(d) and select the Hz in the Frequency in popup.

If you want to see only the $20\log_{10}|H(j\omega)|$ versus ω, rightclick on the mouse in the plot area of the figure 9.2(d), find the Show in a popup, and click the Phase under the Show to see only the magnitude spectrum. Similarly you can view only the phase spectrum $\angle H(j\omega)$ by clicking the Magnitude under the Show.

The default phase in the bode plot is in degree. If you want to turn that to radian, click the Units menu of the figure 9.2(d) and select the radians in the Phase in popup.

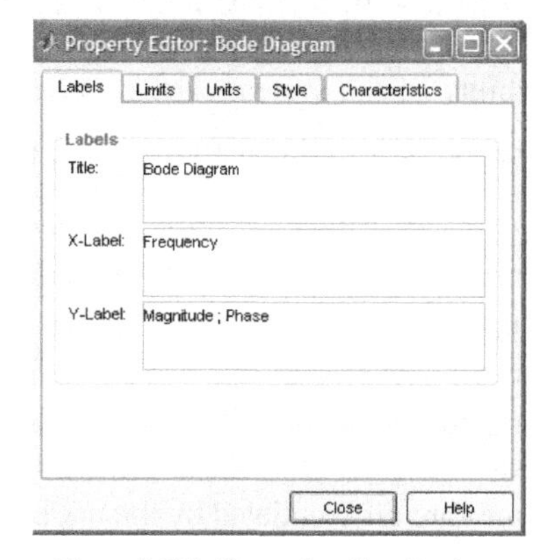

Figure 9.2(d) Property editor for the bode plot

The default title of the figure 9.2(c) is Bode Diagram. If you intend to change that to some other for example System Response, click the Labels menu of the figure 9.2(d) and type System Response from keyboard deleting the Bode Diagram under the Title slot of the Labels. In a similar fashion you can change the horizontal and vertical axes labeling of the figure 9.2(c).

Mouse based access on the graph data:

The mouse pointer helps us find the system function graph data. For this bring your mouse pointer at any point on the bode drawn curve (figure 9.2(c)) and leftclick the mouse. An indicatory box appears on top of the bode plot in which you find the mouse point coordinates of the horizontal and vertical axes quantities of the bode graph.

Adding grid lines to the bode graph:

Figure 9.2(c) shows only the bode graph. If you wish to include grid lines to the horizontal and vertical axes (makes the graph more meaningful) of the graph, execute the command grid at the command prompt.

Graphing multiple system functions:

We have been exercising all along keeping the $H(j\omega)$ in the H. Suppose there is another system function available in the workspace variable H1 (i.e. $H_1(j\omega)$) and we wish to plot the $H(j\omega)$ and $H_1(j\omega)$ over the same ω interval. The bode also keeps the provision for graphing multiple system functions. For the graphing, the command we need is bode(H,H1,{0.1,100}) – the names of assignee holding the system functions and interval description all separated by a comma respectively. In the case of three system functions, we just employ the command bode(H,H1,H2,{0.1,100}) where H2 indicates the third system function over the same interval.

Note: The bode does not function for the negative frequencies unlike the freqresp of the last section and its return is always in magnitude-phase angle form.

9.5 Pole-zero map from a system function

Pole-zero map is widely used in the continuous system analysis to test the stability of the system. The concept of pole-zero is applicable if the given system function is in rational form $\frac{P(s)}{Q(s)}$, the roots of forming the equations $Q(s)=0$ and $P(s)=0$ are called the poles and zeroes of the system respectively. Given a system function $Y(s)$, MATLAB built-in function pzmap (abbreviation for the pole zero map) locates the poles and zeroes of $Y(s)$ in the s plane which is complex in general. Denoted by symbol one writes $s=\sigma+j\omega$ indicating the real and imaginary components in the s plane. The syntax we apply is pzmap(system) where the system is defined in accordance with the section 8.2. In the s plane, the poles and zeroes are differentiated by the markers × and o respectively.

Let us find the pole-zero map for the system function $Y(s)=\frac{2(2s+1)}{s^2+1}$.

The poles (means roots of $s^2+1=0$) and zeroes (means roots of $2(2s+1)=0$) of $Y(s)$ are $s=\pm j$ and $s=-\frac{1}{2}$ respectively. First we define the system by using the tf of section 8.2 as follows:

```
>>Y=tf([4 2],[1 0 1]); ↵
```

In above execution the Y is any user-chosen name which holds the system $Y(s)$. Then we call the pzmap on entered Y as follows:

```
>>pzmap(Y) ↵
```

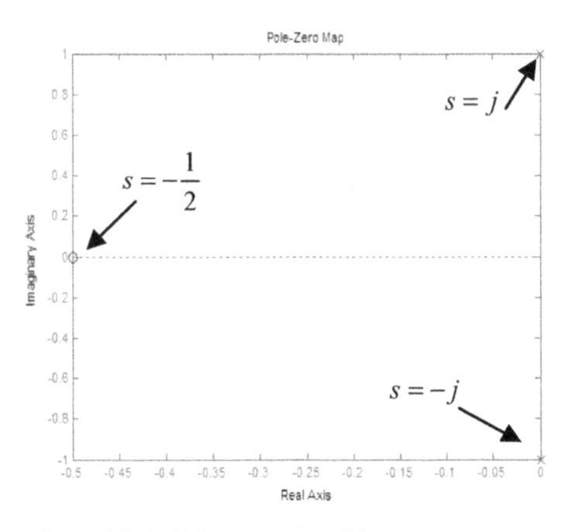

Figure 9.2(e) Pole-zero map of the system $Y(s)$

Response is shown in the figure 9.2(e). Arrows in the figure 9.2(e) indicate the relative positions of the poles and zeroes of $Y(s)$. The horizontal and vertical axes of the figure 9.2(e) refer to the σ and ω respectively. In the case of multiple poles or zeroes, only one is displayed.

The pzmap also keeps the provision for returning the poles and zeroes as column matrix but we need to supply two output arguments (first one for the poles and the second for the zeroes) to pzmap as follows:

```
>>[p,z]=pzmap(Y) ↵   ← Variable p for poles and z for zeroes, p or z can be reader
                       -supplied variable, subsection 1.1.2 for output argument
p =                  ← p holds the poles as a column matrix
     0 + 1.0000i
     0 - 1.0000i
z =                  ← z holds the zero
     -0.5000
```

Let us see the following two examples on different system functions.

Example 1

Find the pole-zero map of the system $Y(s)=\frac{-7s^4+2s^3+24}{s^5}$ along with their values.

Its straightforward implementation is as follows:

>>Y=tf([-7 2 0 0 24], [1 0 0 0 0 0]); ↵ ← Workspace Y holds the system $Y(s)$

>>pzmap(Y) ↵ ← Only for pole-zero map, graph not shown for space reason

>>[p,z]=pzmap(Y); ↵ ← To have the poles and zeroes stored in p and z as a column matrix respectively

Example 2

Analyze the pole-zero map for the system $Y(s)=\frac{2(2s+1)}{s^2+1}+\frac{-7s^4+2s^3+24}{s^5}$.

When the system functions are in addition form, they are treated as parallel system as regards to section 8.3 and the equivalent is obtained by the command parallel as follows:

>>Y1=tf([4 2], [1 0 1]); ↵ ← Workspace Y1 holds the component system $\frac{2(2s+1)}{s^2+1}$

>>Y2=tf([-7 2 0 0 24], [1 0 0 0 0 0]); ↵ ← Workspace Y2 holds the component system $\frac{-7s^4+2s^3+24}{s^5}$

>>Y=parallel(Y1,Y2); ↵ ← Workspace Y holds the system $Y(s)$

The Y, Y1, and Y2 are user-chosen variables. As we conducted in the other examples, execute the command pzmap(Y) or [p,z]=pzmap(Y) for the analysis.

Constant damping ratio and natural angular frequency curves from system function

The variable $s=\sigma+j\omega$ of a system function brings two more parameters in the s plane namely constant damping ratio ζ and constant natural angular frequency ω_n. It is possible to draw these two curves on the existing pole-zero plot by utilizing the command sgrid without any input argument.

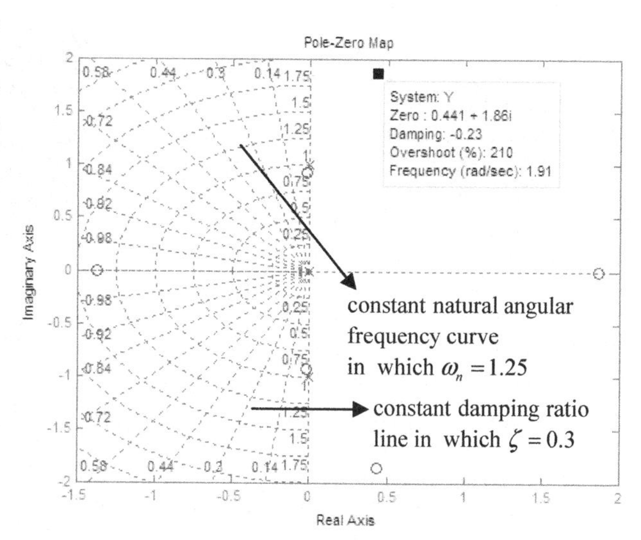

Figure 9.2(f) Pole-zero map with s plane grid

Let us consider the example 2 mentioned system $Y(s)$ which is stored in the workspace Y afterwards we conduct the following:

>>pzmap(Y) ↵ ← Displays the pole-zero map only (not shown for space reason)
>>sgrid ↵ ← Displays the figure 9.2(f) with the s plane grid

In the s plane of figure 9.2(f), two types of grid are drawn – line of constant ζ and round curve of constant ω_n. The arrows in the figure show the constant damping ratio line and constant angular frequency curve for ζ=0.3 and ω_n=1.25 respectively. Also if you bring mouse pointer on any pole or zero and leftclick the mouse pointer, you find the location and other relevant information about the pole or zero as shown in the upper half of the figure 9.2(f). To remove the information box, rightclick the mouse at the pole or zero and click the Delete.

Importance of pole-zero analysis

If any pole of a system is in the right half s plane, the system becomes unstable. In such a system the output quantity (whether voltage, current, or other) shows successively increasing amplitude over time which is not practical at all.

9.6 Modeling step and impulse responses

In section 9.1 we focused step response of a continuous time system in MATLAB. The same step response we can model in SIMULINK. Step response input is always simulated by the Step block of chapter 3. The system can be in any form (like Transfer Fcn, Zero-Pole, or State-Space defined) as discussed in section 8.5. The time domain output we see by using a Scope (subsection 1.2.5).

Impulse response truly speaking can never be modeled in SIMULINK. Implementationally we apply a short existent pulse of high amplitude which serves the purpose of impulse response. Modeling approach is solely applicable for time domain input-output. Let us proceed with the following examples about the step and impulse responses.

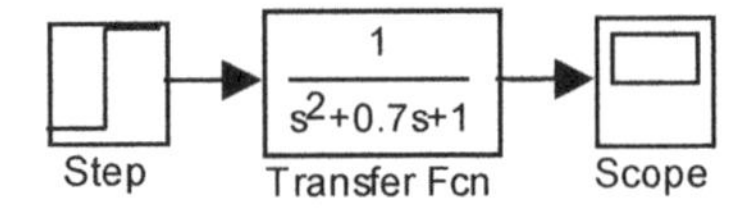

Figure 9.3(a) Modeling step response of a continuous system

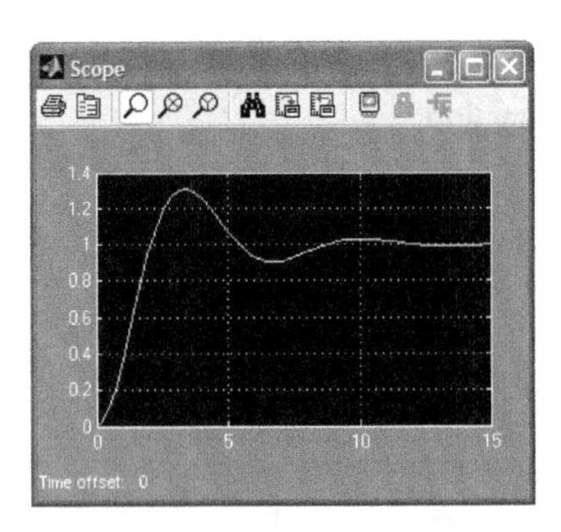

Figure 9.3(b) Step response for the continuous system

Example 1

Referring to figure 9.1(a), we intend to determine the step response $y(t)$ for the system $H(s)=\frac{1}{s^2+0.7s+1}$ over $0 \le t \le 15$ sec in SIMULINK.

This is all about $y(t)$ versus t finding. Open a new SIMULINK model file (subsection 1.2.2), get one Step (for $u(t)$), one Transfer Fcn (for $H(s)$), and one Scope (for viewing $y(t)$) blocks in the model file following the link mentioned in appendix A. Doubleclick the Step to change its Step time from default 1 to 0, doubleclick the Transfer Fcn, enter its Denominator as [1 0.7 1] in the parameter window of the block, and place the blocks relatively and connect them (subsection 1.2.3) according to figure 9.3(a) which is the model for the step response. Change the Stop time of the solver (subsection 1.2.6) from default 10 to 15 for the interval entering, run the model by clicking the ► icon in the menu bar, and doubleclick the Scope. On clicking the autoscale icon (figure 3.1(c)) we find the Scope output as seen in figure 9.3(b). The horizontal and vertical axes of the figure 9.3(b) refer to t and $y(t)$ values respectively.

Example 2

In example 1 the system $H(s)$ is having the state-space representation $\{A, B, C, D\}$ where $A = \begin{bmatrix} 0.1 & 0 \\ 0 & 0.2 \end{bmatrix}$, $B = \begin{bmatrix} 1 \\ 0.2 \end{bmatrix}$, $C = [0.2 \quad 1]$, and $D = [1]$. We intend to obtain the step response over the same interval for this system.

All we need is replace the Transfer Fcn of the figure 9.3(a) by a State-Space block like the figure 9.3(c). Doubleclick the State-Space and enter its A, B, C, and D as [0.1 0;0 0.2], [1;0.2], [0.2 1], and 1 in the parameter window for A, B, C, and D respectively. Run the model by clicking the ► icon in the menu bar and doubleclick the Scope to see the step response $y(t)$ versus t like figure 9.3(d) with autoscale setting.

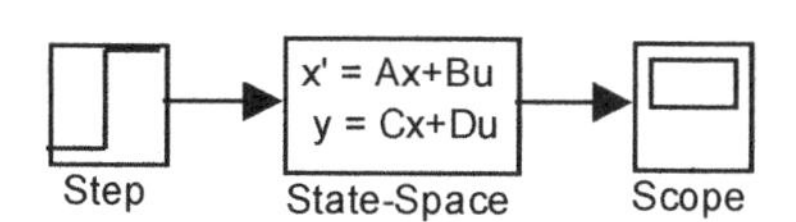

Figure 9.3(c) Step response from state-space model

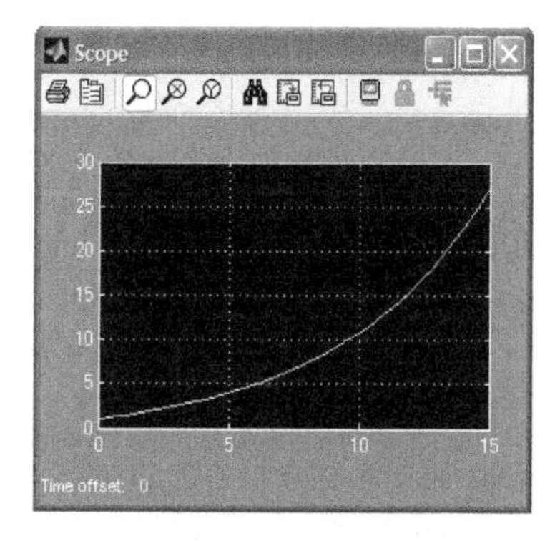

Figure 9.3(d) Step response for the model in figure 9.3(c)

Example 3

This example demonstrates how advantageous SIMULINK modeling can be. Concerning the figure 8.4(b), we wish to see the step response $y(t)$ versus t for the interconnected system over $0 \le t \le 15$ sec in SIMULINK.

We presented complete modeling of the interconnected system in section 8.5 whose model is depicted in figure 8.4(c). We just change the Stop time of the solver from default 10 to 15 for the interval entering, run the model by clicking the ► icon in the menu bar, and doubleclick the Scope. The Scope output with autoscale setting is shown in figure 9.3(e).

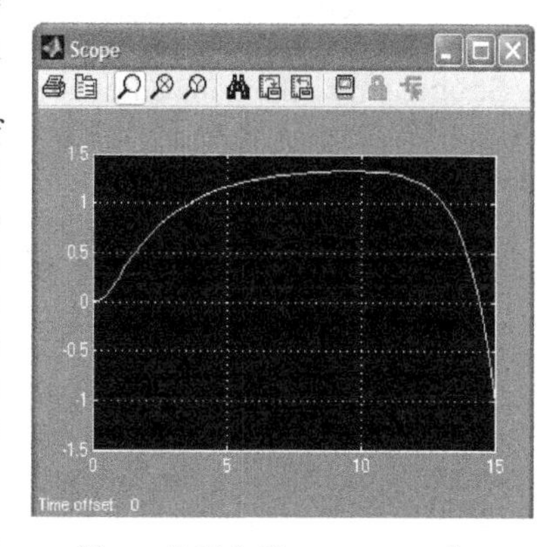

Figure 9.3(e) Step response for the interconnected system of the figure 8.4(b)

Example 4

In this example we illustrate how a practical impulse response we can simulate in SIMULINK. We wish to find the impulse response for the system $H(s) = \frac{1}{s^2 + 0.7s + 1}$ over $0 \le t \le 10$ sec in SIMULINK.

Our concentration is to view the wave shape of $h(t)$ versus t. Section 3.2 explains the modeling of a practical impulse function based on a short duration finite pulse. We assume the short duration is 0.01sec so the amplitude of step function should be 1/0.01 or 100 therefore the practical impulse function has the expression $\delta_{practical}(t) = 100u(t) - 100u(t-0.01)$. However as a whole the modeling procedure is the following:

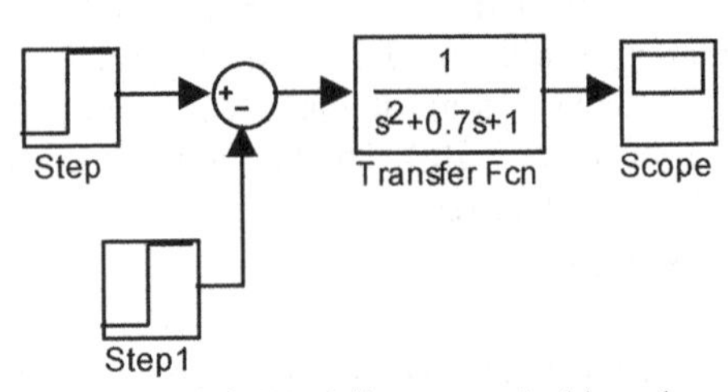

Figure 9.3(f) Modeling a practical impulse response on a continuous system

Open a new SIMULINK model file, get two Step, one Transfer Fcn, one Sum, and one Scope blocks in the model file following the link mentioned in appendix A.

Doubleclick the Step to change its Step time from default 1 to 0, enter its Final Value as 100 (for $100u(t)$), doubleclick the Step1 to change its Step time from default 1 to 0.01, and enter its Final Value as 100 (for $100u(t-0.01)$).

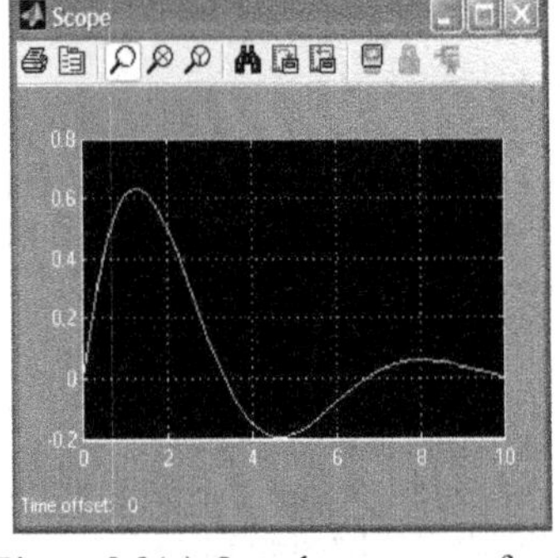

Figure 9.3(g) Impulse response for the continuous system of the example 4

Doubleclick the Transfer Fcn, enter its Denominator as [1 0.7 1] in the parameter window of the block, doubleclick the Sum block, and change its List of Signs from default ++ to +- for the negative step function.

Place the blocks relatively and connect them according to figure 9.3(f) which is the model for the practical impulse response. Since the interval is default in SIMULINK, run the model by clicking the ▶ icon in the menu bar and doubleclick the Scope. On clicking the autoscale icon we see the Scope output as seen in figure 9.3(g). The horizontal and vertical axes of the figure 9.3(g) refer to t and $h(t)$ values respectively.

9.7 Modeling the response of a circuit system

Response from an electric circuit system other than step and impulse is also a matter of interest in the study. In this section our focus is to see the user-required circuit response in time domain by modeling the circuit system in SIMULINK. Since all relevant issues we addressed in previous sections or chapters, let us jump into the implementations through the following examples.

Example 1

We wish to see the circuit response for v_0 and i of the figure 8.6(c) shown RC circuit with $R = 4\ K\Omega$ and $C = 0.5\ mF$ over $0 \le t \le 7$ sec when a ramp input $v_i = 0.2t$ is applied.

The circuit is modeled as example 1 in section 8.8 and the model is seen in figure 8.7(g). We get a Ramp block for v_i input and two Scope blocks for v_0 and i in the model of figure 8.7(g). Doubleclick the Ramp for entering the Slope as 0.2 in the parameter window of the block, rename (subsection 1.2.3) the two Scopes as I and Vo to make sense with the modeling of i and v_0 respectively, set the solver Stop time as 7 (subsection 1.2.6) for the interval, and connect the three blocks like figure 9.4(a) which is the model for this problem.

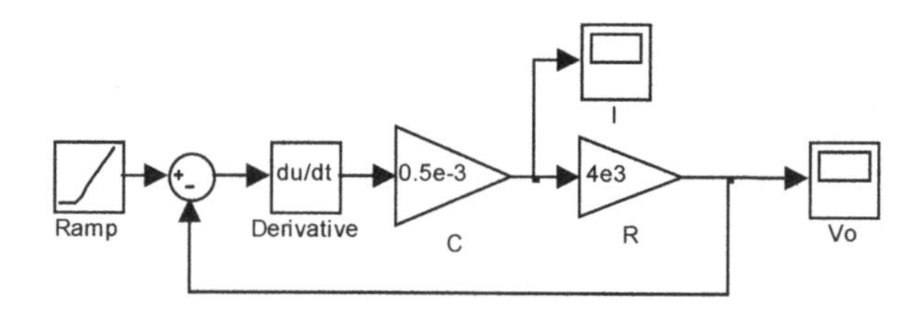

Figure 9.4(a) Circuit response modeling for the circuit of figure 8.6(c) with a ramp input

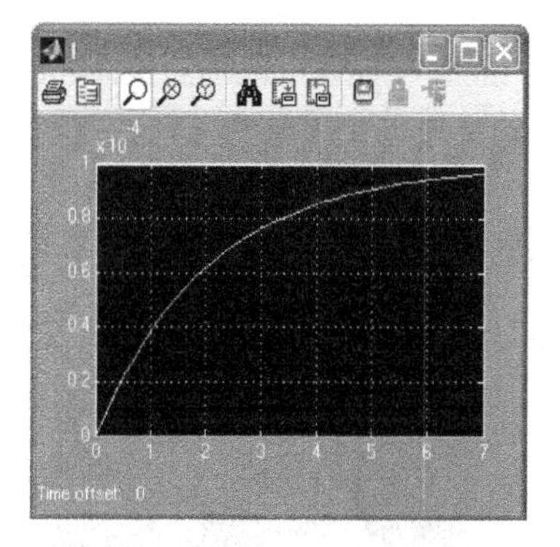

Figure 9.4(b) The i response of the circuit in figure 8.6(c)

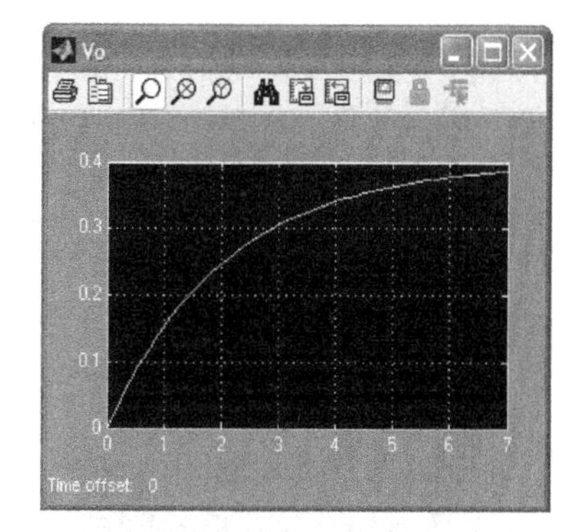

Figure 9.4(c) The v_0 response of the circuit in figure 8.6(c)

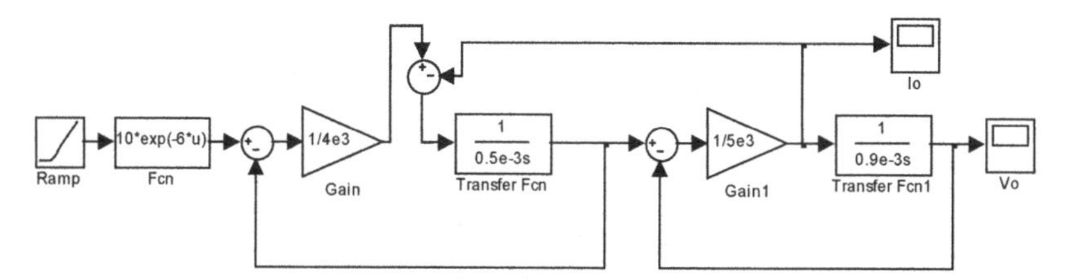

Figure 9.4(d) Model for finding the v_0 and i_0 when an exponential signal is applied to the two stage filtering circuit of figure 8.8(a)

Run the model by clicking the ► icon in the menu bar, doubleclick the I and Vo to see the outputs like the figures 9.4(b) (for i versus t) and 9.4(c) (for v_0 versus t) with the autoscale setting (figure 3.1(c)) respectively.

Example 2

We wish to see the time domain responses of the v_0 and i_0 for the figure 8.8(a) shown two stage RC filtering circuit with R_1=4 $K\Omega$, R_2=5 $K\Omega$, $C_1 = 0.5mF$, and $C_2 = 0.9mF$ over $0 \le t \le 3$ sec when an exponential signal $v_i = 10\, e^{-6t}$ is applied to the input.

Without input and output the circuit is modeled as the example 2 in section 8.8. The model of the two stage filtering circuit is shown in figure 8.8(d). We add the blocks for the input and output to obtain the model of figure 9.4(d) which is the model of this problem. To do so, we get one Ramp, one Fcn, and two Scope blocks in the model of the figure 8.8(d), doubleclick the Fcn, enter its Expression as 10*exp(-6*u) for the exponential source, enlarge the Fcn to see its contents, rename the Scopes as Vo and Io for the v_0 and i_0 respectively, and connect the newly included block as shown in figure 9.4(d).

Lastly set the solver Stop time as 3 for the interval, run the model by clicking the ▶ icon in the menu bar, doubleclick the Vo and Io to see the outputs like the figures 9.4(e) (for v_0 versus t) and 9.4(f) (for i_0 versus t) with the autoscale setting respectively.

Thus the reader can apply any time domain input signal whether periodic or nonperiodic of the chapter 3 to any other electrical circuit and seek the response in time domain in SIMULINK.

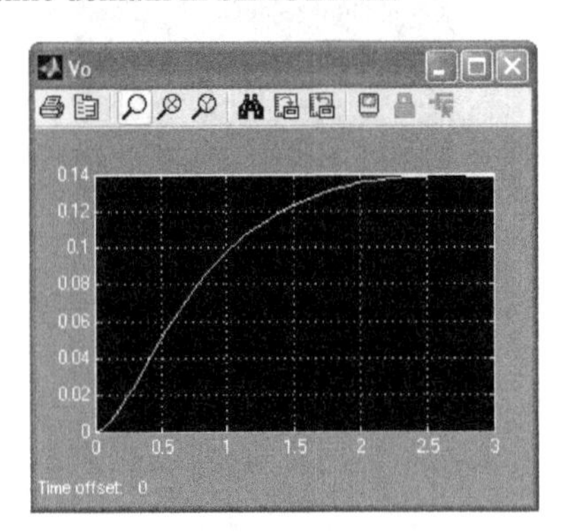

Figure 9.4(e) v_0 response for the circuit of figure 8.8(a)

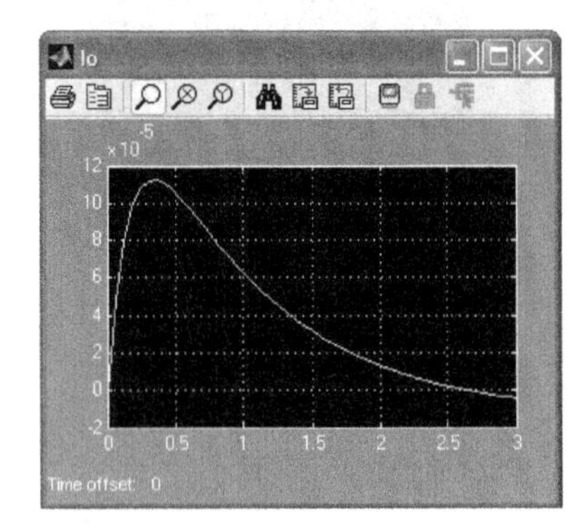

Figure 9.4(f) i_0 response for the circuit of the figure 8.8(a)

9.8 Response on circuit dynamics in time domain

In the last section we concentrated the circuit system solution in terms of wave shapes in SIMULINK whether voltage or current. Expression based solution is also obtainable by exercising the differential equation solver dsolve at MATLAB command prompt. Appendix H describes the input-output format of the built-in function dsolve. In these types of problems mainly we commence with the governing differential equations of the voltage and current in a given circuit and seek for the voltage and current expressions from the governing differential equations. Let us go through the following six examples in this regard.

⬍⬍ Example 1 – first order system and one dependent variable

The capacitor voltage of the circuit shown in figure 9.5(a) is $v_C(t) = 100 - 80e^{-t}$ for $t \geq 0$ coinciding with the polarity of $v_C(0)$ which we wish to determine.

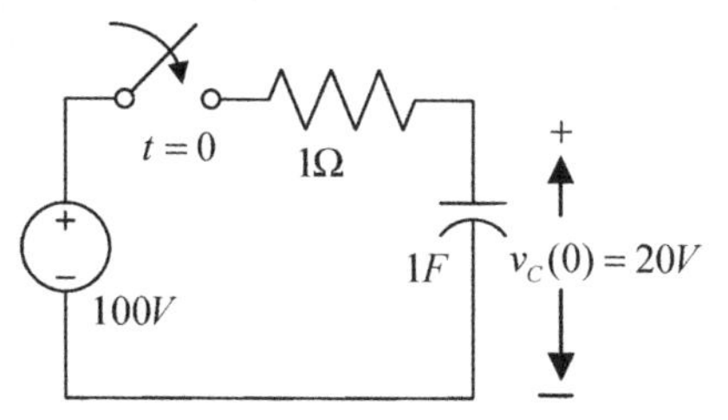

Figure 9.5(a) An R - C circuit system

The instantaneous current in the circuit is $C\frac{dv_C(t)}{dt}$ (current polarity is from +ve to –ve nodes of the $v_C(0)$) and KVL writing around the circuit at $t = 0^+$ provides $RC\frac{dv_C(t)}{dt} + v_C(t) = 100$ or $\frac{dv_C(t)}{dt} + v_C(t) = 100$ by plugging the circuit parameters besides the initial condition $v_C(0) = 20$ utilizing the last two equations we execute the following at the command prompt of MATLAB:

```
>>S=dsolve('DV+V=100','V(0)=20') ↵

S =                    ← Workspace variable S is holding the solution for the vC(t)

100-80*exp(-t)
```

In above execution the S or V is any user-chosen variable name and the variable equivalences are as follows: D⇔$\frac{d}{dt}$, V⇔$v_C(t)$, DV⇔$\frac{dv_C(t)}{dt}$, and V(0)⇔$v_C(0)$.

The solution for the same circuit without the initial capacitor voltage is $v_C(t) = 100(1 - e^{-t})$ which we verify by executing the command S=dsolve('DV+V=100','V(0)=0') at the command prompt.

⬍⬍ Example 2 – second order system and one dependent variable

In the second order circuit of the figure 9.5(b) KVL writing around the circuit loop provides $9i_L + 9\frac{di_L}{dt} + 9\int i_L dt = 70$ (resistor, inductor, and capacitor voltages are $9i_L$, $9\frac{di_L}{dt}$, and $9\int i_L dt$ respectively) for $t \geq 0$. Since we are particularly solving differential equation not the integral one (the component $9\int i_L dt$ is an integral expression), we differentiate the equation which provides the simplified governing differential equation $\frac{d^2 i_L}{dt^2} + \frac{di_L}{dt} + i_L = 0$. Again applying the KVL at $t = 0^+$, the inductor voltage is

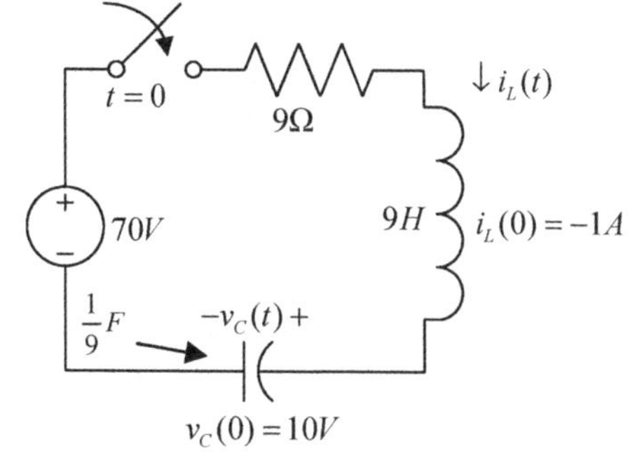

Figure 9.5(b) Second order circuit system with initial conditions

$9\frac{di_L}{dt}=69$ or $\frac{di_L}{dt}=\frac{23}{3}$. Taking the initial condition into account, the $i_L(t)$ is obtained as $-e^{-\frac{t}{2}}\cos\frac{t\sqrt{3}}{2}+\frac{43\sqrt{3}}{9}e^{-\frac{t}{2}}\sin\frac{t\sqrt{3}}{2}$ following the differential equation solving.

Concisely we wish to obtain the inductor current expression $i_L(t)=-e^{-\frac{t}{2}}\cos\frac{t\sqrt{3}}{2}+\frac{43\sqrt{3}}{9}e^{-\frac{t}{2}}\sin\frac{t\sqrt{3}}{2}$ starting from $\frac{d^2i_L}{dt^2}+\frac{di_L}{dt}+i_L=0$, $i_L(0)=-1$, and $\left.\frac{di_L}{dt}\right|_{t=0^+}=\frac{23}{3}$ which needs the following execution at the command prompt:

```
>>S=dsolve('D2I+DI+I=0','I(0)=-1,DI(0)=23/3') ↵

S =                       ← Workspace S is holding the solution for i_L(t) in code form

43/9*3^(1/2)*exp(-1/2*t)*sin(1/2*3^(1/2)*t)-exp(-1/2*t)*cos(1/2*3^(1/2)*t)
```

In above implementation the S or I is any user-chosen variable name maintaining the following variable equivalences: I⇔$i_L(t)$, DI⇔$\frac{di_L(t)}{dt}$, D2⇔$\frac{d^2}{dt^2}$, D2I⇔$\frac{d^2i_L}{dt^2}$, I(0) ⇔$i_L(0)$, and DI(0)⇔$\left.\frac{di_L}{dt}\right|_{t=0^+}$. Usage of the command pretty (appendix C.10) on the codes stored in S displays the following:

```
>>pretty(S) ↵

          1/2                     1/2                           1/2
  43/9 3     exp(-1/2 t) sin(1/2 3    t) - exp(- 1/2 t) cos(1/2 3    t)     ←What we are after
```

Note: Since i or j is the unit imaginary number in MATLAB, we do not use the i or j as the workspace variable for the current for instance Di would return some error.

❖❖ Example 3 – first order system and two dependent variables

In example 2 the prime or dependent variable is $i_L(t)$. What if we need the capacitor voltage $v_C(t)$ of the circuit in figure 9.5(b)? The reader might say $v_C(t)=\frac{1}{C}\int i(t)dt$ is the solution thereby requiring extra integration of $i_L(t)$. Instead of a single differential equation if we write two differential equations relating the $v_C(t)$ and $i_L(t)$ (just like the system of differential equations of appendix H), the solver dsolve yields the solution for both the $v_C(t)$ and $i_L(t)$. For that reason we rewrite the governing differential equations for the circuit in figure 9.5(b) as follows:

Equation 1: $9i_L+9\frac{di_L}{dt}+v_C=70$ from KVL around the circuit loop

Equation 2: $i_L=\frac{1}{9}\frac{dv_C}{dt}$ from KCL at the +ve node of the $v_C(t)$

Initial conditions: $i_L(0)=-1A$ and $v_C(0)=10V$

Solving the two equations and initial conditions, one obtains the solution as $i_L(t) = -e^{-\frac{t}{2}}\cos\frac{t\sqrt{3}}{2} + \frac{43\sqrt{3}}{9}e^{-\frac{t}{2}}\sin\frac{t\sqrt{3}}{2}$ and $v_C(t) = 70 + e^{-\frac{t}{2}}\left(-26\sqrt{3}\sin\frac{t\sqrt{3}}{2} - 60\cos\frac{t\sqrt{3}}{2}\right)$ for $t \geq 0$. Our objective is to get these $i_L(t)$ and $v_C(t)$ expressions from MATLAB by using the differential equations and initial conditions for what reason we go through the following:

```
>>e1='9*I+9*DI+Vc=70'; e2='I=DVc/9'; ↵ ← e1, e2, I, and Vc are user-chosen
                                                       names
>>S=dsolve(e1,e2,'Vc(0)=10,I(0)=-1') ↵  ← S is any user-chosen name

S =
      I: [1x1 sym]
     Vc: [1x1 sym]
```

In the first line of just conducted implementation we assigned the two differential equations to the workspace variables e1 and e2 respectively with the variable equivalence I⇔$i_L(t)$, Vc⇔$v_C(t)$, DI⇔$\frac{di_L(t)}{dt}$, and DVc⇔$\frac{dv_C(t)}{dt}$. The initial condition set has the code equivalence I(0)⇔$i_L(0)$ and Vc(0)⇔$v_C(0)$ and is entered as the third input argument to dsolve in the second command line. As the return says the content of S is a structure array having two members I and Vc presumably the two dependent variables $i_L(t)$ and $v_C(t)$ respectively. In order to extract them from the S, we employ the commands S.I and S.Vc respectively. Obviously the S.I and S.Vc keep the codes for the expressions of $i_L(t)$ and $v_C(t)$ respectively. If we exercise the command pretty on the S.I or S.Vc, we see the solution as follows:

```
>>pretty(S.I) ↵
                                  1/2    1/2                1/2
   - 1/18 exp(- 1/2 t) (-86 sin(1/2 3   t) 3   + 18 cos(1/2 3   t)) ← Wanted expression
                                                                        for i_L(t)

>>pretty(S.Vc) ↵
                               1/2    1/2                1/2
   70 + exp(- 1/2 t) (-26 sin(1/2 3   t) 3   - 60 cos(1/2 3   t))  ← Wanted expression
                                                                       for v_C(t)
```

◆◆ Example 4 – first order system and three dependent variables

Performing the circuit analysis, given $i_{L1}(t) = 2 - \frac{1}{2}e^{-t}(6 - 4\cos t)$, $i_{L2}(t) = 2 + \frac{1}{2}e^{-t}(-6 - 4\sin t)$, and $v_C(t) = e^{-t}(3 + 2\sin t - 2\cos t)$ for $t \geq 0$ with the initial currents and voltage values $i_{L1}(0) = 1A$, $i_{L2}(0) = -1A$, and $v_C(0) = 1V$ are the inductor currents and capacitor voltage expressions

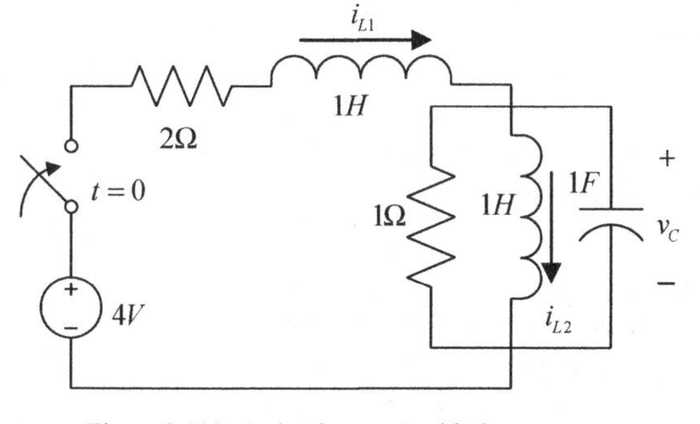

Figure 9.5(c) A circuit system with three states

for the transient circuit in figure 9.5(c). Our objective is to get these expressions from MATLAB.

We write the KVL at $t=0^+$ around the circuit to have $2i_{L1}+\frac{di_{L1}}{dt}+v_C=4$. In the $R-L-C$ parallel circuit, voltages across the inductor and the capacitor are same hence $\frac{di_{L2}}{dt}=v_C$. KCL at $t=0^+$ at the +ve node of the v_C provides $i_{L1}=v_C+i_{L2}+\frac{dv_C}{dt}$ on the fact that the currents through the resistor, inductor, and capacitor in the $R-L-C$ parallel branch are v_C, i_{L2}, and $\frac{dv_C}{dt}$ respectively. Having written the three governing equations, we jump into the execution as follows:

>>e1='2*I1+DI1+Vc=4'; ↵ ← $2i_{L1}+\frac{di_{L1}}{dt}+v_C=4$ is assigned to e1 and e1 is user-chosen name

>>e2='I1=Vc+I2+DVc'; ↵ ← $i_{L1}=v_C+i_{L2}+\frac{dv_C}{dt}$ is assigned to e2 and e2 is user-chosen name

>>e3='Vc=DI2'; ↵ ← $\frac{di_{L2}}{dt}=v_C$ is assigned to e3 and e3 is user-chosen name

>>S=dsolve(e1,e2,e3,'I1(0)=1,I2(0)=-1,Vc(0)=1') ↵ ← Calling the solver for the solution which is in S

S =

I1: [1x1 sym] ← Corresponds to $i_{L1}(t)$

I2: [1x1 sym] ← Corresponds to $i_{L2}(t)$

Vc: [1x1 sym] ← Corresponds to $v_C(t)$

User-chosen variable equivalences are the following: I1⇔$i_{L1}(t)$, I2⇔$i_{L2}(t)$, Vc⇔$v_C(t)$, DI1⇔$\frac{di_{L1}(t)}{dt}$, DI2⇔$\frac{di_{L2}(t)}{dt}$, and DVc⇔$\frac{dv_C(t)}{dt}$ moreover I1(0)⇔$i_{L1}(0)$, I2(0)⇔$i_{L2}(0)$, Vc(0)⇔$v_C(0)$, I1(0)=1⇔$i_{L1}(0)=1A$, I2(0)=-1⇔$i_{L2}(0)=-1A$, and Vc(0)=1⇔$v_C(0)=1V$ for the initial values. The solution for the $i_{L1}(t)$ is made available by calling S.I1 whose mathematics readable form is seen by the following:

>>pretty(S.I1) ↵

2 - 1/2 exp(-t) (6 - 4 cos(t))

Similarly the other two expressions are seen by executing pretty(S.I2) and pretty(S.Vc) for the $i_{L2}(t)$ and $v_C(t)$ respectively.

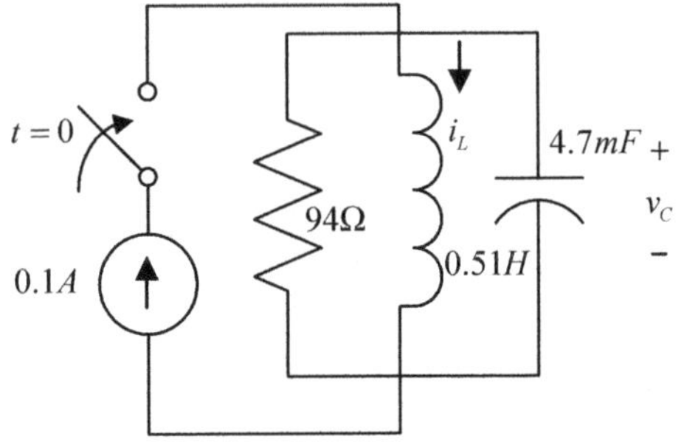

Figure 9.5(d) A current source is to be connected at $t=0$ to a parallel circuit

◆◆ Example 5 – system with fractional unit circuit elements

If we choose the circuit element values simple integers, we obtain simple coefficient differential equations. In the case of fractional unit or electrical engineering unit, the differential equation solution appears in lengthy expression. This example illustrates how we handle circuit elements with fractional or engineering unit. Provided $i_L(t) = \frac{1}{10} + \frac{2209\sqrt{21114867}}{527871675} e^{-\frac{2500}{2209}t} \sin\left(\frac{500\sqrt{21114867}}{112659}t\right)$ and $v_C(t) = -\frac{5}{2209} e^{-\frac{2500}{2209}t} \left\{ \frac{2209\sqrt{21114867}}{2070085} \sin\left(\frac{500\sqrt{21114867}}{112659}t\right) - \frac{2209}{25}\cos\left(\frac{500\sqrt{21114867}}{112659}t\right)\right\}$ for $t \geq 0$ are the inductor current and capacitor voltage of the circuit in figure 9.5(d) with the initial values $i_L(0) = 0.1A$ and $v_C(0) = 0.2V$. We wish to obtain these two expressions in MATLAB.

Two governing equations of the circuit are $0.51\frac{di_L}{dt} = v_C$ (from the equality of the inductor and capacitor voltages at $t = 0^+$) and $0.1 = \frac{v_C}{94} + i_L + 4.7\times10^{-3}\frac{dv_C}{dt}$ (from the KCL at $t = 0^+$). We code and execute the two equations as follows:

```
>>e1='51/100*DI=Vc'; ↵ ← first equation is assigned to e1 and e1 is user-chosen
                                                                          name
>>e2='1/10=I+47/1e4*DVc+Vc/94'; ↵ ← second equation is assigned to e2 and
                                                           e2 is user-chosen name
>>S=dsolve(e1,e2,'I(0)=1/10,Vc(0)=2/10') ↵ ← Calling the solver for the
                                                           solution which is in S
S =

            I: [1x1 sym]      ← Corresponds to iL(t)
           Vc: [1x1 sym]      ← Corresponds to vC(t)
```

In above implementation all circuit parameters are entered in standard unit and in rational form for instance 0.1 as 1/10, $1mF$ as 1/1000 F, etc. The symbology we chose is the following: I⇔$i_L(t)$, Vc⇔$v_C(t)$, DI⇔$\frac{di_L(t)}{dt}$, DVc⇔$\frac{dv_C(t)}{dt}$, I(0)⇔$i_L(0)$, Vc(0)⇔$v_C(0)$, I(0)=1/10⇔$i_L(0) = 0.1A$, and Vc(0)=2/10⇔$v_C(0) = 0.2V$. Having the solution of the ODEs available in the workspace S, calling individual expression takes place as follows:

```
>>pretty(S.I) ↵                    ← for the iL(t) expression

                 2209              2500              1/2        500               1/2
   1/10 + --------------- exp(- ------- t) 21114867    sin(----------- 21114867     t)
              527871675          2209                          112659

>>pretty(S.Vc) ↵                   ← for the vC(t) expression

                     2500     /  2209                   1/2         500              1/2
  - 5/2209 exp(- -------- t) |------------- 21114867    sin(----------- 21114867    t)
                     2209     \2070085                             112659

          2209         500                1/2 \
       - ------ cos(---------- 21114867    t)|
          25        112659                    /
```

♦♦ Example 6 – input is function of time

Foregoing examples demonstrate the computation based on the constant input excitation. That is not the only input the dsolve is designed for. Other input such as exponential, sinusoidal, etc can also be the input excitation to an electric circuit system.

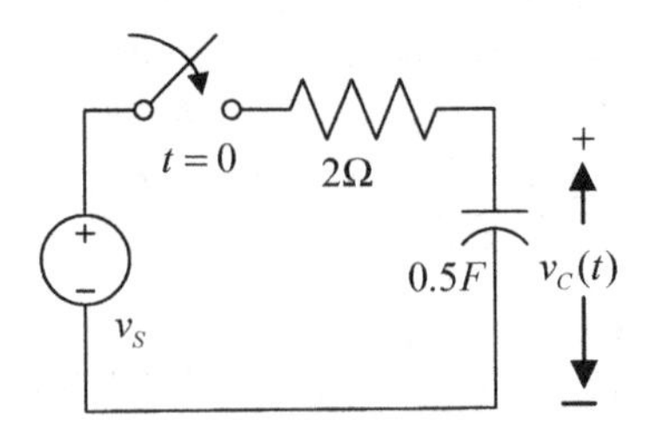

Figure 9.5(e) An R - C transient circuit

The circuit in the figure 9.5(e) has the KCL equation $\frac{v_C - v_S}{2} + 0.5\frac{dv_C}{dt} = 0$ at $t = 0^+$ at the +ve node of the capacitor voltage $v_C(t)$. With the $v_C(0) = 5V$ and $v_S(t) = 3\sin t$, the capacitor voltage of the circuit is found as $v_C(t) = -\frac{3}{2}\cos t + \frac{3}{2}\sin t + \frac{13}{2}e^{-t}$ for $t \geq 0$ which we wish to find and is conducted as follows:

```
>>S=dsolve('(Vc-3*sin(t))/2+1/2*DVc=0','Vc(0)=5'); ↵ ← The solution is stored in S, S is user-chosen name
>>pretty(S) ↵                    ← Just to read out the expression stored in S

      - 3/2 cos(t) + 3/2 sin(t) + 13/2 exp(-t)
```

Applied symbologies are the following: Vc⇔ $v_C(t)$, DVc⇔ $\frac{dv_C(t)}{dt}$, Vc(0)⇔ $v_C(0)$, Vc(0)=5⇔ $v_C(0) = 5V$, and 3*sin(t)⇔ $3\sin t$. Also the derivative coefficient 0.5 is written as 1/2 to make symbolic or rational sense.

If we had $v_S(t) = 3 + 9t$, the command would be S=dsolve('(Vc-3-9*t)/2+ 1/2*DVc=0','Vc(0)=5');.

Again if we had $v_S(t) = 3e^{-3t}$, the command would be S=dsolve('(Vc-3* exp(-3*t))/2+1/2*DVc= 0','Vc(0)=5');.

9.9 Access to circuit system response data

Circuit system solution means mathematical expression obtaining for the user-required voltage or current from the circuit dynamics. Once the expression is found, data of voltage or current might be required from the expression. This section is all about data obtaining from the voltage or current expression coded in MATLAB.

MATLAB built-in function subs helps us get the signal data at any user-defined t value with the syntax subs(functional code under quote, value of t). The reader finds help about the code in appendix B.

For instance the parabolic signal $f(t) = 6t^2 - 5t + 8$ has the functional value $f(1) = 9$ at t =1. In order to see its computation, we execute the following at the command prompt:

```
>>subs('6*t^2-5*t+8',1) ↵

ans =
         9
```

If we wrote f=subs('6*t^2-5*t+8',1), the value 9 would have been assigned to the workspace f where f is any user-chosen name. Let us see how we employ this function for different voltage and current expressions for a circuit system.

Example 1:

It is given that the capacitor voltage $v_C(t)$ of the example 1 in section 9.8 has the functional value $v_C(t)$=34.5015V at t=0.2secs which we wish to compute.

We know that the workspace S is holding the expressional code for the $v_C(t)$ from which we just execute the following at the command prompt:

```
>>subs(S,0.2) ↵

ans =

        34.5015
```

If we used v=subs(S,0.2), the $v_C(t)$ value would be assigned to v where v is any user-chosen name.

For the same capacitor voltage $v_C(t)$ suppose now the values are required for t=0, 1, and 2secs and which should be 20V, 70.5696V, and 89.1732V respectively.

Under this circumstance, the required t points are entered as a three element row matrix as the second input argument to the subs as follows:

```
>>v=subs(S,[0 1 2]) ↵ ← Computed values are assigned to v as a three
                                element row matrix
v =

    20.0000   70.5696   89.1732
```

Example 2:

The capacitor voltage $v_C(t)$ value of the example 1 is now needed for the interval $0 \le t \le 1$sec with the t step 0.01sec.

In a situation like this we first generate the t data by using the given step size and colon operator (subsection 1.1.2) which is t=0:0.01:1; and then call the subs as v=subs(S,t); in which v retains all voltage data as a row matrix. From 0 to 1 and with the step 0.01, there are 101 points in the t and the v also holds the 101 voltage data respectively. If the t is a row/column matrix, so is the vector v.

Example 3:

Referring to the example 4 of the section 9.8, the inductor current $i_{L1}(t)$ has the value 1.12A at t=0.15secs. The expression for the $i_{L1}(t)$ is reached by executing S.I1 so we need to execute the command subs(S.I1,0.15) at the command prompt.

Example 4:

If we wish to see the t and capacitor voltage/inductor current data side by side in the command window, we form a two column matrix and display it. For instance example 4 of section 9.8 cited $i_{L1}(t)$ is to be displayed over $0 \le t \le 0.15$ sec with a step size 0.05sec. We do so as follows:

>>t=[0:0.05:0.15]'; v=subs(S.I1,t); ↵ ← t is a column matrix holding the t data, the return to the **v** is also a column matrix

>>[t v] ↵

```
ans =
         0    1.0000
    0.0500    1.0464
    0.1000    1.0861
    0.1500    1.1200
```

↑ t data ↑ $i_{L1}(t)$ data

With this example we close the system response chapter.

Chapter 10
MISCELLANEOUS SIGNAL TOPICS

Introduction

As the name implies, a mixture of fundamental signal topics are chosen to implement in this chapter. While devising a particular chapter, all topics covering the chapter may not receive evenhanded attention yet the material may bear special importance to the course. Furthermore a certain signal problem may also need the exercise of various section cited functions. This sort of problems is tackled here too. The medley includes the following:

- Two finite time domain signals – $rect(t)$ and $tri(t)$
- Special signals – binary, complex, and statistical
- Input-output on various transform domains

10.1 The rect(t) and tri(t) pulses

In chapters 2 and 3 we explained the sample generation of the unit step $u(t)$ and Dirac delta $\delta(t)$ functions (sections 2.5, 2.8, and 3.2). In signal and system elementary course we come across two more pulses namely $rect(t)$ and $tri(t)$ whose generation we wish to address in this section.

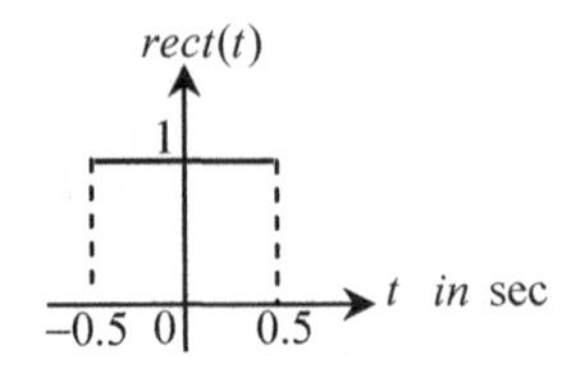

Figure 10.1(a) The $rect(t)$ pulse

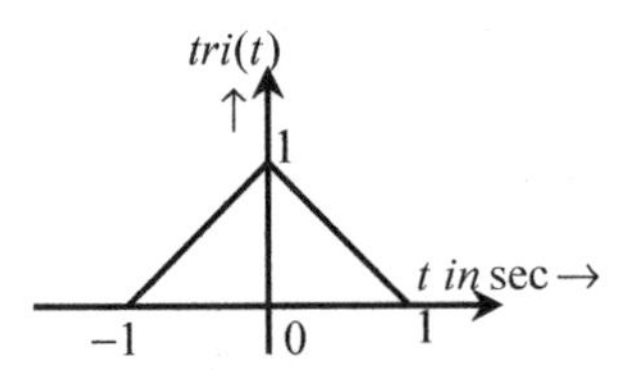

Figure 10.1(b) The $tri(t)$ pulse

```
function y=rec(t,to)
y=[ ];
for k=1:length(t)
    if -0.5<=(t(k)-to)&(t(k)-to)<=0.5
        y=[y 1];
    else
        y=[y 0];
    end
end
```

Figure 10.1(c) M-file code for the generation of the shifted $rect(t)$ function or $rect(t-t_0)$

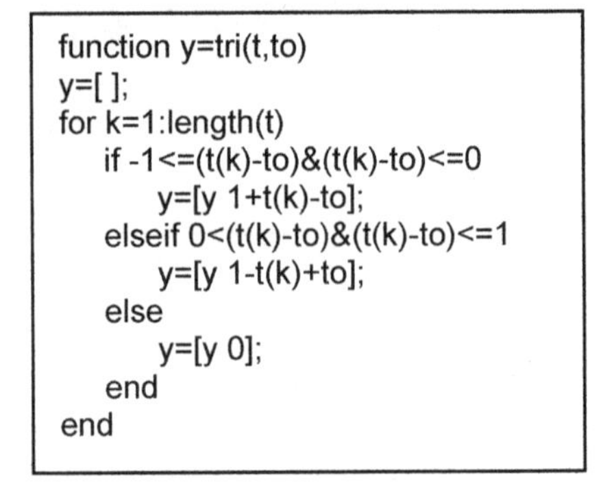

```
function y=tri(t,to)
y=[ ];
for k=1:length(t)
    if -1<=(t(k)-to)&(t(k)-to)<=0
        y=[y 1+t(k)-to];
    elseif 0<(t(k)-to)&(t(k)-to)<=1
        y=[y 1-t(k)+to];
    else
        y=[y 0];
    end
end
```

Figure 10.1(d) M-file code for the generation of the shifted $tri(t)$ function or $tri(t-t_0)$

$rect(t)$ and its derived functions

Figure 10.1(a) presents the $rect(t)$ function which is defined by $rect(t) = \begin{cases} 1 & for\ -0.5 \le t \le 0.5 \\ 0 & elsewhere \end{cases}$ and centered at t_0=0. The $rect(t)$ shifted at t_0 is given by $rect(t-t_0) = \begin{cases} 1 & for\ -0.5 \le (t-t_0) \le 0.5 \\ 0 & elsewhere \end{cases}$ and whose function file based code we provided in figure 10.1(c) following the second approach of section 2.5. The input arguments of rec(t,to) are t and t0 which are the t vector as a row matrix and scalar t_0 value respectively. Type the codes of the figure 10.1(c) in a new M-file (subsection 1.1.2) and save the file by the name rec in your working path.

Example:

Let us generate the samples of the signal $\sin t\ rect(2t-0.3)$ over $0 \le t \le 3$ sec with a step 0.1sec as follows:

>>t=0:0.1:3; ↵ ← t holds the t vector as a row matrix (section 2.3)

>>f=sin(t).*rec(2*t-0.3,0); ↵ ← Functional values of $\sin t\ rect(2t-0.3)$ are assigned to the workspace f as a row matrix where f is user-chosen variable

$tri(t)$ and its derived functions

Another function called $tri(t)$ is also seen in the signal and system text whose graphical representation at t_0=0 is provided in the figure 10.1(b). The shifted version of $tri(t)$ at any $t=t_0$ is formulated by $tri(t-t_0) =$

$\begin{cases} 1+t-t_0 & for\ -1 \le (t-t_0) \le 0 \\ 1-t+t_0 & for\ \ 0 < (t-t_0) \le 1 \\ 0 & elsewhere \end{cases}$ whose function file based code is provided in the figure 10.1(d) following the second approach of section 2.5. The input arguments of tri(t,to) are t and t0 which are the t vector as a row matrix and scalar t_0 value respectively. Type the codes of the figure 10.1(d) in a new M-file and save the file by the name tri in your working path.

Example:

Let us generate the samples of the signal $e^{-2t}\ tri(1.2t-0.7)$ over $0 \le t \le 3$ sec with a step 0.1sec as follows:

```
>>t=0:0.1:3; ↵        ← t holds the t vector as a row matrix
>>f=exp(-2*t).*tri(1.2*t-0.7,0); ↵ ← Functional values of
```

$e^{-2t}\ tri(1.2t-0.7)$ are assigned to the workspace f as a row matrix where f is user-chosen variable

Whether $rect(t)$ or $tri(t)$, we can verify the signal sample generation by looking into the graph which is conducted by the command plot(t,f) (section 2.9 – smaller step size is needed if the reader is not satisfied with the shape).

10.2 Binary signals

In digital signal study we need to generate binary signals. There are two types of binary signals seen in the elementary level, the first of which is 0 and 1 and the second of which is –1 and 1. Mostly the generation needed is in random form and our discussion follows so.

The built-in command randint (abbreviation for random integers) generates random integers based on user-requirement. The function has the syntax randint(user -required row number, user-required column number, integer range as a two element row matrix but first the lower bound and then the upper bound). For a 0-1 binary we write [0 1] as the third input argument of randint and a single scalar has matrix order 1×1 so a single binary random amplitude is generated as follows:

```
>>X=randint(1,1,[0 1]) ↵              ← X is any user-supplied variable

X =

        0
```

We get the return 0 but you may find it as 1. If you say I need 10 random binary numbers as a row matrix, execute the following:

```
>>X=randint(1,10,[0 1]) ↵             ← X is any user-supplied variable

X =

        0    1    0    1    1    1    1    0    0    1
```

Similarly 10 binary random numbers as a column matrix are obtained by the command randint(10,1,[0 1]).

If bipolar binary random amplitudes are needed i.e. –1 and 1, we can not apply just discussed randint for the generation. The reason is usage of randint returns 0 which we do not need. Under this circumstance we apply the built-in function randsrc which generates random elements from a given set with the syntax randsrc(user-required row number, user-required column number, given set as a row matrix).

Let us say we intend to generate a single random amplitude (which has matrix dimension 1×1) from the set of −1 and 1 whose execution is the following:

```
>>X=randsrc(1,1,[-1 1]) ↵              ← X is any user-supplied variable

X =

     -1
```

Again if we wish to form a row matrix of 10 random bipolar amplitudes, we execute the following:

```
>>X=randsrc(1,10,[-1 1]) ↵             ← X is any user-supplied variable

X =

     -1   1  -1   1   1  -1  -1   1  -1   1
```

If column matrix is needed for the above, we execute the command X= randsrc(10,1,[-1 1]).

10.3 Complex signal samples

A complex signal sample can also be generated in MATLAB. The reader is referred to appendix C.12 for complex number basics. The code writing of a complex signal is similar to that of the real one (section 2.3). We present four examples on complex signal sample generation in the following.

Example 1

Let us generate the complex signal samples of $f(t)=3e^{j2t}$ over the time interval $0 \le t \le 2$ sec with a step size 0.1sec.

Even though the signal $f(t)$ is complex, the t variation is real. With the given step size and interval, we generate the t vector as a row matrix as follows:

```
>>t=0:0.1:2; ↵              ← t is any user-supplied variable
```

The scalar code of $3e^{j2t}$ is 3*exp(j*2*t) on having that we perform the following:

```
>>f=3*exp(j*2*t); ↵         ← f is any user-supplied variable
```

The workspace variable f is having the complex signal samples as a row matrix in which every element is a complex number indicating the complex signal value at the indicated t points stored in t respectively. If we wish to extract the real, imaginary, magnitude, and phase angle components of the complex signal samples, following commands can be executed:

```
>>R=real(f); I=imag(f); A=abs(f); P=angle(f); ↵
```

The workspace variables R, I, A, and P (all are user-supplied variables) are now having the four components of the signal samples respectively, each of which is a real identical size row matrix. Note that the domain of the phase angle covers $-\pi \le t \le \pi$.

Example 2

A chirp signal is one whose frequency changes with time. The signal has different forms of representation for example sine, cosine, or exponential. The variation of the frequency may take place linearly, quadratically, or other with respect to time.

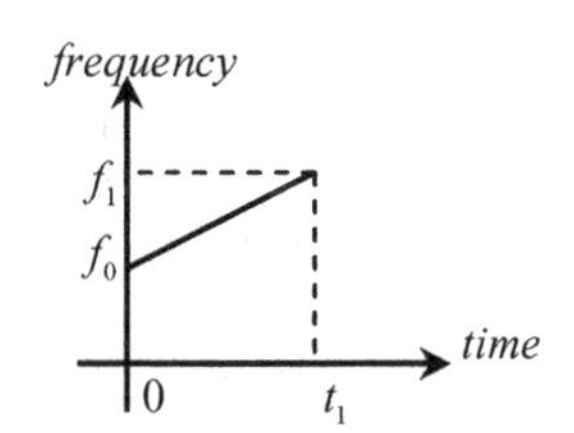

Figure 10.2(a) Frequency versus time variation of a linear chirp signal

Let us consider the chirp signal to be a complex exponential with linear frequency variation and the signal is given by $v(t)=Ae^{-j2\pi(f_0+\beta t)t}$ where f_0 and f_1 are the frequencies (both in Hz) at t=0 and $t=t_1$sec respectively and A is the signal magnitude. The linear variation of the frequency is illustrated in the figure 10.2(a) in which the variation has a line slope $\beta=\frac{f_1-f_0}{t_1}$ and the parameters f_0, f_1, and t_1 are user defined. The instantaneous frequency has the expression $f=f_0+\beta\ t$. Also note that the t_1 is the interval during which the signal should exist.

As an example let us generate a linear exponential chirp signal whose frequency changes from $1\ KHz$ to $3\ KHz$ within $4\ m\sec$ and which has magnitude 3. On given specification, we have A=3, f_0=1 KHz, f_1=3 KHz, and t_1=4 $m\sec$ and let us choose the step size as 0.01 $m\sec$. Procedurally we conduct the following for the generation:

>>t=0:0.01e-3:4e-3; ↵ ← t holds the t variation $0\le t\le 4m\sec$ as a row matrix

>>b=(3e3-1e3)/4e-3; ↵ ← b holds the calculated β value using $\frac{f_1-f_0}{t_1}$, b⇔ β

>>v=3*exp(-j*2*pi*(1e3+b*t).*t); ↵ ← v holds the linear chirp signal samples of $v(t)$ as a row matrix

In above implementation the codes for the $1\ KHz$, $3\ KHz$, $4\ m\sec$, $0.01\ m\sec$, and $3e^{-j2\pi(f_0+\beta t)t}$ are 1e3, 3e3, 4e-3, 0.01e-3, and 3*exp(-j*2*pi*(1e3+b*t).*t) respectively.

Example 3

Referring to the example 2 when an exponential chirp signal frequency variation is quadratic, its instantaneous frequency dependence on time is given by $f=f_0+\beta\ t^2$ where $\beta=\frac{f_1-f_0}{t_1^{\ 2}}$ assuming that the frequency f_0 turns to frequency f_1 within the time t_1. Now the signal expression becomes $v(t)=Ae^{-j2\pi(f_0+\beta t^2)t}$. Considering the same numeric values as those of the example 2, the quadratic chirp signal is generated as follows:

>>t=0:0.01e-3:4e-3; ↵ ← t holds the t variation $0\le t\le 4m\sec$ as a row matrix

>>b=(3e3-1e3)/4e-3^2; ↵ ← b holds calculated β value using $\frac{f_1-f_0}{t_1^{\ 2}}$, b⇔ β

>>v=3*exp(-j*2*pi*(1e3+b*t.^2).*t); ↵ ← v holds the quadratic chirp signal samples of $v(t)$ as a row matrix

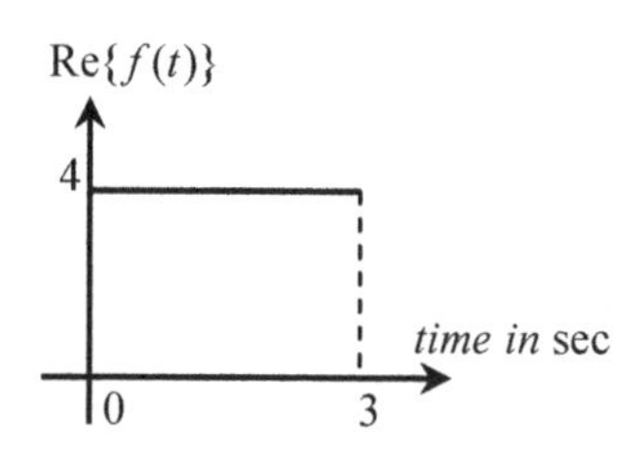

Figure 10.2(b) Real part of a complex signal $f(t)$

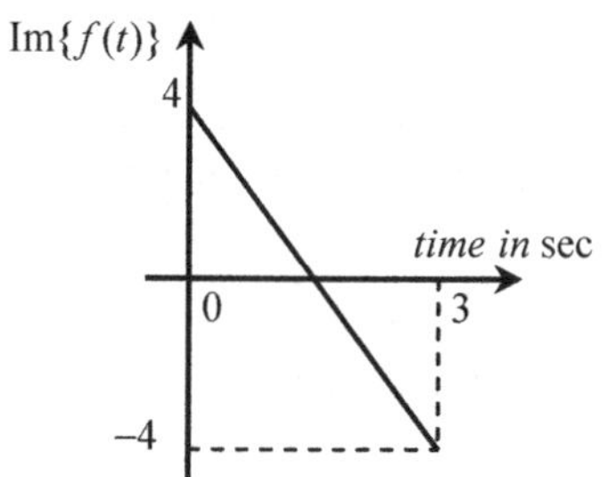

Figure 10.2(c) Imaginary part of a complex signal $f(t)$

Example 4

Complex signal description in terms of graphical figure is no exception. Let us say the real and imaginary parts of a complex signal $f(t)$ are having the characteristics of the figures 10.2(b) and 10.2(c) respectively. Essentially both parts of the signal share common t variation. For the constant value generation we apply the programming technique of section 2.3 (example 2) and generate the complex signal samples by choosing a step size 0.1sec as follows:

```
>>t=0:0.1:3; ↵                ← t holds the t variation 0≤t≤3sec as a row matrix
>>R=4*ones(1,length(t)); ↵ ← R holds the signal Re{f(t)} samples as a row
                                  matrix over 0≤t≤3sec
>>I=-8/3*t+4; ↵    ← I holds the signal Im{f(t)} samples as a row matrix over
                                  0≤t≤3sec
>>f=complex(R,I); ↵ ← f holds the complex signal f(t) samples as a row matrix
                                  over 0≤t≤3sec
```

The imaginary part $\text{Im}\{f(t)\}$ follows a linear characteristic and passes though the points (0,4) and (3,–4) thereby possessing the equation $\text{Im}\{f(t)\} = -\frac{8}{3}t + 4$ (appendix C.6) whose code is used in the computation of I. The command complex has two input arguments (appendix C.12), the first and second of which are the real and imaginary parts of the complex signal respectively, both of which must be identical size matrix. One could have used the command f=R+i*I instead of f=complex(R,I) (syntactically same).

10.4 Statistical signals

Random variable generation comes primarily while discussing about the statistical signal. There are many random variables (at least two dozens) in statistics. Of them, uniform and Gaussian random variables are frequently addressed in preliminary signal and system texts.

Random variable generation might take place in continuous as well as in discrete sense. As the focus has been, we are after either $f(t)$ versus t or $f[n]$ versus n data (section 4.1) generation. Random variable generation means the generation of only $f(t)$ or $f[n]$ at a particular t or n respectively. The type of the variable, uniform or Gaussian, defines the range of the randomness. Or in other words the independent variable is frozen for the random variable generation. Such generation is termed as i.i.d or identically independent distribution. In electrical engineering term one can assume that the random value corresponds to a single voltage or current amplitude not the whole voltage wave.

Uniform random variable generation in continuous sense

MATLAB function unifrnd simulates the continuous uniform variable in continuous sense (truly speaking a perfect continuous can never be simulated by finite digit). A uniform random variable has some range for example the range of $f(t)$ is $-5 \le f(t) \le 3$ indicating any value from –5 to 3 equally likely including fractional. The syntax we apply is unifrnd(lower bound, upper bound). Let us implement the following:

```
>>f=unifrnd(-5,3) ↵

f =                              ← f holds a single random value within –5 to 3, f can be
                                    any variable of your choice
           -0.4574
```

Gaussian random variable generation in continuous sense

MATLAB function normrnd simulates the continuous Gaussian random variable. A Gaussian random variable has two parameters – mean (μ) and standard deviation (σ) but the range of the values is from minus infinity to plus infinity that is $-\infty < f(t) < \infty$. The syntax we apply is normrnd(μ, σ). Let us say we intend to implement a Gaussian random variable with $\mu=-2$ and $\sigma=4$ which we carry out as follows:

```
>>f=normrnd(-2,4) ↵

f =                         ← f holds a single random value within −∞ to ∞, f
                              can be any variable of your choice
         -3.7303
```

A continuous normal random variable is a variant of Gaussian with 0 mean and unity standard deviation that is the command we need is normrnd(0,1).

Uniform random variable generation in discrete sense

In this sort of generation mainly we need the discrete integers and the command we apply is randint with the syntax randint(user-required row number, user-required column number, integer range as a two element row matrix but first the lower bound and then the upper bound).

For instance a uniformly distributed single random integer n within the range $-4 \le n \le 3$ is generated by using the command n=randint(1,1,[-4 3]) where n is user-chosen name which keeps the generated integer. The first two input arguments of randint must be 1 and 1 because a single element has matrix dimension 1×1.

From random variable to statistical signal

If you gather many random variable observation data side by side, collectively these random variables form a discrete statistical signal.

Let us say we intend to generate the continuous statistical signal $f(t)$ samples over the interval $0 \le t \le 3$ sec with a step size 0.5sec in which the $f(t)$ samples are identically and independently distributed (called i.i.d in the literature) uniform random number in the range $-5 \le f(t) \le 3$ at every instant of t. First of all we need to find how many samples there are in the interval which happens by the following:

```
>>t=0:0.5:3; ↵       ← t holds the required time points on the t axis as a row matrix
>>length(t) ↵        ← Finding the total number of time points on t axis by length

ans =

         7
```

From above execution we need to generate 7 samples of $f(t)$ where each one is in the range $-5 \le f(t) \le 3$. Earlier mentioned unifrnd(-5,3) generates just one sample.

For multiple samples (in general as a matrix), we add two more input arguments at the end in the unifrnd. For example the unifrnd(-5,3,3,5) generates a matrix of order 3×5 (last two input arguments) in which every element is in the $-5 \le f(t) \le 3$ (bounds indicated by the first two input arguments). As far as the signal generation is concerned, we have to have a row (means the third input argument is 1) or a column (means the fourth input argument is 1) matrix. Considering a row matrix, let us execute the following:

```
>>f=unifrnd(-5,3,1,7) ↵ ← f holds the random signal samples as a row matrix at
                          the mentioned t points where f is user-chosen
                          variable
f =

         2.6010  -3.1509  -0.1453  -1.1121   2.1304   1.0968  -1.3483
```

This sort of generation is possible for the other types of random variables as well.

Let us say we intend to generate the statistical signal $f(t)$ samples over $0 \le t \le 3$ sec with a step size 0.5sec in which the $f(t)$ samples are i.i.d Gaussian with

μ=–2 and σ=4. The workspace t is also applicable here. Earlier discussed normrnd is applied with added input arguments as follows:

```
>>f=normrnd(-2,4,1,7) ↵
```
← f holds the random signal samples as a row matrix at the mentioned t points where f is user-chosen variable

```
f =
        3.1610   0.6744   2.7634  -6.8098  -2.0792  -2.6269  -8.4163
```

The normrnd has four input arguments, the first two of which are the mean and standard deviation respectively. The third and fourth input arguments indicate the row and column numbers in the required matrix respectively (for the row matrix here the matrix dimension is 1×7).

Random signals with definite power specifications

In communication problems random signals are generated from power specifications. By power we mean the electrical power which is essentially the product of voltage and current waves. In such cases the generation is mostly for earlier mentioned Gaussian but can be for the other as well.

Let us review the definition of average power in a continuous random signal $f(t)$ which exists for T seconds (meaning $0 \le t \le T$ sec). The average power is given by $\frac{1}{T}\int_0^T f^2(t)dt$. In the calculation of average power in a random signal it is assumed that the power is taken through a one ohm resistor hence it does not matter that $f(t)$ represents a voltage or current wave. The instantaneous continuous power is always given by $f^2(t)$. But we take samples of such signal from the continuous one.

Let us say $f[n]$ represents the discrete signal in which the samples are taken from the $f(t)$ with a sampling frequency f_s (hence period $T_s = \frac{1}{f_s}$ sec, section 4.1) on the relationship $f[n] = f(nT_s)$ where n is integer or sample number. Assuming constant value between the consecutive samples (also called the zero order hold), the average power in a time interval T seconds from the discrete counterpart is given by $P_{avg} = \frac{\sum_{n=0}^{N-1} f^2[n]T_s^2}{T}$ where N is the number of samples in the discrete signal.

Often we generate a random signal from a given P_{avg}, T_s, and T.

Example:

Let us generate a random Gaussian signal whose average power is P_{avg}=0.3Watts with a sampling period T_s=0.5secs. The signal existence should be for T=15sec starting from $t=0$ sec.

As the specification requires, our time vector t as a row matrix must be formed from 0 to 15sec with a step 0.5sec that is:

```
>>t=0:0.5:15; ↵
```

The total sample number N in the discrete signal evolves from the number of elements in the time vector t which we find by length(t) as follows:

```
>>N=length(t); ↵
```
← Workspace N holds the N

Previously mentioned normrnd simulates the Gaussian but the user has to decide the noise voltage or current mean and standard deviation. Let us say $\mu=0$ and $\sigma=1$ on that account the signal is generated as follows:

>>f=normrnd(0,1,1,N); ↵ ← Workspace f holds the signal $f(t)$ at the mentioned t points as a row matrix or $f[n]$ at different n where n changes integerwise from 0 to $N-1$

We have no idea about the average power of the signal stored in f. One calculates the P_{avg} for this f as follows:

>>P=0.5^2/15*sum(f.^2); ↵ ← Workspace P (any user-hosen variable) holds the P_{avg} for the last f, not the one we need, appendix B and C.11

In above command the f.^2 squares every sample in f and the sum adds all samples. Whatever be the power P for sure it is a scalar or single value. If we divide P by P, we get 1. If we multiply the 1 by specified power 0.3Watts, we get the required signal power. We are not after power but after the signal $f[n]$ whence multiplying every sample in the last f by $\sqrt{\frac{0.3}{P}}$ (where $P \Leftrightarrow$P) provides the signal with specified power as follows:

>>f=sqrt(0.3/P)*f; ↵ ← We assigned the signal f again to f (can be any variable of your choice)

Since power across a one ohm resistor is square of the voltage (assuming the signal is a voltage wave), we applied the square root while multiplying the signal. With all these, the last assignee f is holding the signal with specified power, sampling period, and interval. You can even recalculate to check the power by using 0.5^2/15*sum(f.^2) whether it is 0.3Watts.

10.5 Input-output on Fourier system

Sometimes for a continuous system we apply the signal in time domain and wish to see the output in time or in frequency domain. All relevant functions and implementations are addressed before. Just to be specific with some example, let us go through the following.

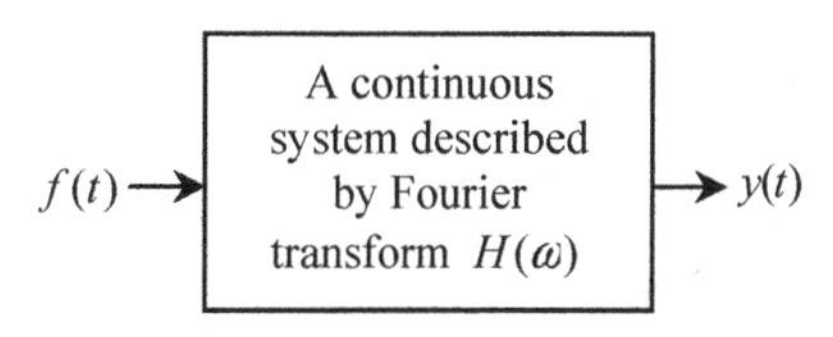

Figure 10.3(a) A single input - single output Fourier system

Example 1

Suppose a continuous system is characterized by the Fourier transform $H(\omega)=\frac{3}{3+j4\omega}$. When an input $f(t)=2\,rect(t)$ is applied to the system in time domain, what should be the output $y(t)$ in time domain as well as in frequency domain?

The $f(t)$ can be written (section 10.1) as $f(t)=2[\,u(t+0.5)-u(t-0.5)\,]$ and its Fourier transform (section 5.3) we obtain as follows:

>>syms t w ↵ ← Declaring related variable by the syms

>>f=2*(heaviside(t+0.5)-heaviside(t-0.5)); ↵ ← f holds the code of $f(t)$

>>F=fourier(f); ↵ ← F holds the Fourier transform of $f(t)$ or $F(\omega)$

>>H=3/(3+j*4*w); ↵ ← H holds the code of $H(\omega)$, appendix B

From the Fourier theory we know that the output in frequency domain $Y(\omega)$ is the product of $F(\omega)$ and $H(\omega)$ which we obtain as follows:

```
>>Y=F*H; ↵                    ← Y holds the code of Y(ω)
>>pretty(Y) ↵                 ← Appendix C.10 for the pretty
          sin(1/2 w)
     12 ----------------
        w (3 + 4 I w)
```

From the above return we can say that the $Y(\omega)$ is $\dfrac{12\sin\frac{\omega}{2}}{\omega(3+j4\omega)}$ and its inverse Fourier transform we obtain as follows:

```
>>y=ifourier(Y,w,t); ↵        ← Subsection 5.3.2 for the inverse transform
>>pretty(y) ↵
    2 (exp(- 3/4 t + 3/8) - 1) heaviside(t - 1/2)

       + 2 (-exp(- 3/4 t - 3/8) + 1) heaviside(t + 1/2)
```

From above execution we have the time domain output as $y(t) = 2(e^{-\frac{3t}{4}+\frac{3}{8}} - 1)u(t-0.5)$ $+ 2(-e^{-\frac{3t}{4}-\frac{3}{8}} + 1)u(t+0.5)$.

If you wish to plot the $y(t)$ or $Y(\omega)$, apply the appendix F cited ezplot on some interval to see the graphical output for example $|Y(\omega)|$ over $-20 \le \omega \le 20 rad/\sec$ by ezplot(abs(Y),[-20 20]) where Y is holding the code of $Y(\omega)$ from earlier executions.

Example 2

The series R - L circuit of the figure 10.3(b) has R=3Ω and L=4 H. When v_i = $2\,rect(t)$ is applied to the circuit, what should be the output v_0?

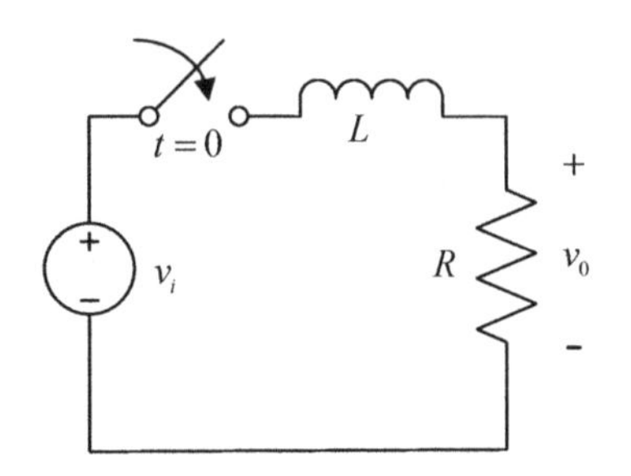

Figure 10.3(b) A series R - L circuit

For some angular frequency ω, the input to output $H(\omega)$ is $\dfrac{V_0(\omega)}{V_i(\omega)} = \dfrac{3}{3+j4\omega}$ from voltage division. Then it is simply the problem of the example 1.

Example 3

Now the input to the circuit of figure 10.3(b) is the half cycle voltage sine wave of the figure 10.3(c). We wish to see the v_0 versus t over $0 \le t \le 20\sec$.

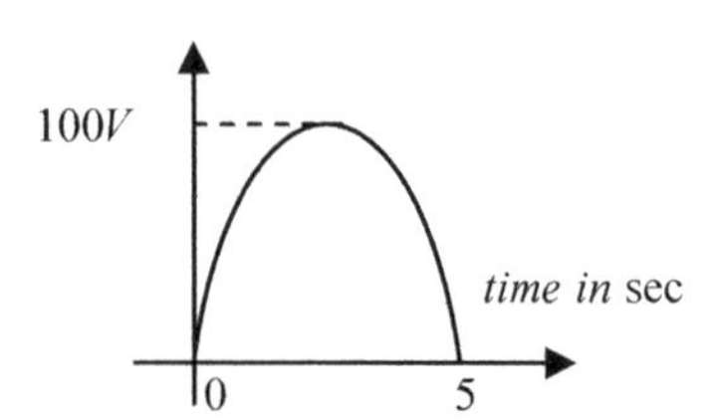

Figure 10.3(c) A half cycle sine wave

Even though the information is in half cycle, we need the full cycle duration or time period T to write the expression for the input v_i which is 10sec so the frequency of the input sine wave is $f = \dfrac{1}{T}$ =0.1sec therefore the v_i can be written in functional form as v_i=100[$u(t) - u(t-5)$] $\sin 2\pi 0.1t$ whose Fourier transform we find as follows:

```
>>syms t w ↵                  ← Declaring related variable by the syms
>>f=100*(heaviside(t)-heaviside(t-5))*sin(2*pi*0.1*t); ↵ ← f holds the code of
                                                            f(t) or v_i
>>F=fourier(f); ↵             ← F holds the Fourier transform of v_i or F(ω)
```

```
>>H=3/(3+j*4*w); ↵          ← H holds the code of H(ω) with example 2 values
>>Y=F*H; ↵                  ← Y holds the code of Y(ω)
>>y=ifourier(Y,w,t); ↵      ← Inverse transform provides y or v0
```

Should the reader need the expression for the output v_0, the command pretty(y) can be executed (needless to mention which is a lengthy one). However the last y is holding the time domain code of the v_0 and we see the v_0 versus t by dint of the appendix F cited command ezplot as follows:

```
>>ezplot(y,[0 20]) ↵
```

Figure 10.3(d) shows the output in which the horizontal and vertical axes correspond to t and v_0 respectively. Note that although the input is a half cycle sine wave, the output is not so. On top of the figure 10.3(d) you see some texts that is basically some part of the code shown automatically by the plotter ezplot. If you do not wish to see that, click the Edit plot icon (figure 1.2(c)), select the title text of figure 10.3(d), click the Cut icon, and click again the Edit plot icon to deactivate the selection.

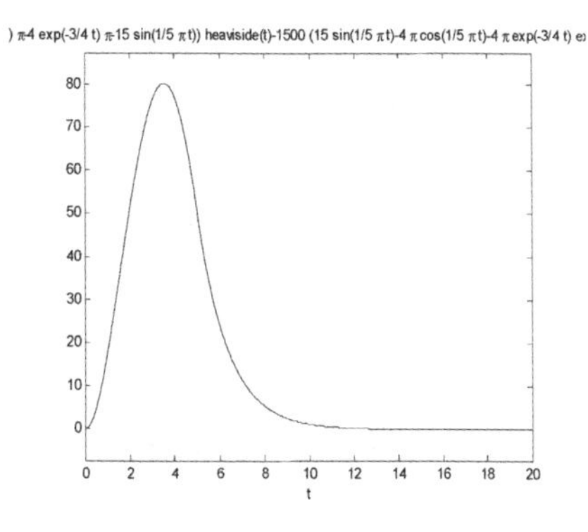

Figure 10.3(d) The response from the circuit in figure 10.3(b) when a half cycle sine wave is applied

10.6 Input-output on Laplace system

As a matter of fact greater part discussion on the signal and system text is based on transform domain or Laplace system which we see in chapters 6, 8, and 9. Just to maintain parallel treatment on the subject we quote some examples. In examples 1, 2, and 3 of the last section we utilized fourier and ifourier now that will be laplace and ilaplace for the forward and inverse counterparts respectively. In the system function now we will have s related expression instead of $j\omega$. We present some example in the following.

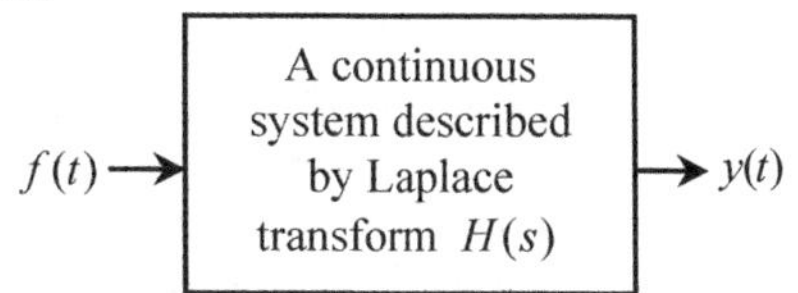

Figure 10.3(e) A single input - single output Laplace system

Example

Suppose a continuous system is characterized by the Laplace transform $H(s)=\dfrac{3}{3+4s}$ as seen in figure 10.3(e). When an input $f(t)=2\,rect(t)$ is applied to the system in time domain, what should be the output $y(t)$ in time domain as well as in s domain?

The $f(t)$ is written as $f(t)=2[\,u(t+0.5)-u(t-0.5)\,]$ and its Laplace transform (section 6.3) we obtain as follows:

```
>>syms t s ↵                 ← Declaring related variable by the syms
>>f=2*(heaviside(t+0.5)-heaviside(t-0.5)); ↵  ← f holds the code of f(t)
>>F=laplace(f); ↵            ← F holds the Laplace transform of f(t) or F(s)
>>H=3/(3+4*s); ↵             ← H holds the code of H(s), appendix B
```

From the Laplace theory we know that the output in s domain $Y(s)$ is the product of $F(s)$ and $H(s)$ which we obtain as follows:

```
>>Y=F*H; ↵                  ← Y holds the code of Y(s)
>>pretty(simple(Y)) ↵       ← Applying appendix C.10 cited pretty after
                              simplification by the command simple

       -1 + exp(- 1/2 s)
    -6 -----------------
          s (3 + 4 s)
```

We read off the expression for the output as $Y(s)=-6\frac{-1+e^{-\frac{s}{2}}}{s(3+4s)}$ and its inverse Laplace transform we obtain as follows:

```
>>y=ilaplace(Y,s,t); ↵      ← Section 6.8 for the inverse Laplace transform
>>pretty(y) ↵

      4 exp(- 3/8 t) sinh(3/8 t)

         - 4 heaviside(t - 1/2) exp(- 3/8 t + 3/16) sinh(3/8 t - 3/16)
```

On above return, we have the time domain expression for the output as $y(t)=4e^{-\frac{3t}{8}}\sinh\frac{3t}{8}-4u(t-0.5)e^{-\frac{3t}{8}+\frac{3}{16}}\sinh\left(\frac{3t}{8}-\frac{3}{16}\right)$. If you wish to plot the $y(t)$, apply the appendix F cited ezplot on some interval for example $y(t)$ over $0\le t\le 20\sec$ by ezplot(y,[0 20]) where y is holding the code of $y(t)$ from earlier executions.

Example 2 of last section which presents a circuit is also handled in a similar way. This time the circuit has $H(s)=\frac{3}{3+4s}$.

For more circuit based examples see section 8.7. For a long or complicated circuit you can obtain the expression for the $H(s)$ system first and apply input afterwards as conducted in this example.

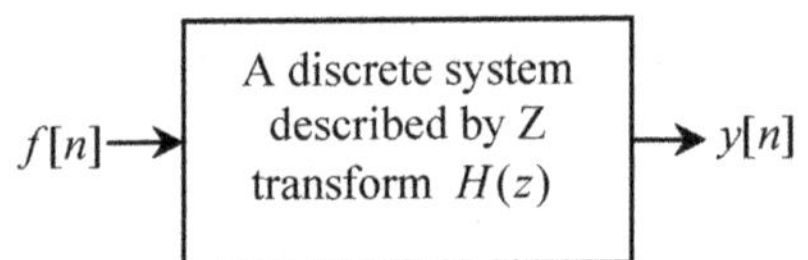

Figure 10.4(a) A single input - single output Z transform system

10.7 Input-output on Z transform system

Parallel strategy like the Fourier and Laplace transform systems (last two sections) certainly exists for the discrete system as shown in figure 10.4(a). Now the Z transform forward and inverse counterparts are ztrans and iztrans respectively (chapter 7). Both the input $f[n]$ and output $y[n]$ are purely function of integer index n. The output finding is very similar to that of the last two sections therefore we go through the following example.

Suppose we apply a discrete signal $f[n]=2^{-n}$ to the Z transform system of the figure 10.4(a) with $H(z)=\frac{2z}{z-2}$. We wish to see the output in transform z as well as in index n domain. Maintaining ongoing symbology and function we execute the following at the command prompt:

```
>>syms n z ↵                ← Declaring related variable by the syms
>>f=2^(-n); ↵               ← f holds the code of f[n]
>>F=ztrans(f); ↵            ← F holds the Z transform of f[n] or F(z)
>>H=2*z/(z-2); ↵            ← H holds the code of H(z)
>>Y=F*H; ↵                  ← Y holds the code of Y(z) from F(z)×H(z)
```

```
>>pretty(simple(Y)) ↵
```
← Appendix C.10 said pretty after simplification by the command simple

```
            2
           z
    4 -----------------
      (2 z - 1) (z - 2)
```

As the return says we have the $Y(z)$ as $\frac{4z^2}{(2z-1)(z-2)}$ from which the $y[n]$ is obtained by the following:

```
>>y=iztrans(Y,z,n); ↵
```
← Section 7.4 for the inverse Z transform where y holds the code of $y[n]$

```
>>pretty(y) ↵
            n          n
     8/3 2   -  2/3 (1/2)
```

By expression we write it as $y[n] = \frac{8\times 2^n}{3} - \frac{2}{3}\left(\frac{1}{2}\right)^n$. If you say I need to see the output graph of the $y[n]$ versus n, employ the function stem or scatter of the appendix F.

Sample or data based input-output

Just cited example presents expression based input-output on the Z transform system. When the input is sample or data based, the Z transform discrete system needs convolution considering that the transform system is in index domain and which we addressed in section 4.5.

This time we have the system given in transform domain as in figure 10.4(a). The input and output of the system in figure 10.4(a) are simply samples or data. For this type of computation we apply the built-in function filter with the syntax filter(B,A,f) where the input arguments of the filter implement any Z transform rational system function which has the form $H(z) = \frac{\sum_{k=0}^{M} b_k z^{-k}}{1 - \sum_{k=1}^{N} a_k z^{-k}}$ and that evolves from the general difference equation $y[n] - \sum_{k=1}^{N} a_k y[n-k] = \sum_{k=0}^{M} b_k x[n-k]$.

In the filter input argument, we feed the coefficients of the numerator and denominator of $H(z)$ as a row matrix but in ascending power of z^{-1} to the B and A respectively. Any missing coefficient is set to 0 and normalization is necessary if $H(z)$ is not given in proper form. The f is the input discrete signal $f[n]$ of the figure 10.4(a) as a row matrix. The return from the filter is the discrete signal $y[n]$ as a row matrix.

For example $H(z) = \frac{3+2z^{-1}}{1-2z^{-1}+z^{-2}}$ has B=[3 2] and A=[1 –2 1].

For example $H(z) = \frac{3+2z^{-1}}{3-2z^{-1}+z^{-2}}$ has B=[3 2]/3 and A=[3 –2 1]/3 after normalization by 3 both the denominator and numerator.

For example $H(z) = \frac{3z+2z^2}{2+3z^4}$ needs normalization by z^4 (highest power in numerator and denominator) first to have $\frac{3z^{-3}+2z^{-2}}{2z^{-4}+3}$ or

$\dfrac{2z^{-2}+3z^{-3}}{3+2z^{-4}}$ and by 3 afterwards to have B=[0 0 2 3]/3 and A=[3 0 0 0 2]/3.

With all these definitions of the discrete system now we wish to present one example.

Example:

In the figure 10.4(a) suppose the discrete signal $f[n]$ is [1 1 1 1 1] and the Z transform system is $H(z)=\dfrac{3+2z^{-1}}{1-2z^{-1}+z^{-2}}$. It is given that the output $y[n]$ is [3 11 24 42 65] – which we wish to obtain.

Just mentioned explanation prompts us to proceed with the following:

```
>>f=[1 1 1 1 1]; ↵            ← Entering the f[n] to f as a row matrix
>>B=[3 2]; ↵     ← Entering the numerator coefficients of H(z) to B
>>A=[1 -2 1]; ↵ ← Entering the denominator coefficients of H(z) to A
>>y=filter(B,A,f) ↵            ← Calling the filter for solution

y =
         3   11   24   42   65
```

10.8 Residue and steady state value of a system

The concept of a pole comes first (section 9.5) while discussing about the residue because residue computation happens for a particular pole. Again the concept of a pole is applicable only for the rational system function (means numerator and denominator polynomial expressible form). Given the rational system function $Y(s)$ with m poles at $s=p$, the residuals are expressed by $R_n = \dfrac{1}{(n-1)!}\dfrac{d^{n-1}(s-p)^m Y(s)}{ds^{n-1}}\Bigg|_{s=p}$ where n can vary integerwise from 1 to m.

Symbolic way of finding residue

For repetitive poles:

Let us consider the system function $Y(s)=\dfrac{s^2+4}{(s-1)^3}$ which has the pole at $s=1$ and of multiplicity 3 indicating $p=1$ and $m=3$. There must be three residuals because $m=3$ (n changes from 1 to 3) and we calculate them as follows: $R_1=(s-1)^3Y(s)\Big|_{s=1}$ =5, $R_2=\dfrac{1}{1!}\dfrac{d(s-1)^3Y(s)}{ds}\Bigg|_{s=1}$ =2, and $R_3=\dfrac{1}{2!}\dfrac{d^2(s-1)^3Y(s)}{ds^2}\Bigg|_{s=1}$ =1. Undoubtedly the computation is symbolic and we carry out the symbolic computation by first declaring the related variable by syms as follows:

```
>>syms s ↵                    ← s⇔ s
```

Then we assign the code (vector code, appendix B) of $Y(s)$ to Y (where Y is user-chosen name) as follows:

```
>>Y=(s^2+4)/(s-1)^3; ↵
```

Next we form $(s-1)^3Y(s)$ and assign that to E (user-chosen name):

```
>>E=(s-1)^3*Y; ↵
```

For example the R_3 (say as variable R3) we execute as follows:

```
>>R3=diff(E,2)/factorial(2) ↵

R3 =

1
```

The code factorial(n) is equivalent to $n!$. The command diff(E,n) performs n times differentiation of the string stored in E with respect to the concern variable (here it is s). As a whole the command diff(E,2)/factorial(2) implements $\frac{1}{2!}\frac{d^2(s-1)^3Y(s)}{ds^2}$. Similarly E/factorial(0) and diff(E,1)/ factorial(1) implement the expressions for the R_1 and R_2 respectively as follows:

```
>>R1=E/factorial(0);R2=diff(E,1)/factorial(1); ↵ ← R1⇔ R1, R2⇔ R2
```

For the R_1 and R_2 we will have the expressions in terms of s so we need substitution of the value of s and that happens by the command subs as subs(R1,1) and subs(R2,1) respectively.

For non-repetitive poles:

If $Y(s)$ does not hold repetitive poles, the derivative term in the residue expression disappears for example $Y(s)=\frac{3}{(2s+3)(4s-5)}$ which has the poles at $s=-\frac{3}{2}$ and $s=\frac{5}{4}$. The residue at the first pole is obtained as R_1 $=\left(s+\frac{3}{2}\right)Y(s)\Big|_{s=-\frac{3}{2}}=-\frac{3}{22}$ similarly second residue $R_2=\left(s-\frac{5}{4}\right)Y(s)\Big|_{s=\frac{5}{4}}=\frac{3}{22}$.

Under this type of circumstance we need to find the limiting value of the $Y(s)$ which happens through the function limit with the syntax limit(function,value). Let us implement the two residues as follows:

```
>>syms s ↵     ← Declaring the independent variable of Y(s) as symbolic
>>Y=3/(2*s+3)/(4*s-5); ↵   ← Assigning code of Y(s) to Y, Y⇔ Y(s)
>>E=(s+3/2)*Y; ↵ ← Forming (s+3/2) Y(s) and assigned to workspace
                                      E, E⇔ (s+3/2) Y(s)
>>R1=limit(E,-3/2) ↵          ←R1⇔ R1

R1 =

-3/22
```

Similarly the other residue we obtain by E=(s-5/4)*Y; R2=limit(E,5/4).

How if we deal with the complex poles for example $Y(s)=\frac{2}{(s+1)(s^2+1)}$ which has the poles at $s=-1$, $s=j$, and $s=-j$? The residue R_1 at the pole $s=-j$ is computed as follows:

```
>>syms s ↵ ← Declaring the independent variable of Y(s) as symbolic
>>Y=2/(s+1)/(s^2+1); ↵ ← Assigning the code of Y(s) to workspace Y,
                                         Y⇔ Y(s)
>>E=(s+j)*Y; ↵ ← Forming (s+j) Y(s) and assigned to workspace E,
                                         E⇔ (s+j) Y(s)
>>R1=limit(E,-j) ↵          ← Appendix C.12 for complex number

R1 =

-1/2+1/2*i                  ← It means the residue is -1/2 + j/2
```

Numeric way of finding residue

If you need decimal value of any rational expression, use the command double for example the last residue in decimal is seen by using double(R1).

MATLAB function residue finds all residuals from the system $Y(s)$ defined in terms of numerator-denominator polynomial coefficients with the syntax [user-supplied variable for residue, user-supplied variable for pole]=residue(numerator coefficient as a row matrix, denominator coefficient as a row matrix).

For example the system $Y(s)=\frac{2s+6}{(s+1)(s^2+1)}$ has the residuals 2, $-1-j2$, and $-1+j2$ at $s=-1$, j, and $-j$ respectively. We implement it as follows:

```
>>N=[2 6]; ↵
>>D=conv([1 1],[1 0 1]); ↵          ← Section 8.2 for conv
>>[R,P]=residue(N,D); ↵             ← Calling the residue
>>[R P] ↵                         ← To see the residue and pole side by side

ans =
        2.0000              -1.0000            ← Residue 2 at s =-1
       -1.0000 - 2.0000i    -0.0000 + 1.0000i
       -1.0000 + 2.0000i    -0.0000 - 1.0000i
```

The N, D, R, and P are all user-chosen names.

Steady state value

The steady state value of a system function $Y(s)$ is defined as the limiting value of $sY(s)$ at $s=0$ or symbolically $\underset{s\to 0}{Lt}\ sY(s)$. As an example the system function $Y(s)=\frac{2(2s+1)}{s^2+1}$ has the steady state value 0 and we implement it by earlier mentioned limit as follows: syms s, Y=2*(2*s+1)/(s^2+1); limit(s*Y,0).

That brings an end to our miscellaneous signal topic chapter.

Chapter 11

Mini Problems on Computation/Simulation

In this chapter basic signal and system problems from all chapter headings are randomly accumulated. Instead of organizing the signal and system problems on chapter specific term, we treat every signal and system problem as a mini project because each one needs MATLAB computation or SIMULINK modeling. Attached answer and associated clue in every mini project would provide the reader some level of confidence on signal analysis through the wing of MATLAB or SIMULINK. Although the mini projects seem simple, they pave the way for insight of much more complicated signal and system design problems for instance in signal processing chip design by using the delay elements. However the mini project collection is as follows:

◆◆ Project 1 (on signal sample generation from expression)

Write the code for each of the following signal expressions to generate its sample values in standard unit over the given interval and step size:

(a) Voltage signal $f(t) = -0.7t + 6\ V$ over $-1 \le t \le 3$ sec with step 0.1sec

(b) Voltage signal $f(t) = |-0.7t + 6|\ V$ over $-1 \le t \le 3$ sec with step 0.1sec

(c) Current signal $i(t) = 4.4\ mA$ over $0 \le t \le 2\ m$sec with step 0.1 $m$sec

(d) Parabolic current signal $i(t) = 2 - 0.2t + 0.7t^2\ A$ over $0 \le t \le 2$ sec with step 0.1sec

(e) Damped sine signal $f(t) = 3.1e^{-0.5t}\sin(1.5t + 34^0)$ over $0 \le t \le 3$ sec with step 0.1sec

(f) A sine signal with harmonics which is composed of fundamental 60 Hz along with the 3rd and 5th harmonics and their amplitudes are 100 V, 20 V, and 1 V respectively over $0 \le t \le \frac{1}{30}$ sec with step 0.0015sec

Hint: section 2.3

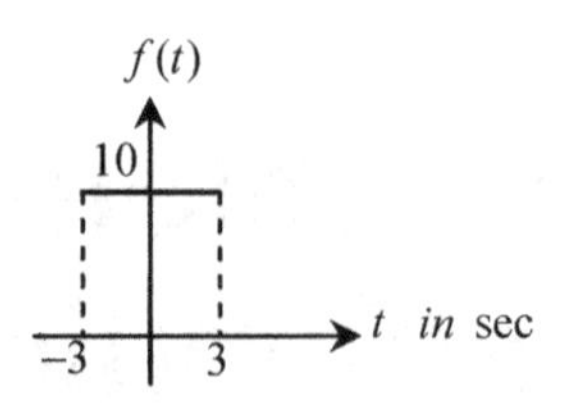

Figure 11.1(a) A finite duration symmetric rectangular voltage pulse

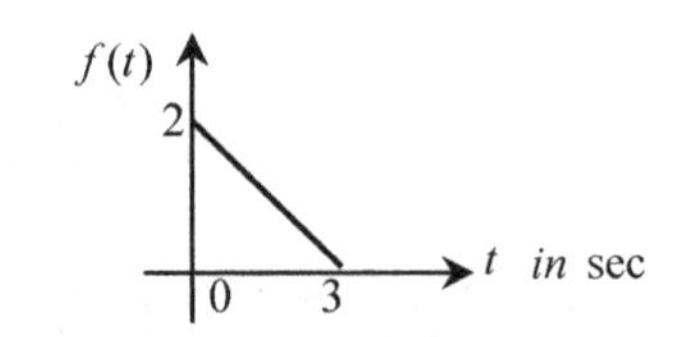

Figure 11.1(b) A finite duration ramp voltage pulse

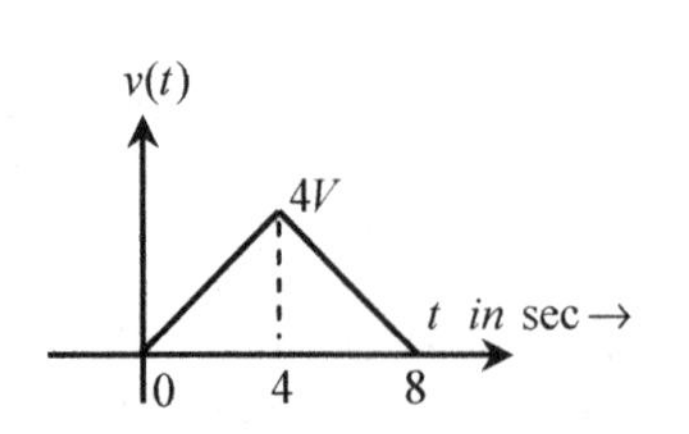

Figure 11.1(c) A symmetric triangular pulse

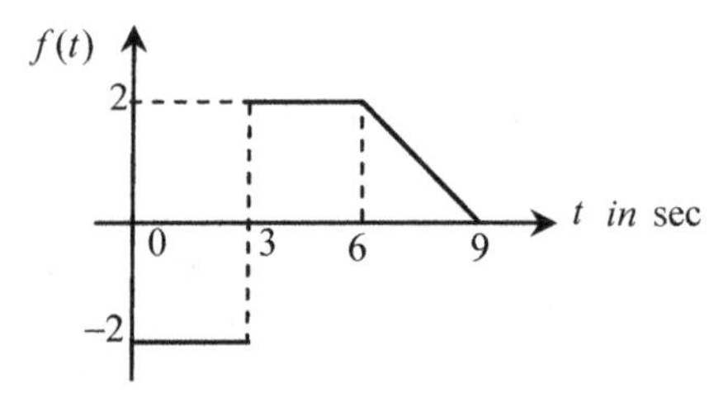

Figure 11.1(d) A piecewise continuous signal

◆◆ Project 2 (on signal sample generation from graphical representation)

Write the code for each of the following graphically represented signals to generate its sample values over the given interval and step size and after the coding verify the signal shape by section 2.9 cited plot command:

(a) Finite duration symmetric rectangular voltage pulse $f(t)$ of the figure 11.1(a) over $-5 \le t \le 5$ sec with step 0.01sec

(b) Finite duration ramp voltage pulse $f(t)$ of the figure 11.1(b) over $-1 \le t \le 4$ sec with step 0.01sec

(c) Shifted symmetric triangular voltage pulse $v(t)$ of the figure 11.1(c) over $0 \le t \le 10$ sec with step 0.01sec

(d) Piecewise continuous signal of the figure 11.1(d) over $-1 \le t \le 10$ sec with step 0.01sec

Hint: section 2.4

◆◆ Project 3 (on plotting a signal from its expression)

Apply the ezplot of section 2.9 on each of the following expressions to view the signal variation over the given interval:

(a) Parabolic current signal $i(t) = 2 - 0.2t + 0.7t^2$ A over $0 \le t \le 2$ sec

(b) Shifted symmetric voltage signal $f(t) = |2(t-1)|$ V over $-1 \le t \le 3$ sec

Answers: (a) the graph should look like the figure 11.1(e)
(b) the graph should look like the figure 11.1(f)

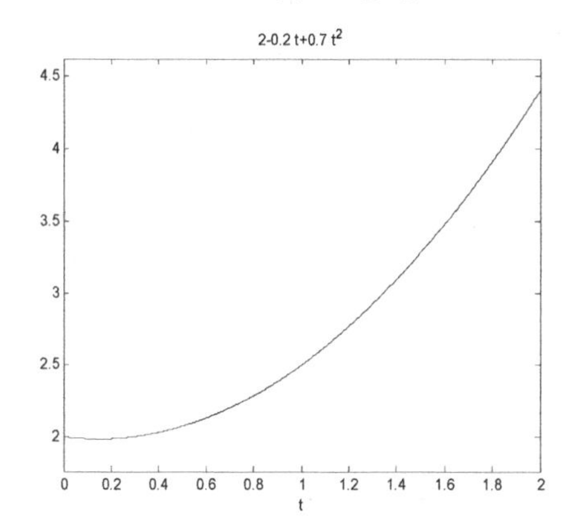

Figure 11.1(e) Parabolic current signal of project 3a

Figure 11.1(f) Shifted symmetric voltage signal of project 3b

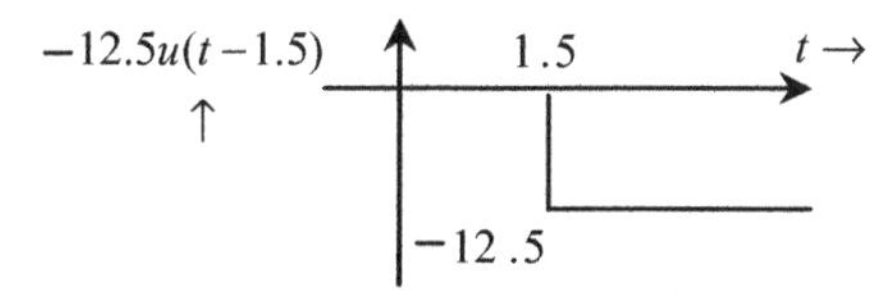

Figure 11.1(g) Shifted and scaled unit step function

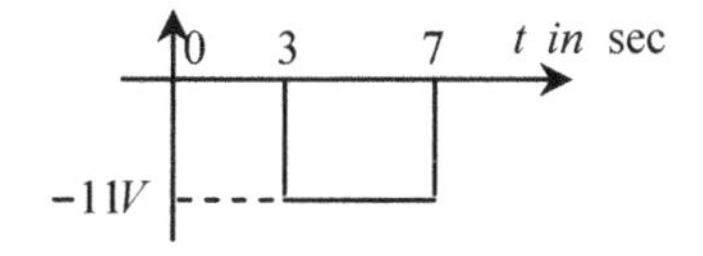

Figure 11.1(h) A finite duration voltage pulse of negative amplitude

♦♦ Project 4 (on modeling on unit step derived functions)

Apply SIMULINK approach to model each of the following signals and verify the modeled signal shape by a Scope:

(a) Shifted and scaled unit step function of the figure 11.1(g) over $0 \le t \le 2$ sec

(b) Finite duration voltage pulse of the figure 11.1(h) over $0 \le t \le 10$ sec

(c) Ideal Dirac delta function of figure 11.1(i) in numerical sense over $0 \le t \le 5$ sec

Hint: section 3.2

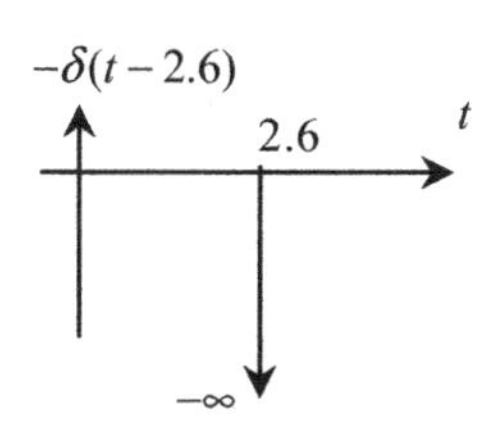

Figure 11.1(i) Ideal –ve Dirac delta function

♦♦ Project 5 (on modeling on ramp derived functions)

Apply SIMULINK approach to model each of the following signals and verify the modeled signal shape by a Scope:

(a) The ramp signal of the figure 11.2(a) which has a slope over $0 \le t \le 1.9$ sec

(b) The ramp signal of the figure 11.2(b) which has a vertical axis intercept over $0 \le t \le 2$ sec
(c) The ramp signal of the figure 11.2(c) which has both axes intercept over $0 \le t \le 10$ sec
(d) The ramp signal of the figure 11.2(d) which forms a triangle with a negative slope over $0 \le t \le 10$ sec
(e) The ramp signal of the figure 11.2(e) which has a horizontal axis intercept over $0 \le t \le 10$ sec
(f) The ramp signal of the figure 11.2(f) which forms a triangle with a positive slope over $0 \le t \le 10$ sec

Hint: section 3.3

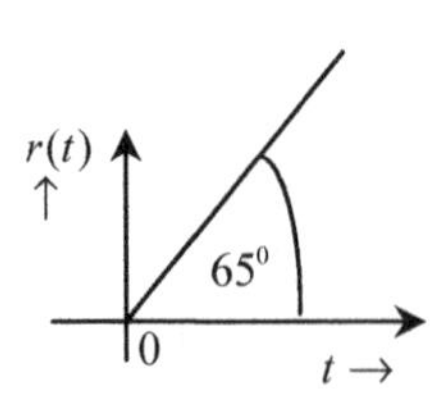

Figure 11.2(a) A ramp signal with a slope

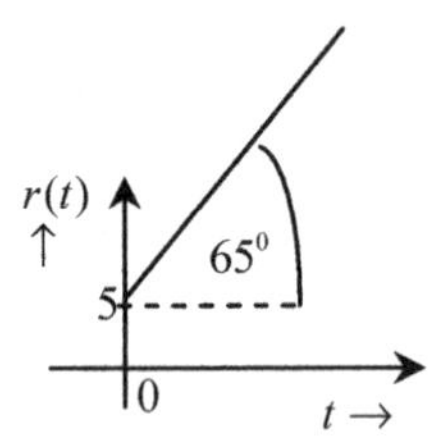

Figure 11.2(b) A ramp signal with vertical axis intercept

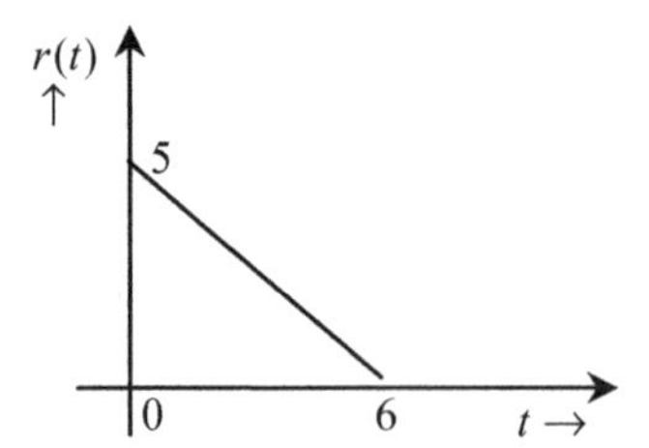

Figure 11.2(c) A ramp function with both axes intercept

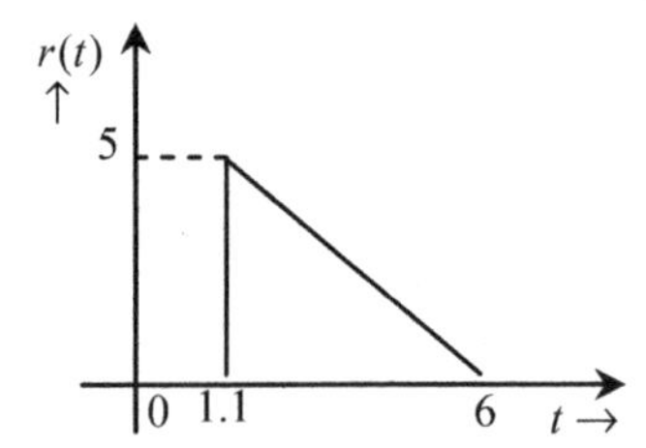

Figure 11.2(d) A ramp signal forms a single triangle with –ve slope

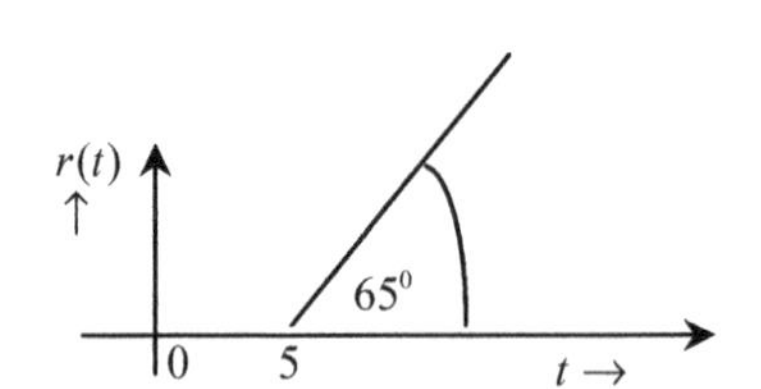

Figure 11.2(e) A ramp signal with horizontal axis intercept

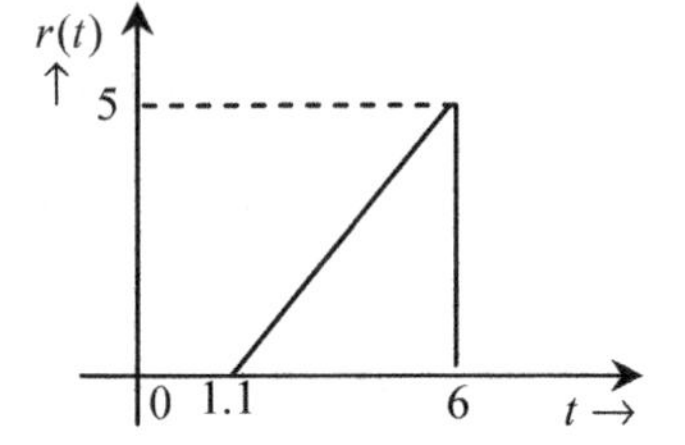

Figure 11.2(f) A ramp signal forms a single triangle with +ve slope

◆◆ Project 6 (on Fourier series coefficients of a continuous periodic signal)

(a) Determine the real form 1 Fourier series coefficients for the continuous periodic signal $f(t)$ of the figure 11.2(g) in standard time unit in MATLAB

(b) Determine the real form 2 Fourier series coefficients for the continuous periodic signal $f(t)$ of the figure 11.2(g) in standard time unit in MATLAB

(c) Determine the complex form Fourier series coefficients for the continuous periodic signal $f(t)$ of the figure 11.2(g) in standard time unit in MATLAB

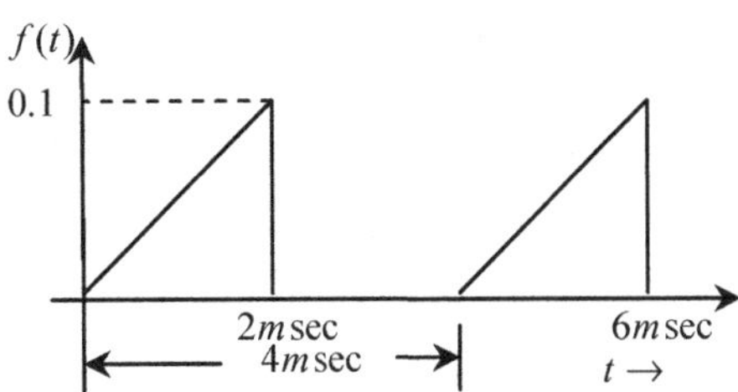

Figure 11.2(g) A triangular periodic wave with some off interval

Hint: section 5.2

Answers: (a) $A_n = \dfrac{(-1)^n - 1}{10n^2\pi^2}$ and $B_n = -\dfrac{(-1)^n}{10\pi n}$

(b) $C_n = \dfrac{\sqrt{2 + 2(-1)^{n+1} + n^2\pi^2}}{10n^2\pi^2}$ and $\varphi_n = -\tan^{-1}\left\{\dfrac{n\pi(-1)^n}{(-1)^n - 1}\right\}$

(c) $C_n = \dfrac{(-1)^n - 1 + (-1)^n \pi n j}{20n^2\pi^2}$

◆◆ Project 7 (on forward Laplace transform)

Apply the MATLAB built-in function laplace to determine the forward Laplace transform for each of the following continuous signals (obviously $t \geq 0$):

(a) $f(t) = \sin 2t + 3\cos 2t$ (b) $f(t) = 1 + e^{-2t}\sin 7t$

(c) $f(t) = (t^3 + 7)e^{-2t}$ (d) $f(t) = 4\sinh 3t + 3\cosh 2t$

(e) $f(t) = 7u(t-3)$ (f) $f(t) = 4\delta(t-3)$

(g) $f(t) = 20 - 4t + 9t^3$ (h) $f(t) = 9t^2 u(t-3)$

(i) $f(t) = 9t^2[u(t) - u(t-4)]$ (j) $f(t) = 4tJ_0(4t)$

Hint: section 6.2

Answers: (a) $F(s) = \dfrac{3s + 2}{s^2 + 4}$ (b) $F(s) = \dfrac{s^2 + 11s + 53}{s(s^2 + 4s + 53)}$

(c) $F(s) = \dfrac{7s^3 + 42s^2 + 84s + 62}{(s+2)^4}$

(d) $F(s) = \dfrac{3s^3 + 12s^2 - 27s - 48}{(s^2 - 9)(s^2 - 4)}$

(e) $F(s) = \dfrac{7e^{-3s}}{s}$ (f) $F(s) = 4e^{-3s}$

(g) $F(s) = \dfrac{20s^3 - 4s^2 + 54}{s^4}$ (h) $F(s) = \dfrac{9e^{-3s}(9s^2 + 6s + 2)}{s^3}$

(i) $F(s) = \dfrac{-18(8s^2 e^{-4s} + 4se^{-4s} + e^{-4s} - 1)}{s^3}$

(j) $F(s)=\dfrac{4s}{(s^2+16)^{\frac{3}{2}}}$

◆◆ Project 8 (on inverse Laplace transform)

Apply the MATLAB built-in function ilaplace to determine the inverse Laplace transform for each of the following s domain functions:

(a) $F(s)=\dfrac{5s+2}{s^2+18}$ (b) $F(s)=\dfrac{s^3-27s-48}{(s^2-1)(s^2+4)}$

(c) $F(s)=\dfrac{e^{-2s}s}{s^2+4}$ (d) $F(s)=\dfrac{2s^5-4s^2+4}{s^7}$

(e) $F(s)=\ln\left(\dfrac{3s}{s^2+16}\right)$ (f) $F(s)=\dfrac{s^3}{(s^2+4)^2}$

Hint: section 6.8

Answers (clearly $t\ge 0$ for the following functions):

(a) $f(t)=5\cos(3t\sqrt{2})+\dfrac{\sqrt{2}\sin(3t\sqrt{2})}{3}$

(b) $f(t)=\dfrac{11e^{-t}}{5}+\dfrac{31\cos 2t}{5}+\dfrac{24\sin 2t}{5}-\dfrac{37e^{t}}{5}$

(c) $f(t)=u(t-2)\cos(2t-4)$

(d) $f(t)=\dfrac{t^6}{180}-\dfrac{t^4}{6}+2t$

(e) $f(t)=\delta(t)\ln 3-\dfrac{1}{t}+\dfrac{2\cos 4t}{t}$

(f) $f(t)=-t\sin 2t+\cos 2t$

◆◆ Project 9 (on system function from a differential equation)

Apply MATLAB built-in function laplace as well as solve to find the Laplace transform system function from each of the following differential equations:

(a) $5\dfrac{dy}{dt}+3y=\sin t$ with initial condition $y(0)=9$

(b) $2\dfrac{d^2y}{dt^2}+2\dfrac{dy}{dt}+7y=t+3$ with initial conditions $y(0)=9$ and $y'(0)=-1$

(c) $16\dfrac{d^4y}{dt^4}-\dfrac{dy}{dt}=e^{-3t}$ with initial conditions $y(0)=3$, $y'(0)=0$, $y''(0)=-1$, and $y'''(0)=-6$

(d) $12\dfrac{d^3y}{dt^3}-15\dfrac{dy}{dt}+3y=\cos 2t$ when $y(0)=9$, $y'(0)=3$, and $y''(0)=2$

Hint: section 6.4

Answers: (a) $F(s)=\dfrac{45s^2+46}{(s^2+1)(5s+3)}$

(b) $F(s)=\dfrac{18s^3+16s^2+3s+1}{s^2(2s^2+2s+7)}$

(c) $F(s)=\dfrac{48s^4+144s^3-16s^2-147s-296}{s(s+3)(16s^3-1)}$

(d) $F(s)=\dfrac{108s^4+36s^3+321s^2+145s-444}{3(s^2+4)(4s^3-5s+1)}$

♦♦ Project 10 (on Fourier series coefficients of a continuous periodic signal)

When we connect our mobile charger to a $110V$ power supply, a full wave rectification takes place like the figure 11.2(h). For the continuous periodic signal $f(t)$ of the figure 11.2(h), determine the following in standard time unit in MATLAB:

(a) real form 1 Fourier series coefficients
(b) real form 2 Fourier series coefficients
(c) complex form Fourier series coefficients

Hint: section 5.2

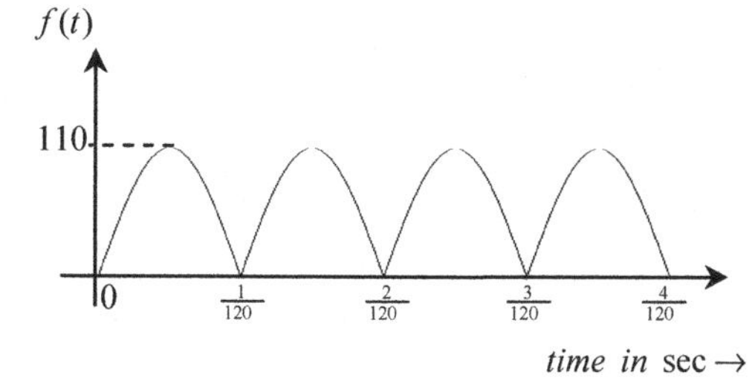

Figure 11.2(h) A full rectified sine wave

Answers: (a) $A_n=-\dfrac{440}{\pi(4n^2-1)}$ and B_n=0

(b) $C_n=\dfrac{440}{\pi(4n^2-1)}$ and φ_n=0

(c) $C_n=-\dfrac{220}{\pi(4n^2-1)}$

♦♦ Project 11 (on Laplace transform of graphical functions)

Write appropriate mathematical expression for each of the following graphical functions and apply MATLAB built-in function laplace to find its forward Laplace transform:

(a) For the finite ramp function of the figure 11.2(c)
(b) For the finite duration pulse of the figure 11.1(h)
(c) For the finite triangular pulse of the figure 11.1(c)
(d) For the piecewise continuous signal of the figure 11.1(d)

Hint: section 6.3

Answers: (a) $F(s)=\dfrac{5(6s-1+e^{-6s})}{6s^2}$ (b) $F(s)=-\dfrac{11(e^{-3s}-e^{-7s})}{s}$

(c) $F(s)=\dfrac{1-2e^{-4s}+e^{-8s}}{s^2}$

(d) $F(s)=\dfrac{2(-3s+6se^{-3s}-e^{-6s}+e^{-9s})}{3s^2}$

♦♦ Project 12 (on Fourier transform graphing)

Graph the magnitude ($|F(\omega)|$ versus ω), phase ($\angle F(\omega)$ versus ω), real ($\text{Re}\{F(\omega)\}$ versus ω), and imaginary ($\text{Im}\{F(\omega)\}$ versus ω) spectra of the Fourier transform $F(\omega)=\dfrac{j200\omega}{2j\omega-7\omega^2-2000}$ over the interval $-200\le\omega\le 200$ rad/sec.

Hint: Subsection 5.3.3

Answers: (a) figure 11.3(a), (b) figure 11.3(b), (c) figure 11.3(c), and (d) figure 11.3(d)

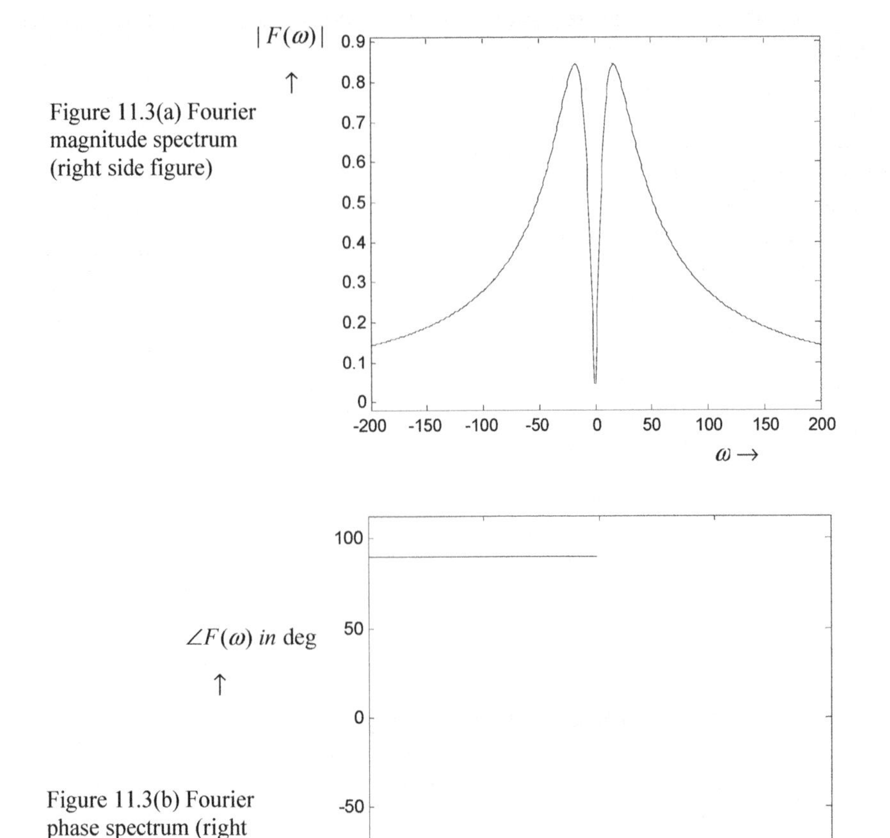

Figure 11.3(a) Fourier magnitude spectrum (right side figure)

Figure 11.3(b) Fourier phase spectrum (right side figure)

⬍⬍ Project 13 (on forward Z transform)

Apply the MATLAB built-in function ztrans to determine the unilateral forward Z transform for each of the following discrete signals:

(a) $f[n] = \sin 2n - \cos 2n$ (b) $f[n] = 4^{-n} \cos n$

(c) $f[n] = 2^n - 3^{-3n}$ (d) $f[n] = n^3 3^{-n}$

(e) $f[n] = 4\delta[n-3]$ (f) $f[n] = 2^{-n}(u[n] - u[n-3])$

Hint: section 7.2

Answers: (a) $F(z) = \dfrac{z\sin 2 + z\cos 2 - z^2}{1 - 2z\cos 2 + z^2}$ (b) $F(z) = \dfrac{4z(4z - \cos 1)}{16z^2 - 8z\cos 1 + 1}$

(c) $F(z) = \dfrac{53z}{(z-2)(27z-1)}$ (d) $F(z) = \dfrac{3z(9z^2 + 12z + 1)}{(3z-1)^4}$

(e) $F(z)=\dfrac{4}{z^3}$ (f) $F(z)=1+\dfrac{1}{2z}+\dfrac{1}{4z^2}$

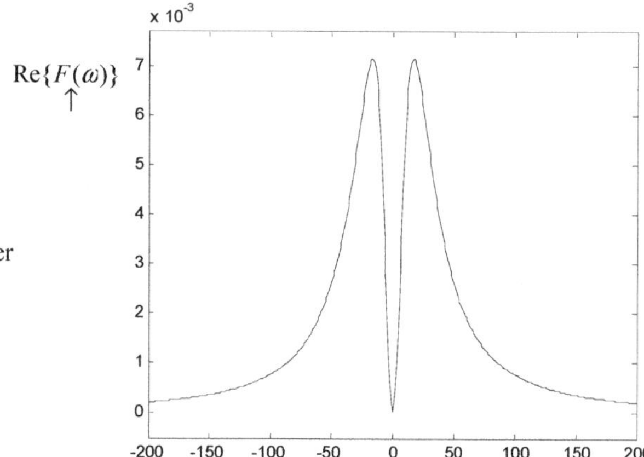

Figure 11.3(c) Fourier real spectrum (right side figure)

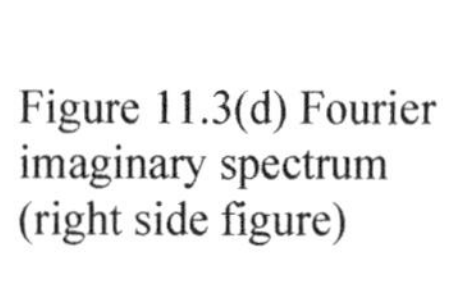

Figure 11.3(d) Fourier imaginary spectrum (right side figure)

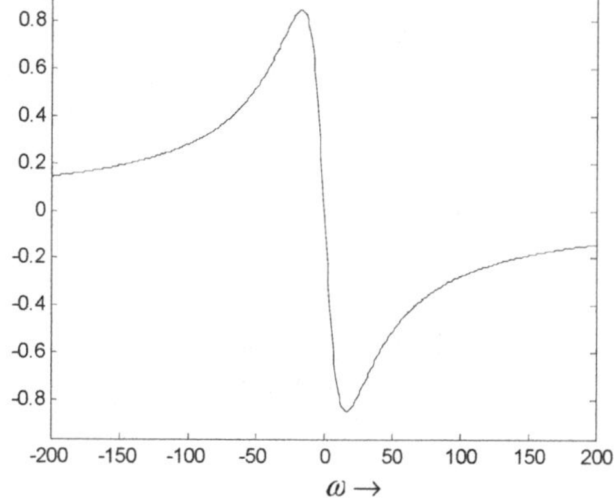

◆◆ Project 14 (on continuous system formation)

Form a continuous system from each of the following representations by applying appropriate MATLAB built-in function:

(a) $H(s)=\dfrac{8.4s-1}{s^3-6s-9}$ (b) $H(s)=\dfrac{8.4s-1}{(s^3-6s-9)(s^2+3)}$

(c) $H(s)=\dfrac{1}{(s^2+3)^3}$ (d) $H(s)=\dfrac{4(s+1)}{(s+4)(s-5)(s+6)}$

(e) $H(s)$ with $\begin{Bmatrix} \text{zeroes: -6.6, 2.3, 9} \\ \text{poles: 3, 0,–2,6} \\ \text{gain: 2.9} \end{Bmatrix}$

(f) $H(s)$ which has the pole-zero map of figure 11.3(e) assuming unity gain

Hint: section 8.2

Answers:

(a) tf([8.4 -1],[1 0 -6 -9])

(b) tf([8.4 -1],conv([1 0 -6 -9],[1 0 3]))
(c) tf(1,conv([1 0 3],conv([1 0 3],[1 0 3])))
(d) zpk(-1,[-4 5 -6],4)
(e) zpk([-6.6 2.3 9],[3 0 -2 6],2.9)
(f) zpk([-2 -5.25],[i*2 -i*2 -3.75+i*2 -3.75-i*2 -3.75],1)

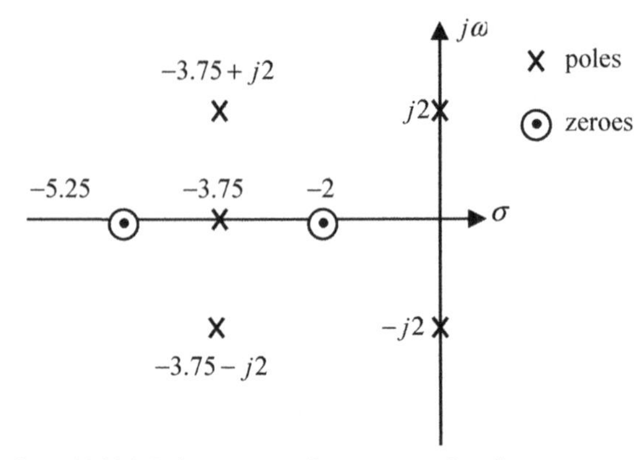

Figure 11.3(e) Pole-zero map of some system function

♦♦ Project 15 (on discrete Fourier transform)

Compute the forward discrete Fourier transform $F[k]$ for each of the following discrete signals:

(a) $f[n]=[9 \quad 3 \quad 5 \quad 0 \quad 1 \quad 7 \quad -12]$
(b) $f[n]=3\,e^{-2n}$ over $0 \le n \le 5$
(c) $f[n]=3\sin\frac{2\pi n}{5}+5\cos\frac{2\pi n}{5}$ over $0 \le n \le 4$
(d) $f[n]$ of (b) in magnitude-angle form
(e) $f[n]$ of (c) only for $F[1]$ in magnitude-angle form

Hint: section 5.4

Answers:

(a) $F[k]=[13 \quad -0.1826-j\,9.3437 \quad 0.8146-j\,16.2735 \quad 24.3681-j\,7.0970 \quad 24.3681+j\,7.0970 \quad 0.8146+j\,16.2735 \quad -0.1826+j\,9.3437]$
(b) $F[k]=[3.4695 \quad 3.1677-j\,0.3982 \quad 2.7764-j\,0.3048 \quad 2.6424 \quad 2.7764+j\,0.3048 \quad 3.1677+j\,0.3982]$
(c) $F[k]=[0 \quad 12.5-j\,7.5 \quad 0 \quad 0 \quad 12.5+j\,7.5]$
(d) $F[k]=[3.4695\angle 0^0 \quad 3.1926\angle -7.1651^0 \quad 2.7931\angle -6.2646^0 \quad 2.6424\angle 0^0 \quad 2.7931\angle 6.2646^0 \quad 3.1926\angle 7.1651^0\,]$
(e) $14.5774\angle -30.9638^0$, use abs(F(2)) and 180*angle(F(2))/pi after F=fft(f)

Figure 11.3(f) A discrete signal $f[n]$ is downsampled by a factor M

Figure 11.3(g) A discrete signal $f[n]$ is upsampled by a factor M

♦♦ Project 16 (on downsampling of a discrete signal)

Figure 11.3(f) shows the downsampling of a discrete signal $f[n]$ by a factor M. Employ the built-in MATLAB function to determine the output $g[n]$ for each of the following discrete signals:

(a) $f[n]=3n+2$ over $0 \le n \le 5$ when M=3

(b) $f[n]=3\,e^{-n/10}$ over $0 \le n \le 19$ when M=5

(c) $f[n]=3\sin\frac{2\pi(n-1)}{3}$ over $1 \le n \le 10$ when M=2

Hint: section 4.2

Answers:

(a) $g[n]$=[2 11] (b) $g[n]$=[3 1.8196 1.1036 0.6694]

(c) $g[n]$=[0 −2.5981 2.5981 0 −2.5981]

♦♦ Project 17 (on upsampling of a discrete signal)

Figure 11.3(g) shows the upsampling of a discrete signal $f[n]$ by a factor M. Employ the built-in MATLAB function to determine the output $g[n]$ for each of the following discrete signals:

(a) $f[n]=n^2-2$ over $0 \le n \le 2$ when M=3

(b) $f[n]=6\,e^{-n}$ over $0 \le n \le 3$ when M=2

(c) $f[n]=3\sin\frac{2\pi(n-1)}{3}$ over $5 \le n \le 9$ when M=2

Hint: section 4.2

Answers:

(a) $g[n]$=[−2 0 0 −1 0 0 2 0 0]

(b) $g[n]$=[6 0 2.2073 0 0.8120 0 0.2987 0]

(c) $g[n]$=[2.5981 0 −2.5981 0 0 0 2.5981 0 −2.5981 0]

$f(t)=2e^{-2t}\cos(2t+60^0)$ → C/D → $f[n]$

60 Hz

Figure 11.3(h) A damped cosine signal is sampled for digitization over $0 \le t \le \frac{4}{15}$sec

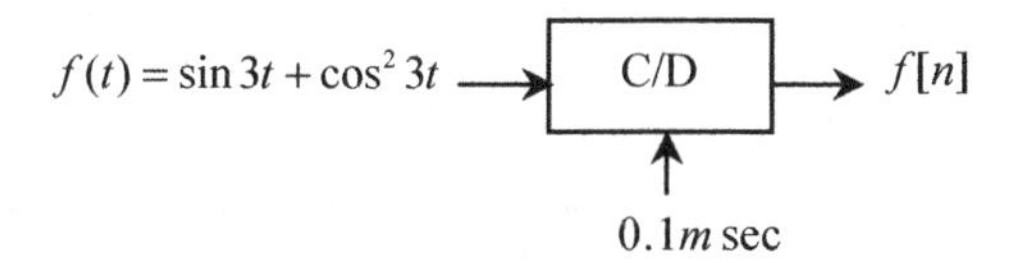

Figure 11.3(i) A continuous signal is sampled for digitization over $0 \le t \le 0.2$sec

♦♦ Project 18 (on continuous to discrete conversion of a signal)

(a) Obtain the discrete signal $f[n]$ as a row matrix by sampling the damped cosine signal $f(t)$ of the figure 11.3(h)

(b) Obtain the discrete signal $f[n]$ as a row matrix by sampling the two frequency signal $f(t)$ of the figure 11.3(i)

(c) How many samples are in $f[n]$ in part (a)?

(d) How many samples are in $f[n]$ in part (b)?

(e) Find $f[4]$ in part (a)

(f) Find $f[7]$, $f[8]$, and $f[9]$ as a row matrix in part (b)

Hint: section 4.2

Answers:

(c) 17 (d) 2001

(e) $f[4]$=0.6659 (f) [1.0021 1.0024 1.0027]

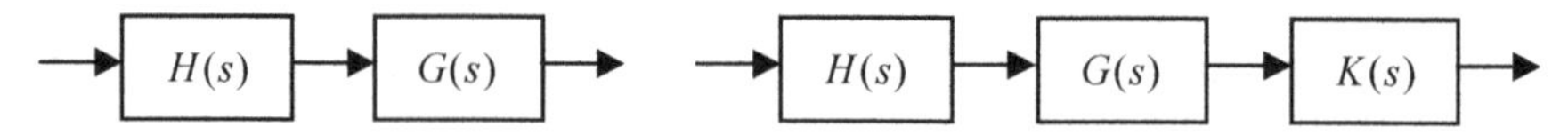

Figure 11.4(a) Two system functions connected in series

Figure 11.4(b) Three system functions connected in series

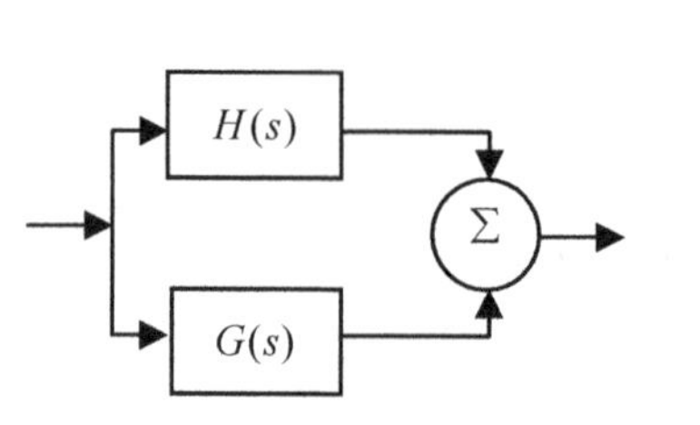

Figure 11.4(c) Two system functions connected in parallel

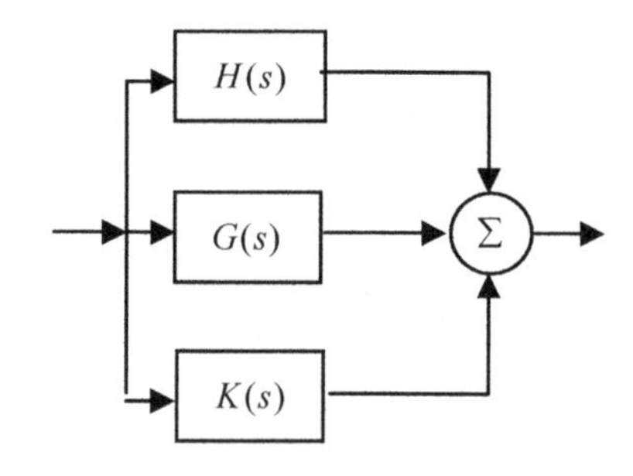

Figure 11.4(d) Three system functions connected in parallel

♦♦ Project 19 (on equivalent system function)

(a) Figure 11.4(a) shows two system functions connected in series where $H(s)=\dfrac{3}{4.6s+2}$ and $G(s)=\dfrac{1.4}{3s^2+2.5}$. Verify that their equivalent system function is $H_{eq}(s)=\dfrac{4.2}{13.8s^3+6s^2+11.5s+5}$.

(b) Figure 11.4(c) shows two system functions connected in parallel where $H(s)=\dfrac{14s-8}{(s+2)(s-3)}$ and $G(s)=\dfrac{1.4}{3s^2+2.5}$. Verify that their equivalent system function is $H_{eq}(s)=\dfrac{42s^3-22.6s^2+33.6s-28.4}{3s^4-3s^3-15.5s^2-2.5s-15}$.

(c) Figure 11.4(b) shows three system functions connected in series where $H(s)=\frac{14s-8}{(s+2)(s-3)}$, $G(s)=\frac{1.4}{3s^2+2.5}$, and $K(s)=\frac{8}{3s(s+2.5)}$. Verify that their equivalent system function is $H_{eq}(s)=\frac{156.8s-89.6}{9s^6+13.5s^5-69s^4-123.8s^3-63.75s^2-112.5s}$.

(d) Figure 11.4(d) shows three system functions connected in parallel where $H(s)=\frac{14(s-5)}{s+2}$, $G(s)=\frac{s-5}{s^2+2.5}$, and $K(s)=\frac{1}{3s}$. Verify that their equivalent system function is $H_{eq}(s)=\frac{14s^4-68.67s^3+32.67s^2-184.2s+1.667}{s^4+2s^3+2.5s^2+5s}$.

Hint: section 8.3

⬍⬍ Project 20 (on modeling a continuous system)

Model each of the following system functions in a SIMULINK model by employing suitable blocks and you may verify the modeling by looking into the expression inside the block:

(a) $H(s)=\frac{156.8s-89.6}{9s^6+13.5s^5-69s^4-123.8s^3-63.75s^2-112.5s}$

(b) $H(s)=\frac{42s^3-22.6s^2+33.6s-28.4}{3s^4-3s^3-15.5s^2-2.5s-15}$

(c) $H(s)=\frac{5.5(s+1.1)}{s(s+6)(s-2.2)}$

(d) $H(s)=\frac{7.9(s+2)(s-4)}{s(s+6)(s-2.2)(s+5)(s+2.1)}$

(e) $H(s)=\frac{5(s+11)(s-6)}{s(s^4+6)(s^2+9)}$

(f) $H(s)=\frac{156.8s-89.6}{(9s^6-69s^4-63.75s^2-112.5)(s+4)(s+2)}$

(g) $H(s)=\{A,B,C,D\}$ in state space form where $A=\begin{bmatrix}-7 & -1\\ 9 & 3\end{bmatrix}$, $B=\begin{bmatrix}-3\\ -1\end{bmatrix}$, $C=[-3 \quad 5]$, and $D=[7]$

(h) $H(s)=\{A,B,C,D\}$ in state space form where $A=\begin{bmatrix}-7 & -1 & 9\\ 9 & 3 & 7\\ 3 & -2 & 2\end{bmatrix}$, $B=\begin{bmatrix}-3\\ -1\\ 5\end{bmatrix}$, $C=[-3 \quad 5 \quad 5]$, and $D=[12]$

Hint: section 8.5

⬍⬍ Project 21 (on forward Z transform of finite sequences)

Determine the forward Z transform for each of the following finite sequences by using MATLAB built-in functions:

(a) $f[n]=\begin{Bmatrix} 3^{-n} & for\ 0\le n<4 \\ 0 & elsewhere \end{Bmatrix}$

(b) $\begin{Bmatrix} f[n] & \rightarrow & -2 & -4 & 3 & 0 & 12 & 16 \\ n & \rightarrow & 2 & 3 & 4 & 5 & 6 & 7 \end{Bmatrix}$

(c) $f[n]=\begin{Bmatrix} e^{-n}\sin\dfrac{\pi(n-1)}{5} & for\ 1\le n<5 \\ 0 & elsewhere \end{Bmatrix}$

(d) The finite sequence envelope of the figure 11.4(e) follows $f[n]=100-n^2$ over $0\le n\le 7$

Hint: section 7.3

Answers:

(a) $F(z)=1+\dfrac{1}{3z}+\dfrac{1}{9z^2}+\dfrac{1}{27z^3}$

(b) $F(z)=\dfrac{-2z^5-4z^4+3z^3+12z+16}{z^7}$

(c) $F(z)=\dfrac{e^{-2}\left(z^2\sin\dfrac{\pi}{5}+e^{-1}z\sin\dfrac{2\pi}{5}+e^{-2}\sin\dfrac{2\pi}{5}\right)}{z^4}$

(d) $F(z)=100+\dfrac{99}{z}+\dfrac{96}{z^2}+\dfrac{91}{z^3}+\dfrac{84}{z^4}+\dfrac{75}{z^5}+\dfrac{64}{z^6}+\dfrac{51}{z^7}$

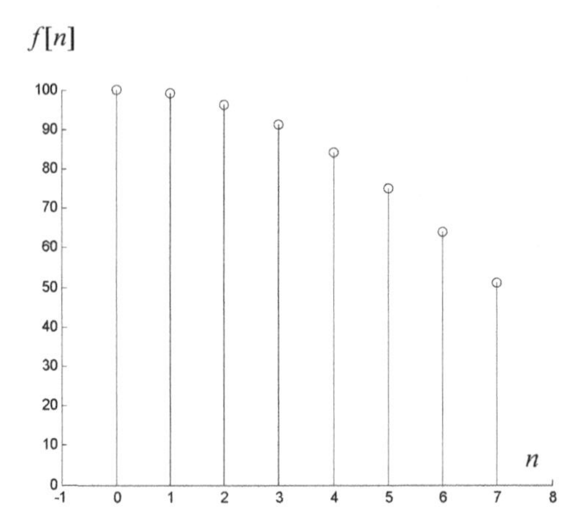

Figure 11.4(e) A finite sequence follows parabolic envelope

♦♦ Project 22 (on signal sample generation)

Generate the signal samples as a row matrix for each of the following continuous signals over given interval and the step size adjacent to the signal:

(a) $f(t)=3-2\sin^2 2\pi t$ over $-2\le t\le 5$ sec with a step 0.25sec

(b) $f(t)=(t+1)^2\cos(\pi t-67^0)$ over $0\le t\le 5$ sec with a step 0.005sec

(c) $x(t)=10e^{-2t}\sin(3t+45^0)\sin c(t)$ over $-4\le t\le 4$ sec with a step 0.05sec

(d) $x(t)=10e^{-|2t-3|}\,|\sin(3t+45^0)\operatorname{sin}c(t)|$ over $-4 \le t \le 4$ sec with a step 0.1sec

Hint: section 2.3

Answers:

(a) 3-2*sin(2*pi*t).^2 (b) (t+1).^2.*cos(pi*t-67/180*pi)

(c) 10*exp(-2*t).*sin(3*t+pi/4).*sinc(t)

(d) 10*exp(-abs(2*t-3)).*abs(sin(3*t+pi/4).*sinc(t))

⬍⬍ Project 23 (on discrete Fourier transform implications)

A composite voltage signal $x(t)$ is composed of three sinusoidal frequencies $\begin{Bmatrix} 4Hz \\ 16Hz \\ 32Hz \end{Bmatrix}$ with identical phase whose amplitudes are $\begin{Bmatrix} \pm 20V \\ \pm 18V \\ \pm 16V \end{Bmatrix}$ respectively. The $x(t)$ exists for $0 \le t \le 0.5\text{sec}$. Considering a sampling frequency of f_s=200 Hz, prove that three strong peaks exist exactly at the given frequencies in the magnitude DFT versus discrete frequency plot.

Hint: section 5.4

⬍⬍ Project 24 (on signal sample generation)

Charging and discharging of electrical charge through a capacitor in conjunction with a series resistance follow the wave shape of the figure 11.4(f). Both the charging and discharging portions have the time constant $\frac{1}{2}$ sec and steady state voltage value for the charging portion is $10V$. Generate the wave shape samples over the interval $0 \le t \le 10$ sec with a step 0.05sec and verify the shape by graphing.

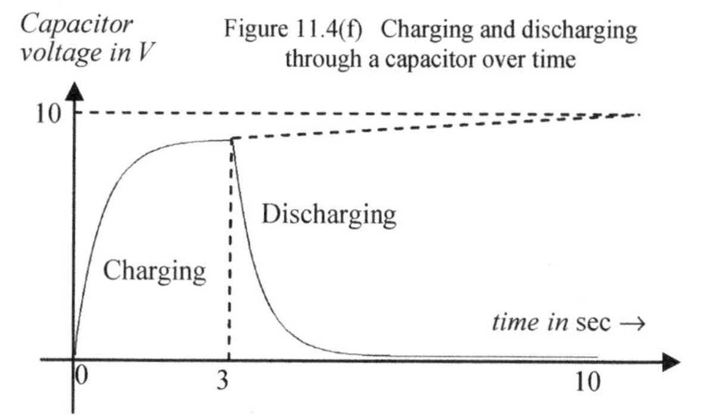

Figure 11.4(f) Charging and discharging through a capacitor over time

Hint: sections 2.3 and 2.9, capacitor voltage $v(t)=10(1-e^{-2t})\ V$ over $0 \le t < 3\text{sec}$ and $v(t)=v_3 e^{-2(t-3)}\ V$ over $3 \le t < 10\text{sec}$ where $v_3=10(1-e^{-6})\ V$.

⬍⬍ Project 25 (on Fourier transform of continuous signals)

Determine the Fourier transform for each of the following continuous signals by using MATLAB built-in functions:

(a) $f(t)=e^{-\frac{(t-1)^2}{4}}$ (b) $f(t)=\frac{1}{9+4t^2}$ (c) $f(t)=4|8t+25|$

(d) $rect(t)$ of the figure 10.1(a) (e) $tri(t)$ of the figure 10.1(b)

Hint: section 5.3 and use heaviside(t) for $rect(t)$ and $tri(t)$

Answers:

(a) $F(\omega)=2\sqrt{\pi}\ e^{-j\omega-\omega^2}$ (b) $F(\omega)=\frac{\pi}{6}[e^{\frac{3\omega}{2}}u(-\omega)+e^{-\frac{3\omega}{2}}u(\omega)]$

(c) $F(\omega) = -\dfrac{64e^{j\frac{25}{8}\omega}}{\omega^2}$ (d) $F(\omega) = \dfrac{2\sin\frac{\omega}{2}}{\omega}$

(e) $F(\omega) = \dfrac{2(1-\cos\omega)}{\omega^2}$

♦♦ Project 26 (on step response of a continuous system)

(a) Verify that the system $H(s) = \dfrac{5.5(s+1.1)}{(s+6)(s+2.2)}$ has the figure 11.4(g) shown step response over the interval $0 \le t < 4\,\text{sec}$

(b) Verify that the system $H(s) = \dfrac{6s+4}{9s^3+4s^2+50s+9}$ has the figure 11.4(h) shown step response over the interval $0 \le t < 25\,\text{sec}$

Hint: section 9.1

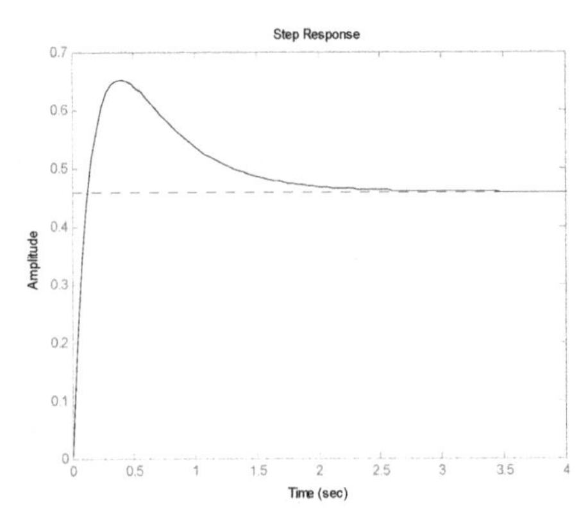

Figure 11.4(g) Step response for the system in project 26(a)

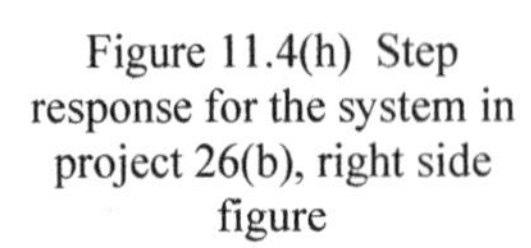
Figure 11.4(h) Step response for the system in project 26(b), right side figure

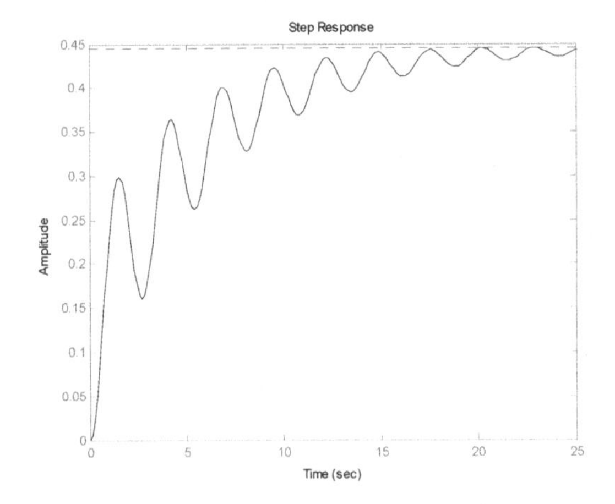

♦♦ Project 27 (on feedback systems)

Determine the equivalent system function $\dfrac{Y(s)}{R(s)}$ for each of the following feedback connected systems:

(a) Negative feedback system of the figure 8.2(e) when $G(s) = \dfrac{5.5(s+1.1)}{(s+6)(s+2.2)}$ and $H(s) = \dfrac{3}{s+2}$

(b) Positive feedback system of the figure 8.2(f) when $G(s) = \dfrac{2.5(s-4)}{(s-3.1)(s-1)}$ and $H(s) = \dfrac{3}{s+2}$

(c) Negative feedback system of the figure 8.2(e) when $G(s) = \dfrac{6s+4}{9s^3+4s^2+50s+9}$ and $H(s) = \dfrac{3}{s(s+2)}$

(d) Positive feedback system of the figure 8.2(f) when $G(s) = \dfrac{6s^2+4}{9s^4-4s^2-5s-9}$ and $H(s) = \dfrac{3(s-5)}{s(s+2)}$

Hint: section 8.4

Answers:

(a) $\dfrac{Y(s)}{R(s)} = \dfrac{5.5s^2+17.05s+12.1}{s^3+10.2s^2+46.1s+44.55}$ (b) $\dfrac{Y(s)}{R(s)} = \dfrac{2.5s^2-5s-20}{s^3-2.1s^2-12.6s+36.2}$

(c) $\dfrac{Y(s)}{R(s)} = \dfrac{6s^3+16s^2+8s}{9s^5+22s^4+58s^3+109s^2+36s+12}$

(d) $\dfrac{Y(s)}{R(s)} = \dfrac{6s^4+12s^3+4s^2+8s}{9s^6+18s^5-4s^4-31s^3+71s^2-30s+60}$

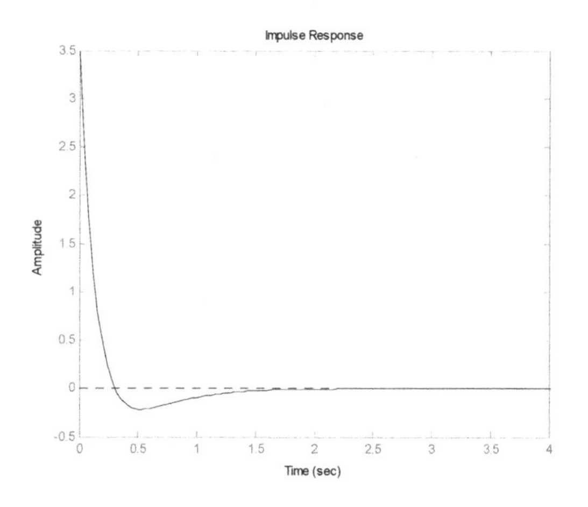

Figure 11.5(a) Impulse response for the system in project 28(a), right side figure

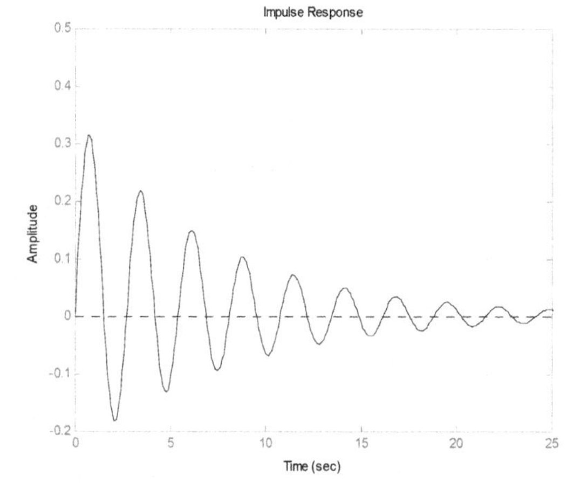

Figure 11.5(b) Impulse response for the system in project 28(b), right side figure

⬍⬍ Project 28 (on impulse response of a continuous system)

(a) Verify that the system $H(s)=\dfrac{3.5(s+1.1)}{(s+3)(s+6.2)}$ has the figure 11.5(a) shown impulse response over the interval $0 \le t < 4$sec

(b) Verify that the system $H(s)=\dfrac{6s+4}{9s^3+4s^2+50s+9}$ has the figure 11.5(b) shown impulse response over the interval $0 \le t < 25$sec

Hint: section 9.2

⬍⬍ Project 29 (on step and impulse responses of a continuous system)

(a) Verify that the system $H(s)=\dfrac{3.5(s+1.1)}{(s+3)(s+6.2)}$ has the step response

$$y(t)=\frac{77}{372}-\frac{1785}{1984}e^{-\frac{31}{5}t}+\frac{133}{192}e^{-3t}$$

(b) Verify that the system $H(s)=\dfrac{6s+4}{2s^2-2s-4}$ has the step response $y(t)=$

$$-1+e^{\frac{t}{2}}\left(\cosh\frac{3t}{2}+\frac{5}{3}\sinh\frac{3t}{2}\right)$$

(c) Verify that the system $H(s)=\dfrac{3.5(s+1.1)}{(s+3)(s+6.2)}$ has the impulse response

$$h(t)=\frac{357}{64}e^{-\frac{31}{5}t}-\frac{133}{64}e^{-3t}$$

(d) Verify that the system $H(s)=\dfrac{6s+4}{2s^2-2s-4}$ has the impulse response

$$h(t)=e^{\frac{t}{2}}\left(3\cosh\frac{3t}{2}+\frac{7}{3}\sinh\frac{3t}{2}\right)$$

Hint: sections 9.1 and 9.2 (clearly all solution is for $t \ge 0$)

⬍⬍ Project 30 (on step and impulse responses of continuous systems)

(a) Verify that the system $H(s)=\dfrac{3.5(s+1.1)}{(s+3)(s+6.2)}$ has the step response functional value $y(t)=0.3403$ at t =0.4sec

(b) Verify that the system $H(s)=\dfrac{3.5(s+1.1)}{(s+3)(s+6.2)}$ has the step response functional values $y(t)=0$, 0.2362, 0.3268, 0.3486, and 0.3403 at t =0, 0.1, 0.2, 0.3, and 0.4sec respectively

(c) Verify that the system $H(s)=\dfrac{3.5(s+1.1)}{(s+3)(s+6.2)}$ has the impulse response functional value $h(t)$ =0.4737 at t =0.2sec

(d) Verify that the system $H(s)=\dfrac{3.5(s+1.1)}{(s+3)(s+6.2)}$ has the impulse response functional values $h(t)=3.5$, 1.4612, 0.4737, 0.0235, and –0.1588 at t =0, 0.1, 0.2, 0.3, and 0.4sec respectively

(e) Verify that the system $H(s)=\dfrac{3s+8}{s^3-7s+3}$ has the step response functional value $y(t)=69.5027$ at t =2sec

(f) Verify that the system $H(s)=\dfrac{3s+8}{s^3-7s+3}$ has the step response functional values $y(t)=4.3470$, 5.9588, 8.0596, 10.7841, and 14.3037 at t =1, 1.1, 1.2, 1.3, and 1.4sec respectively

(g) Verify that the system $H(s)=\dfrac{3s+8}{s^3-7s+3}$ has the impulse response functional value $h(t)$ =3.099 at t =0.5sec

(h) Verify that the system $H(s)=\dfrac{3s+8}{s^3-7s+3}$ has the impulse response functional values $h(t)$ =14.0474, 18.3644, 23.8739, 30.9, and 39.8549 at t =1, 1.1, 1.2, 1.3, and 1.4sec respectively

Hint: sections 9.1 and 9.2

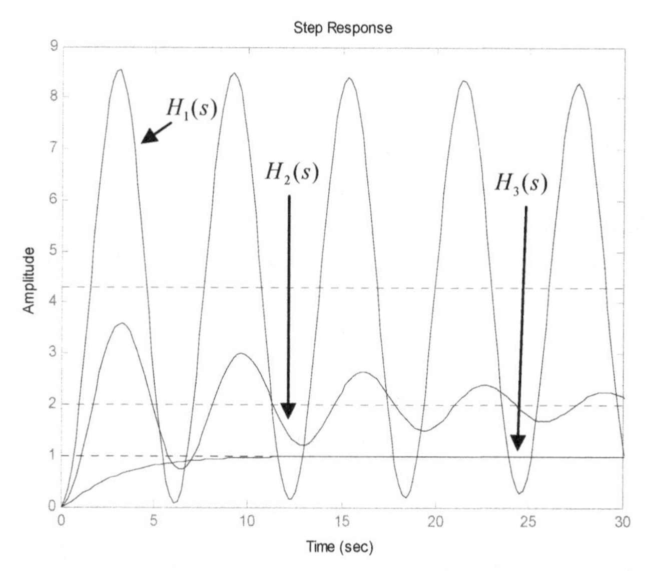

Figure 11.5(c) Step responses of three systems over common time variation

♦♦ Project 31 (on step and impulse responses of multiple systems)

(a) Verify that the three systems $H_1(s)=\dfrac{9}{2s^2+0.01s+2.1}$, $H_2(s)=\dfrac{4}{2.1s^2+0.3s+2}$, and $H_3(s)=\dfrac{3}{3s^2+9.5s+3}$ have the step responses over common time interval $0 \le t < 30\text{sec}$ as displayed in figure 11.5(c)

(b) Verify that the three systems of part (a) have the impulse responses over common time interval $0 \le t < 30\text{sec}$ as displayed in figure 11.5(d)

Hint: sections 9.1 and 9.2

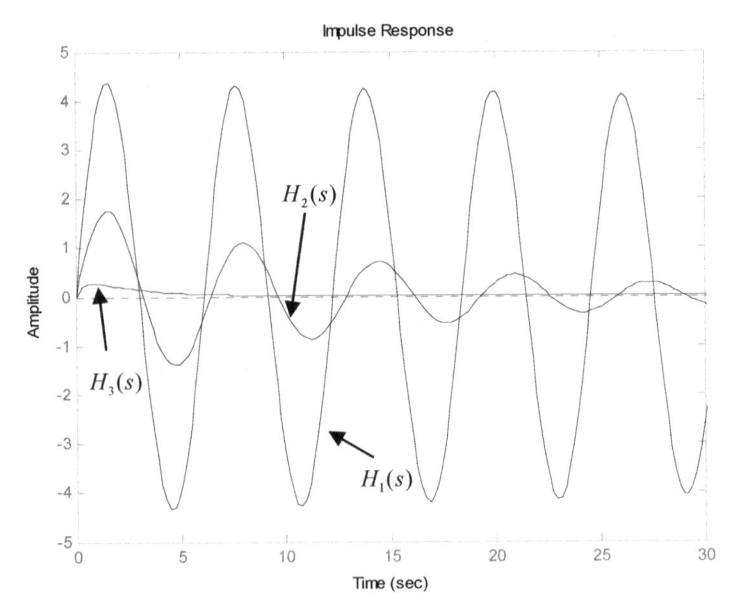

Figure 11.5(d) Impulse responses of three systems over common time variation

◆◆ Project 32 (on sequence solution of a difference equation)

Determine the sequence solution (for k or $n \geq 0$) for each of the following difference equations by applying built-in MATLAB function:

(a) $y_{k+1} = -4y_{k-1} - 4y_k$, $y_0 = 4$, and $y_1 = 0$

(b) $7y[n+2] - 6y[n+1] - y[n] = 4 - n$, $y[0]=2$, and $y[1]=3$

(c) $y[n-3] - 2y[n-2] + y[n-1] = n+1$, $y[0]=0$, and $y[1]=-1$

(d) $3x[n] - 2x[n-1] = 5\delta[n]$ and $x[0]=0$

Hint: section 7.6

Answers:

(a) $y_k = 8(-2)^k + (-4k-4)(-2)^k$

(b) $y[n] = \frac{1199}{512} - \frac{175}{512}\left(-\frac{1}{7}\right)^n + \frac{43n}{64} - \frac{n^2}{16}$

(c) $y[n] = \frac{n(n^2+9n-16)}{6}$

(d) $y[n] = \frac{5}{3}\sum_{m=1}^{m=n}\left(\frac{2}{3}\right)^{n-m}\delta[m]$

◆◆ Project 33 (on convolution of continuous signals)

Apply convolution in t-domain to determine the continuous $r(t)$ signal response of the system in figure 11.5(e) for each of the following cases ($t \geq 0$):

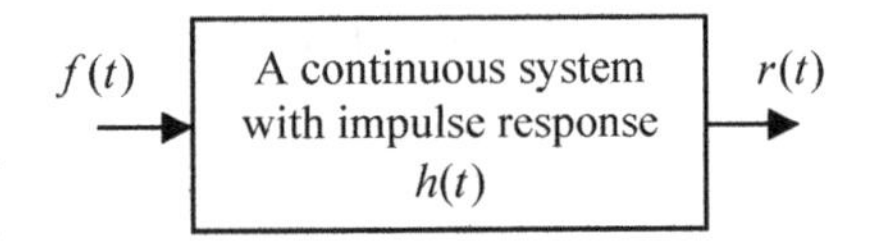

Figure 11.5(e) A continuous system has input $f(t)$ and output $r(t)$

(a) when $h(t) = 2t - 9$ and $f(t) = \sin 2t$

(b) when $h(t)=2e^{-3t}$ and $f(t)=2t^2-78$

(c) when the system in figure 11.5(e) represents a filter with $h(t)=\dfrac{\sin t}{t}$ and $f(t)=2$

(d) graph the $h(t)$, $f(t)$, and $r(t)$ of the part c in a single graph over $-3\pi \le t \le 3\pi$, $-3\pi \le t \le 3\pi$, and $-6\pi \le t \le 6\pi$ respectively

Hint: section 4.4

Answers:

(a) $r(t)=\dfrac{9\cos 2t}{2}-\dfrac{\sin 2t}{2}+t-\dfrac{9}{2}$ (b) $r(t)=\dfrac{1396}{27}e^{-3t}+\dfrac{4t^2}{3}-\dfrac{1396}{27}-\dfrac{8t}{9}$

(c) $r(t)=2\,Si(t)$ where $Si(t)$ is the sine integral with definition

$$Si(t)=\int_{w=0}^{w=t}\frac{\sin w}{w}dw$$

(d) figure 11.5(f)

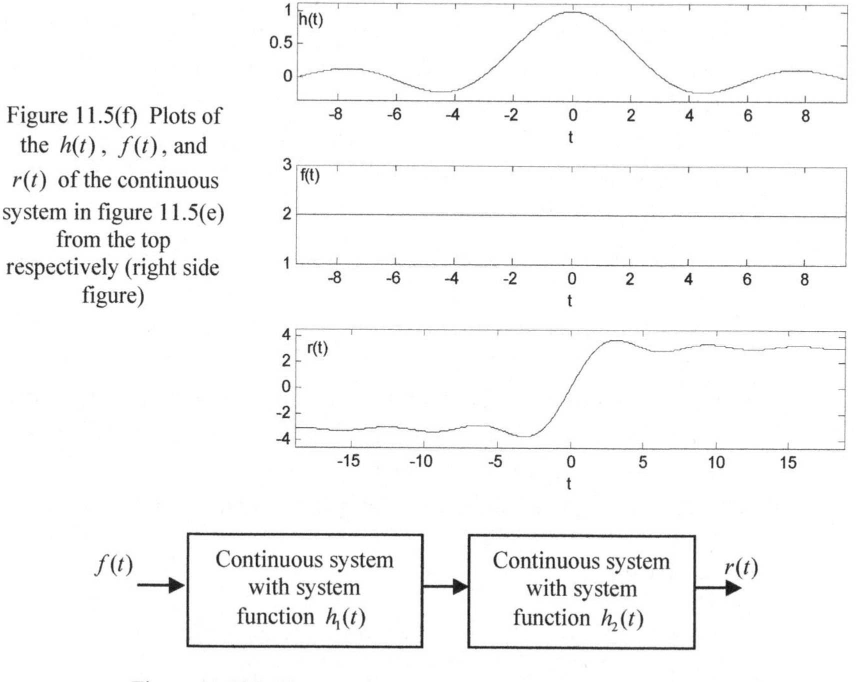

Figure 11.5(f) Plots of the $h(t)$, $f(t)$, and $r(t)$ of the continuous system in figure 11.5(e) from the top respectively (right side figure)

Figure 11.5(g) Two continuous systems connected in series

♦♦ Project 34 (on convolution of continuous signals)

Figure 11.5(g) shows two continuous time systems connected in series. Apply convolution in t-domain to determine the following ($t \ge 0$):

(a) when $h_1(t)=5t+7$ and $h_2(t)=2^{-t}$, what is the equivalent t-domain system function from $f(t)$ to $r(t)$?

(b) when $h_1(t) = e^{-2t}$ and $h_2(t) = 2t - 5$, what is the equivalent t-domain system function from $f(t)$ to $r(t)$?

(c) suppose an input $f(t) = \sin t$ is applied to the system of the figure 11.5(g), what is the output $r(t)$ for $h_1(t) = 5t + 7$ and $h_2(t) = \cos t$?

(d) suppose an input $f(t) = 7e^{-2t}$ is applied to the system of the figure 11.5(g), what is the output $r(t)$ for $h_1(t) = 3t - 7$ and $h_2(t) = e^{-t}$?

Hint: section 4.4

Answers:

(a) $\dfrac{7\times 2^t \ln 2 - 5\times 2^t + 5t2^t \ln 2 - 7\ln 2 + 5}{2^t (\ln 2)^2}$ (b) $3e^{-2t} + t - 3$

(c) $r(t) = 5 - 5\cos t - \dfrac{5t\sin t}{2} - \dfrac{7t\cos t}{2} + \dfrac{7\sin t}{2}$

(d) $r(t) = \dfrac{7(-23e^{2t} + 6te^{2t} + 40e^t - 17)}{4e^{2t}}$

◆◆ Project 35 (on convolution of discrete signals)

(a) In figure 4.4(e) a discrete signal $f[n]$=[3 3 0 5 2 4] is applied to the discrete system with $h[n]$=[2 2 –3 8 9 2], what is the output signal $r[n]$ from the system?

(b) In figure 4.4(e) a discrete binary signal $f[n]$=[1 0 1 0 1 1 0] is applied to the discrete system with $h[n]$=[1 1 1 1], what is the output signal $r[n]$ from the system?

(c) In figure 4.4(e) a discrete signal $f[n]=23\sin\dfrac{2n\pi}{7}$ over $0 \le n \le 4$ is applied to the discrete system with $h[n] = 12e^{-\frac{n}{9}}$ over $0 \le n \le 4$, what is the output signal $r[n]$ from the system?

(d) In figure 4.4(e) a discrete signal $f[n] = 2n - 3$ over $1 \le n \le 4$ is applied to the discrete system with $h[n] = 12\ln n$ over $3 \le n \le 6$, what is the output signal $r[n]$ from the system?

(e) In part d, graph all three discrete signals $f[n]$, $h[n]$, and $r[n]$ in a single plot

Hint: section 4.5

Answers:

(a) $r[n]$=[6 12 –3 25 65 30 48 49 60 40 8]

(b) $r[n]$=[1 1 2 2 2 3 2 2 1 0]

(c) $r[n]$=[0 215.7855 462.1734 533.3229 357.4864 319.8929 162.4450 –9.0234 –76.7826] where $0 \le n \le 8$ assuming the 0 starting

(d) $r[n]$=[–13.1833 –3.4522 36.8723 113.6355 162.6185 161.0696 107.5056] where $1 \le n \le 7$ assuming starting from $n = 1$

(e) figure 11.5(h)

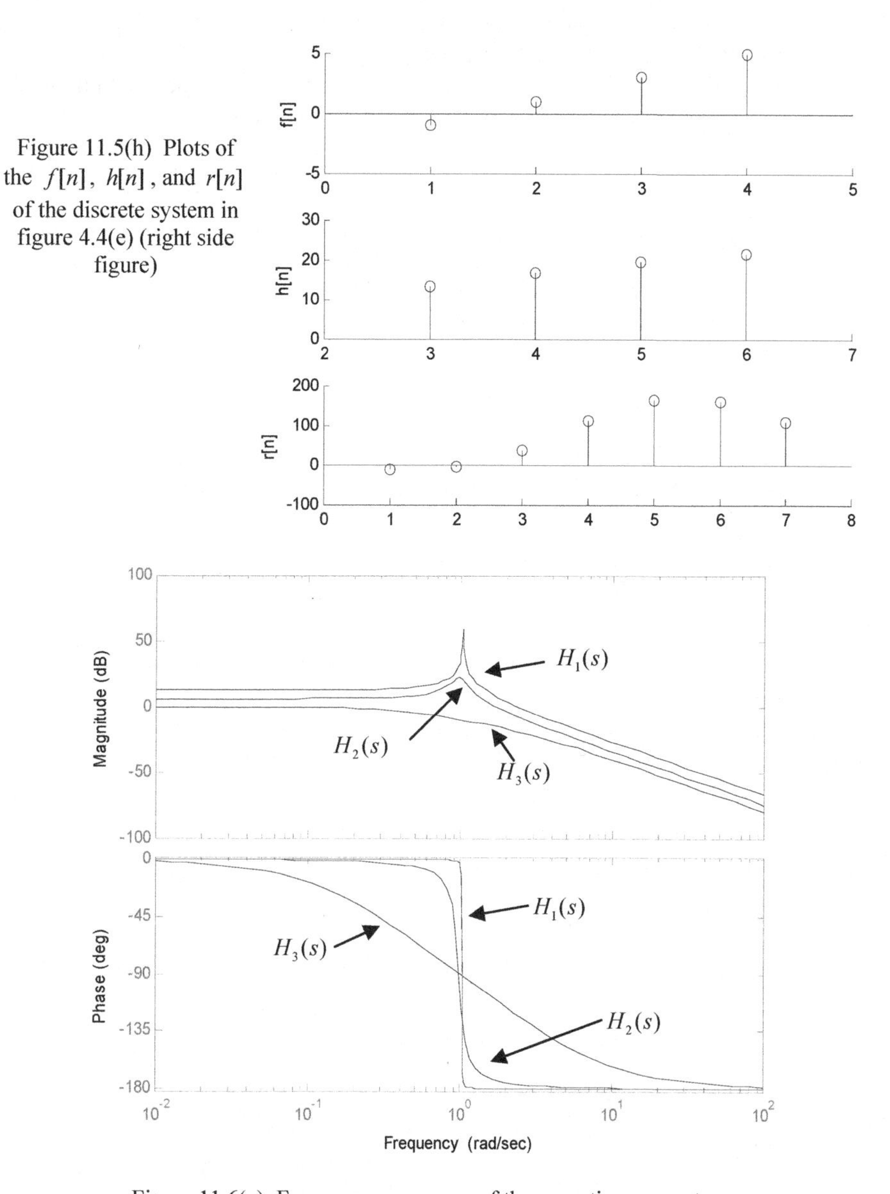

Figure 11.5(h) Plots of the $f[n]$, $h[n]$, and $r[n]$ of the discrete system in figure 4.4(e) (right side figure)

Figure 11.6(a) Frequency responses of three continuous systems

◆◆ Project 36 (on frequency response of a continuous system)

(a) Determine the $H(j2)$ value for the system $H(s)=\dfrac{9}{2s^2+0.01s+2.1}$ by using the freqresp

(b) Determine the $H(-j0.1)$, $H(j0.1)$, $H(j0.5)$, and $H(j1)$ values for the system $H(s)=\dfrac{9}{2s^2+0.01s+2.1}$ by using the freqresp

(c) in part a in magnitude and phase angle form
(d) in part b in magnitude and phase angle form
(e) Determine the $H(j\omega)$ values of the system in part a as a column matrix over $-25 \le \omega \le 25$ $rad/\sec$ with a ω step 0.1 $rad/\sec$
(f) Verify that the three systems $H_1(s) = \frac{9}{2s^2 + 0.01s + 2.1}$, $H_2(s) = \frac{4}{2.1s^2 + 0.3s + 2}$, and $H_3(s) = \frac{3}{3s^2 + 9.5s + 3}$ have the frequency responses (magnitude-phase angle form) over common angular frequency interval $0.01 \le \omega \le 100$ $rad/\sec$ as displayed in figure 11.6(a)

Hint: section 9.3

Answers:

(a) $H(j2)$ =–1.5254– j 0.0052
(b) $H(-j0.1)$ =4.3269+ j 0.0021, $H(j0.1)$=4.3269– j 0.0021, $H(j0.5)$ =5.6249– j 0.0176, and $H(j1)$ =89.1089– j 8.9109
(c) $H(j2) = 1.5254\angle -179.81^0$
(d) $H(-j0.1) = 4.3269\angle -179.97^0$, $H(j0.1) = 4.3269\angle 179.97^0$, $H(j0.5) = 5.625\angle -0.179^0$, and $H(j1) = 89.5533\angle -5.7106^0$

◆◆ Project 37 (on Z transform system function from difference equations)

Determine the unilateral (before $n \ge 0$ the signal value is zero) Z transform system function for each of the following difference equations:

(a) The difference equation $y_{k+2} - 3y_{k+1} = 7y_k$ with initial values $y_0 = -1$ and $y_1 = -2$
(b) The difference equation $y[n+2] - 6y[n+1] + 11y[n] = 4 - n + n^2$ with initial values $y[0]$ =3 and $y[1]$ =–3
(c) The difference equation $y[n-2] - 2y[n-1] + 3y[n] = 4\sin\frac{\pi n}{2}$
(d) The system of difference equations $\begin{Bmatrix} 3y[n+1] - 2x[n-1] = 5^n & \\ & y[0] = -9 \\ 2x[n] + 7y[n] = 3u[n] & \end{Bmatrix}$

Hint: section 7.5

Answers:

(a) $Y(z) = -\frac{z(z-1)}{z^2 - 3z - 7}$
(b) $Y(z) = \frac{z(3z^4 - 30z^3 + 76z^2 - 74z + 27)}{(z-1)^3(z^2 - 6z + 11)}$
(c) $Y(z) = \frac{4z^3}{(z^2+1)(3z^2 - 2z + 1)}$
(d) $X(z) = \frac{(99z^2 - 593z + 476)z^2}{3z^4 - 18z^3 + 22z^2 - 42z + 35}$ and $Y(z) = -\frac{z(-163z^2 + 133z + 27z^3 + 15)}{3z^4 - 18z^3 + 22z^2 - 42z + 35}$

◆◆ Project 38 (on inverse Z transform from discrete system function)

Determine the unilateral inverse Z transform for each of the following discrete system functions:

(a) $F(z)=\dfrac{z}{z-1}+\dfrac{z}{z-4}$ (b) $F(z)=\dfrac{z^2}{(z-3)(z-8)}$

(c) $F(z)=\dfrac{4z^{-1}+5z^{-2}}{\left(1-\frac{1}{3}z^{-1}\right)\left(1+\frac{1}{4}z^{-1}\right)}$ (d) $F(z)=\dfrac{2z^{-1}+9z^{-2}}{(1-\frac{9}{7}z^{-1})(1-\frac{2}{5}z^{-1})(5-\frac{1}{7}z^{-1})}$

(e) $F(z)=\dfrac{1}{(1-z^{-1})^3(1-z^{-2})}$

Hint: section 7.4

Answers (obviously for $n\geq 0$):

(a) $f[n]=1+4^n$ (b) $f[n]=\dfrac{8^{n+1}-3^{n+1}}{5}$

(c) $f[n]=-60\delta[n]+\dfrac{192}{7}\left(-\dfrac{1}{4}\right)^n+\dfrac{228}{7}\left(\dfrac{1}{3}\right)^n$

(d) $f[n]=\dfrac{2835}{1364}\left(\dfrac{9}{7}\right)^n-\dfrac{2401}{403}\left(\dfrac{2}{5}\right)^n+\dfrac{2219}{572}\left(\dfrac{1}{35}\right)^n$

(e) $f[n]=\dfrac{1}{16}(-1)^n+\dfrac{15}{16}+\dfrac{17n}{12}+\dfrac{5n^2}{8}+\dfrac{n^3}{12}$

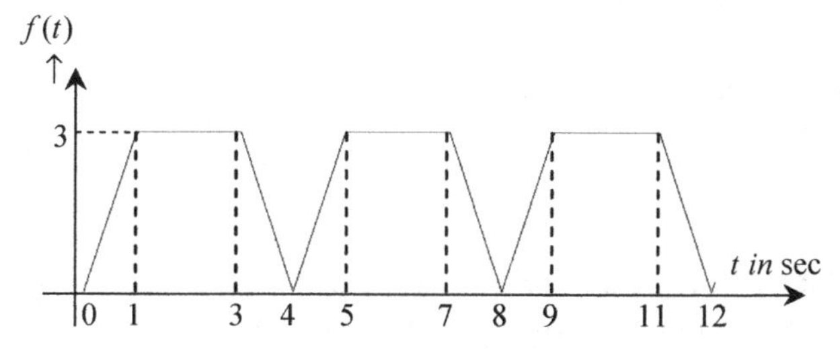

Figure 11.6(b) A trapezoidal periodic wave

♦♦ Project 39 (on Fourier series coefficients of a continuous periodic signal)

(a) Determine the real form 1 Fourier series coefficients for the continuous periodic signal $f(t)$ of the figure 11.6(b) in MATLAB

(b) Determine the real form 2 Fourier series coefficients for the continuous periodic signal $f(t)$ of the figure 11.6(b) in MATLAB

(c) Determine the complex form Fourier series coefficients for the continuous periodic signal $f(t)$ of the figure 11.6(b) in MATLAB

Hint: section 5.2

Answers: (a) $A_n=\dfrac{6\left(\cos\frac{n\pi}{2}-2+\cos\frac{3n\pi}{2}\right)}{n^2\pi^2}$ and $B_n=\dfrac{6\left(\sin\frac{n\pi}{2}+\sin\frac{3n\pi}{2}\right)}{n^2\pi^2}$

(b) $C_n=\dfrac{6\sqrt{-4\cos\frac{n\pi}{2}+2(-1)^n+6-4\cos\frac{3n\pi}{2}}}{n^2\pi^2}$ and

$$\varphi_n = \tan^{-1}\left\{\frac{\sin\frac{n\pi}{2}+\sin\frac{3n\pi}{2}}{\cos\frac{n\pi}{2}-2+\cos\frac{3n\pi}{2}}\right\}$$

(c) $C_n = 3\dfrac{(-1)^{\frac{n}{2}} - 2 + (-1)^{\frac{3n}{2}}}{n^2\pi^2}$

◆◆ Project 40 (on statistical signal samples)

(a) Simulate a uniformly distributed random voltage $v(t)$ in continuous sense over the range $3 \le v(t) \le 10V$

(b) Generate a Gaussian distributed single voltage with a mean $-5\,V$ and standard deviation $2\,V$

(c) Generate a random binary voltage whose magnitude has only two levels, $-10\,V$ and $0\,V$

(d) Generate a random binary voltage whose magnitude has only two levels, $-5\,V$ and $5\,V$

(e) Generate one hundred binary random voltages (0 or 1) as a column matrix

(f) Generate a Gaussian signal whose average power is P_{avg} =0.8Watts with a sampling period T_s=0.25secs. The signal existence should be over T =10secs starting from $t=0$ sec

Hint: sections 10.2 and 10.4

Answers: (a) unifrnd(3,10) (b) normrnd(-5,2)
(c) randsrc(1,1,[-10 0]) (d) randsrc(1,1,[-5 5])
(e) randint(100,1,[0 1])

◆◆ Project 41 (on quantization of a signal)

Apply built-in function of MATLAB to each of the following:

(a) quantize the single voltage value 6.8 V on a Δ =0.3 V

(b) quantize the single current value 22.5 mA on a Δ =0.5 mA

(c) quantize the discrete signal $f[n]$=[−9.3 7.3 0 6.8] in V on a Δ =0.6 V

(d) discretize the continuous signal $f(t)=12e^{-2t}\sin 2t$ with a sampling period T_s=0.05sec over the interval $0 \le t \le \frac{2}{\pi}$ sec and then quantize the discrete counterpart $f[n]$ on a Δ =0.5

(e) discretize the continuous signal $f(t)=3t-5$ with a sampling period T_s=0.1sec over the interval $0 \le t \le 10$ sec and then quantize the discrete counterpart $f[n]$ on a Δ =0.25

Hint: section 4.7

Answers: (a) 6.9 V (b) 22.5 mA
(c) $q[n]$=[-9.6 7.2 0 6.6] in V
(d) t=0:0.05:2/pi; f=12*exp(-2*t).*sin(2*t); Q=quant(f,0.5);
(e) t=0:0.1:10; f=3*t-5; Q=quant(f,0.25);

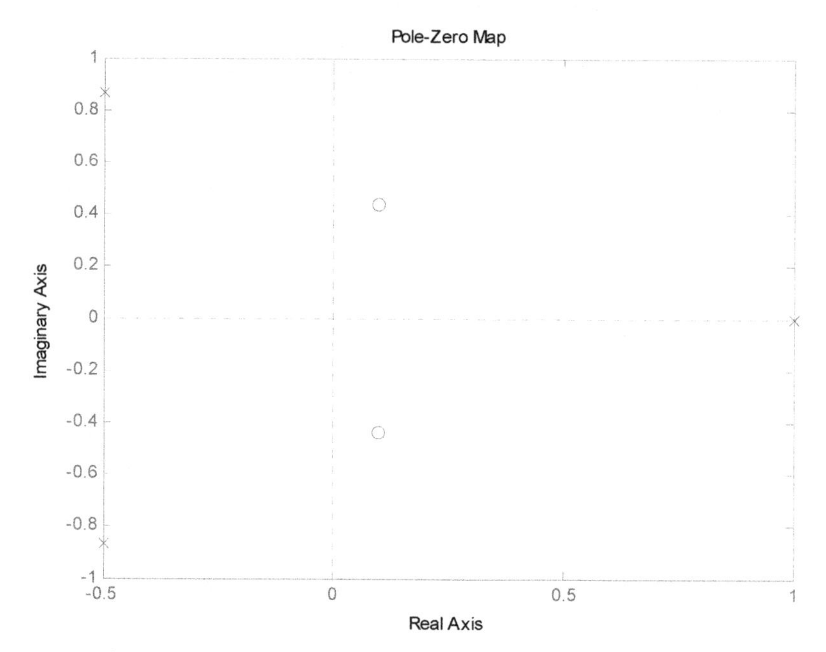

Figure 11.6(c) Pole-zero map for the system function of example a in project 42

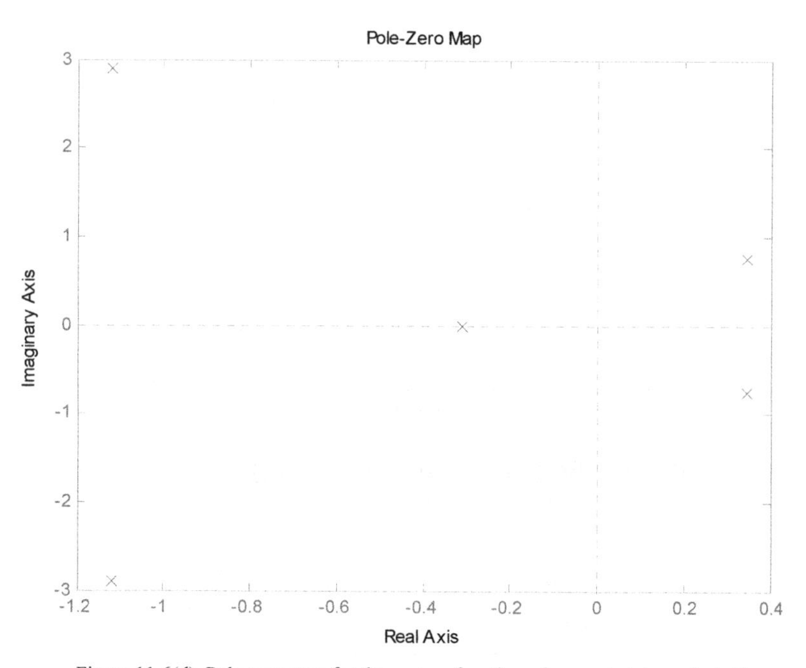

Figure 11.6(d) Pole-zero map for the system function of example b in project 42

♦♦ Project 42 (on pole-zero map of a system function)

In the following some system function is given. Verify the pole-zero map of the system stated beside it by using the MATLAB built-in function.

(a) $H(s)=\dfrac{5s^2-s+1}{s^3-1}$ has the pole-zero map of the figure 11.6(c)

(b) $H(s)=\dfrac{3}{4.3s^5+8s^4+40s^3-10s^2+22s+9}$ has the pole-zero map of the figure 11.6(d)

(c) $H(s)=\dfrac{1.3s^4+6s^3+7s^2+8s+1}{7s^7+10s^3+2.1s^2+9.9s}$ has the pole-zero map of the figure 11.6(e)

Hint: section 9.5

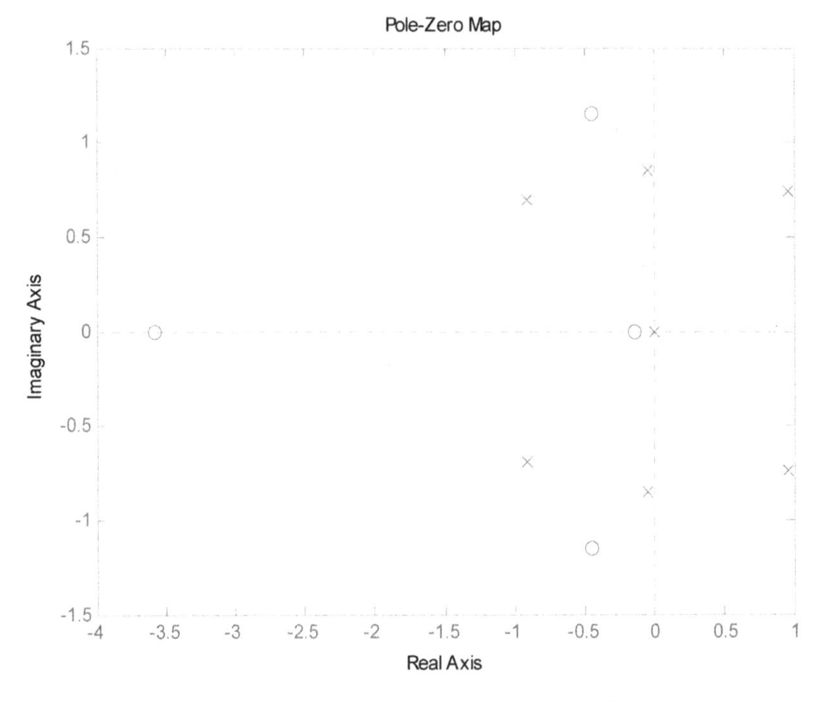

Figure 11.6(e) Pole-zero map for the system function of example c in project 42

♦♦ Project 43 (on differential equation solving)

Determine the solution to each of the following differential equations by using the **dsolve** function of MATLAB:

(a) $6\dfrac{dy}{dt}+7y=t$ subject to $y(0)=-1$

(b) $\dfrac{d^2y}{dt^2}+9y=50$ subject to $y(0)=1$ and $y'(0)=0$

(c) $\frac{d^3y}{dt^3}+\frac{d^2y}{dt^2}+\frac{dy}{dt}+y=100$ subject to $y(0)=0$, $y'(0)=0$, and $y''(0)=\pi$

(d) $t^2\frac{d^2y}{dt^2}+2t\frac{dy}{dt}+y=0$ subject to $y(1)=0$ and $y'(1)=6$

Answers:

(a) $y(t)=\frac{t}{7}-\frac{6}{49}-\frac{43}{49}e^{-\frac{7}{6}t}$ (b) $y(t)=\frac{50}{9}-\frac{41}{9}\cos 3t$

(c) $y(t)=100+\left(\frac{\pi}{2}-50\right)\sin t+\left(-\frac{\pi}{2}-50\right)\cos t+\left(\frac{\pi}{2}-50\right)e^{-t}$

(d) $y(t)=\frac{4\sqrt{3}\sin\left(\frac{\sqrt{3}}{2}\ln t\right)}{\sqrt{t}}$

Hint: appendix H

♦♦ Project 44 (on system of differential equations)

Determine the solution to each of the following systems of differential equations by using the dsolve function of MATLAB:

(a) $8\frac{dy}{dt}+2x=t$ and $\frac{dx}{dt}+y=23$ subject to $y(0)=-1$ and $x(0)=-1$

(b) $5\frac{dy}{dt}+5x=t$ and $\frac{d^2x}{dt^2}+\frac{dy}{dt}=0$ subject to $y(0)=-1$, $x(0)=0$, and $x'(0)=0$

(c) $3\frac{d^2i_1}{dt^2}+2\frac{di_1}{dt}=100\cos t$ and $-2\frac{di_1}{dt}+4\frac{di_2}{dt}+3i_2=0$ subject to $i_1(0)=0$, $i_1'(0)=-1$, and $i_2(0)=0$

Answers:

(a) $x(t)=-24e^{-\frac{1}{2}t}+23e^{\frac{t}{2}}+\frac{t}{2}$ and $y(t)=-12e^{-\frac{t}{2}}-\frac{23}{2}e^{\frac{t}{2}}+\frac{45}{2}$

(b) $x(t)=\frac{1}{10}e^{-t}-\frac{1}{10}e^{t}+\frac{t}{5}$ and $y(t)=\frac{1}{10}e^{-t}+\frac{1}{10}e^{t}-\frac{6}{5}$

(c) $i_1(t)=\frac{200\sin t}{13}-\frac{300\cos t}{13}+\frac{639}{26}e^{-\frac{2t}{3}}-\frac{3}{2}$ and

$i_2(t)=-\frac{48\cos t}{13}+\frac{136\sin t}{13}-\frac{1278}{13}e^{-\frac{2t}{3}}+102e^{-\frac{3}{4}t}$

Hint: appendix H

♦♦ Project 45 (on system formation from integrodifferential equations)

Determine the system function in transform domain from each of the following integro-differential equations:

(a) $3y'+4\int_{x=0}^{x=t}y(x)dx=6u(t-3)$ subject to $y(0)=-3$

(b) $6y''-2y'-y-\int_{x=0}^{x=t}y(x)dx=2u(t)$ with $y(0)=1$ and $y'(0)=-1$

Answers:

(a) $Y(s) = -3\dfrac{3s - 2e^{-3s}}{3s^2 + 4}$ (b) $Y(s) = 2\dfrac{3s^2 - 4s + 1}{6s^3 - 2s^2 - s - 1}$

Hint: section 6.6

◆◆ Project 46 (on system formation from a system of differential equations)

Determine the system functions from each of the following systems of differential equations:

(a) $\begin{Bmatrix} x' = -5x + 2y, \ x(0) = 0 \\ y' = 2x - 3y, \ y(0) = -3 \end{Bmatrix}$ (b) $\begin{Bmatrix} x' = 2x - 3y + \cos t, \ x(0) = 2 \\ y' = 9t - 3x - 5y, \ y(0) = 1 \end{Bmatrix}$

(c) $2\dfrac{dx}{dt} - 7\dfrac{dz}{dt} = 0$, $2\dfrac{d^2z}{dt^2} + 5\dfrac{dy}{dt} = 1 - t$, $\dfrac{dy}{dt} = e^{-t}$, $x(0) = 2$, $y(0) = 1$, $z'(0) = 5$, and $z(0) = -3$

(d) $8\dfrac{dy}{dt} - 6\int_{p=0}^{p=t} x(p)dp = 2e^{-3t}$, $3\dfrac{dy}{dt} - \dfrac{dx}{dt} = \sin t$, $x(0) = -2$, and $y(0) = 2$

Answers:

(a) $X(s) = \dfrac{-6}{s^2 + 8s + 11}$ and $Y(s) = -\dfrac{3(s+5)}{s^2 + 8s + 11}$

(b) $X(s) = \dfrac{2s^5 + 8s^4 + 7s^3 - 20s^2 - 27}{s^2(s^4 + 3s^3 - 18s^2 + 3s - 19)}$ and $Y(s) = \dfrac{-8s^4 + 7s^3 - 26s^2 + 9s - 18}{s^2(s^4 + 3s^3 - 18s^2 + 3s - 19)}$

(c) $X(s) = \dfrac{8s^4 + 78s^3 + 42s^2 - 7}{4s^4(s+1)}$, $Y(s) = \dfrac{s+2}{s(s+1)}$, and $Z(s) = -\dfrac{6s^4 - 4s^3 - 6s^2 + 1}{2s^4(s+1)}$

(d) $X(s) = -\dfrac{(8s^3 + 21s^2 + 12s + 33)s}{4s^5 + 12s^4 - 5s^3 - 15s^2 - 9s - 27}$ and $Y(s) = \dfrac{8s^5 + 25s^4 - 16s^3 - 47s^2 - 27s - 81}{s(4s^5 + 12s^4 - 5s^3 - 15s^2 - 9s - 27)}$

Hint: section 6.7

◆◆ Project 47 (on Fourier series coefficients)

For the trapezoidal periodic signal of the figure 11.6(b), determine the following:

(a) real form 1 Fourier series coefficients in symbolic and in decimal forms for the fundamental frequency
(b) real form 1 Fourier series coefficients in symbolic and in decimal forms for the 2nd, 3rd, and 4th harmonics
(c) real form 2 Fourier series coefficients in symbolic and in decimal forms for the fundamental frequency
(d) real form 2 Fourier series coefficients in symbolic and in decimal forms for the 2nd, 3rd, and 4th harmonics
(e) complex Fourier series coefficients in symbolic and in decimal forms for the fundamental frequency

(f) complex Fourier series coefficients in symbolic and in decimal forms for the 2nd, 3rd, and 4th harmonics

(g) graph the real form 2 Fourier series coefficients (magnitude and phase angle in degree) for the first 10 harmonics

(h) average or DC value of the signal by using the real form 1 coefficient

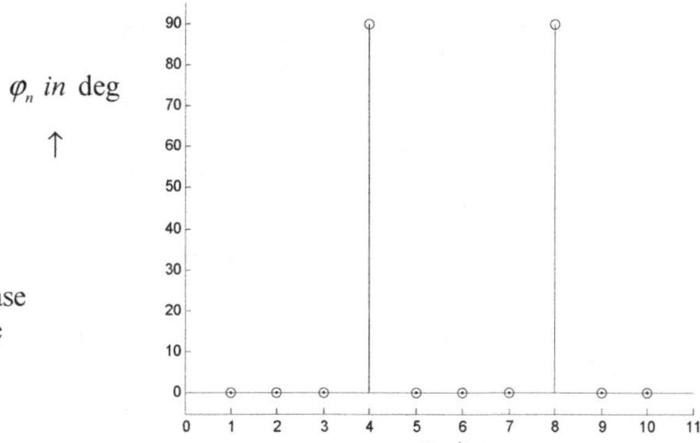

Figure 11.6(f) Fourier series phase spectra (right side figure)

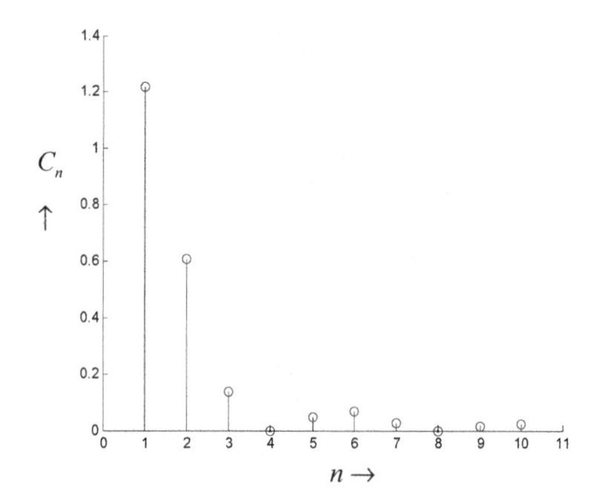

Figure 11.6(g) Fourier series magnitude spectra (right side figure)

Answers:

(a) $A_1 = -\frac{12}{\pi^2}$=–1.2159 and B_1=0

(b) $[A_2 \quad A_3 \quad A_4]=[-\frac{6}{\pi^2} \quad -\frac{4}{3\pi^2} \quad 0]$= [–0.6079 –0.1351 0] and B_2, B_3, and B_4=0

(c) $C_1 = \frac{12}{\pi^2}$=1.2159 and $\varphi_1 = 0$

(d) $[C_2 \quad C_3 \quad C_4]=[\frac{6}{\pi^2} \quad \frac{4}{3\pi^2} \quad 0]$=[0.6079 0.1351 0] and φ_2, φ_3, and $\varphi_4 = \frac{\pi}{2}$ (0/0 appears in the 4th harmonic so the limiting value is taken)

(e) $C_1 = -\frac{6}{\pi^2}$=–0.6079

(f) $[C_2 \quad C_3 \quad C_4]$=$[-\frac{3}{\pi^2} \quad -\frac{2}{3\pi^2} \quad 0]$=[−0.3040 −0.0675 0]

(g) figure 11.6(g) for magnitude and figure 11.6(f) for phase

(h) $A_0/2=\frac{9}{4}$=2.25 (take limit by the function limit at 0)

Hint: section 5.2

♦♦ Project 48 (on signal samples and graphing)

Generate the samples for each of the following continuous signals:

(a) $f(t)=5te^{-4|t|}[u(t-5)-u(t+5)]$ over the interval $-8 \le t \le 8$ sec with the step 0.1sec

(b) $f(t)$=$2\cos 3t \; rect(3t-5)$ over the interval $0 \le t \le 8$ sec with step 0.01sec

(c) $f(t)$=$2\,u(t+5)+5\cos t-9\,rect(t+3)$ over $-5 \le t \le 10$ sec with the step 0.1sec

Graph each of the following continuous signals by first generating the signal samples and then using the built-in command plot:

(a) $f(t)=3te^{-2|t|}[u(t-3)-u(t+3)]$ over the interval $-5 \le t \le 5$ sec with the step 0.01sec

(b) $f(t)$=$2\cos 3t \; rect(3t-5)$ over the interval $0 \le t \le 8$ sec with the step 0.01 sec

(c) $f(t)$=$2\,u(t+5)+5\,tri(t+3)-9\,rect(t+3)$ over $-7 \le t \le 2$ sec with the step 0.01sec

Graph each of the following discrete signals by first generating the signal samples and then using the built-in command stem or scatter:

(a) $f[n]=ne^{-3n}$ over the interval $0 \le n \le 8$

(b) $f[n]=5^{-n}-3^{-n}$ over the interval $0 \le n \le 5$

(c) $f[n]=\frac{5^{-|3n|}}{1+|n|}$ over the interval $-3 \le n \le 5$

Hint: sections 2.3, 2.8, and 10.1 and appendix F

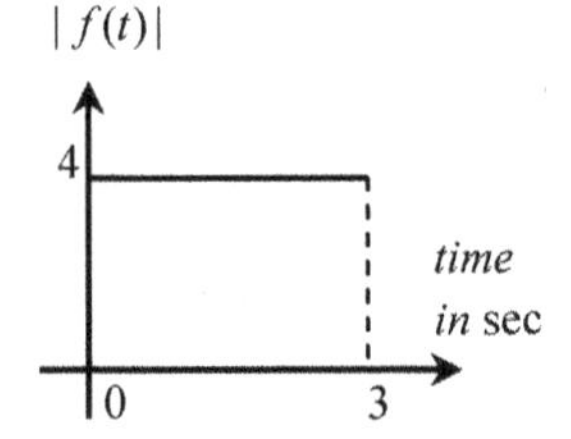

Figure 11.7(a) Magnitude part of a complex signal $f(t)$

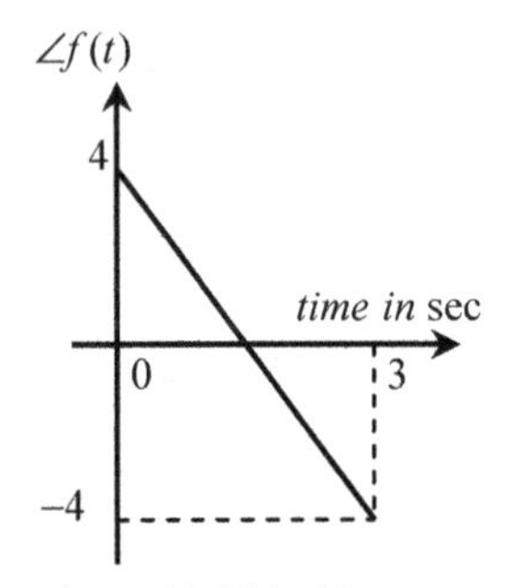

Figure 11.7(b) Phase angle part of a complex signal $f(t)$

♦♦ Project 49 (on complex signal samples)

Generate the samples for each of the following complex signals:

(a) $f(t)=100e^{-j4t}$ over the interval $-1 \le t \le 1$ sec with the step 0.1sec

(b) $f(t)=(1+t)+jt^2$ over the interval $-1 \le t \le 1$ sec with the step 0.1sec

(c) A linear exponential chirp signal frequency changes from 3 KHz to 7 KHz within 10 $m$sec, choose a step size 0.01 $m$sec

(d) A quadratic exponential chirp signal

frequency changes from 3.3 KHz to 7.2 KHz within 9 $m$sec, choose a step size 0.01 $m$sec

(e) The complex $f(t)$ which has the magnitude and radian phase angle variations as shown in the figures 11.7(a) and 11.7(b) respectively, choose a step size 0.01 sec

(f) The frequency domain expression $F(\omega)=\frac{3+j\omega}{5-j\omega}$ over the interval $-10\le\omega\le10$ $rad/\sec$ with the ω step 0.1 $rad/\sec$

Hint: section 10.3

♦♦ Project 50 (on input-output on Z transform system)

(a) The discrete system of the figure 10.4(a) is characterized by $H(z)=\frac{3-2z^{-1}}{3+2z^{-1}-z^{-2}}$. Determine the output $y[n]$ when the input is $f[n]$=[2 2 2 2]

(b) In part (a) now $H(z)=\frac{3z+2z^2}{2+3z^4}$

(c) In part (a) now $f[n]=3^{-n}$ over $0\le n\le5$

(d) In part (a) now $f[n]=e^{-n}\sin n$ over $0\le n\le7$

Hint: section 10.7

Answers:

(a) $y[n]$=[2 −0.6667 1.7778 −0.7407]

(b) $y[n]$=[0 0 1.3333 3.3333]

(c) $y[n]$=[1 −1 0.8889 −0.9630 0.9259 −0.9424]

(d) $y[n]$=[0 0.3096 −0.2897 0.2213 −0.2626 0.2516 −0.2517 0.2527]

♦♦ Project 51 (on pole-zero map of a continuous system)

Graph the pole-zero map for each of the following system functions:

A. $Y(s)=\frac{s-5}{(s^2-3s)^2+4s-12}$

B. $Y(s)=\frac{s^2+5}{(s^2+4)(s^2-36)}$

C. $Y(s)=\frac{3s^2+65}{[(s+3)^2+4][(s-4)^2+36]}$

Also in each pole-zero map, include the constant damping ratio line and constant natural frequency curve.

Hint: section 9.5

♦♦ Project 52 (on residue and steady state value of a system)

Compute the residue in symbolic as well as in decimal form for each of the following system functions:

A. $Y(s)=\frac{s-5}{s^2(s+2)}$

B. $Y(s)=\frac{s+5}{(s^2+4)(s-6)}$

C. $Y(s)=\frac{5s}{(4s+3)(7s+6)}$

Answers:

(a) Symbolic: $-\frac{7}{4}$ at s=−2 and $\frac{7}{4}$ and $-\frac{5}{2}$ at s=0 and Numeric: −1.75, 1.75, and −2.5 respectively

(b) Symbolic: $\frac{11}{40}$, $-\frac{11}{80}-j\frac{13}{80}$, and $-\frac{11}{80}+j\frac{13}{80}$ at $s=6$, $s=-2j$, and $s=2j$ and Numeric: 0.275, $-0.1375-j\,0.1625$, and $-0.1375+j\,0.1625$ respectively

(c) Symbolic: $-\frac{5}{4}$ and $\frac{10}{7}$ at $s=-\frac{3}{4}$ and $s=-\frac{6}{7}$ and Numeric: -1.25 and 1.4286 respectively

Compute the steady state value for each of the following systems:

A. $Y(s)=\dfrac{s^2-5s}{s^2(s+2)}$ B. $Y(s)=\dfrac{s+5}{(s^2+4)(s-6)}$

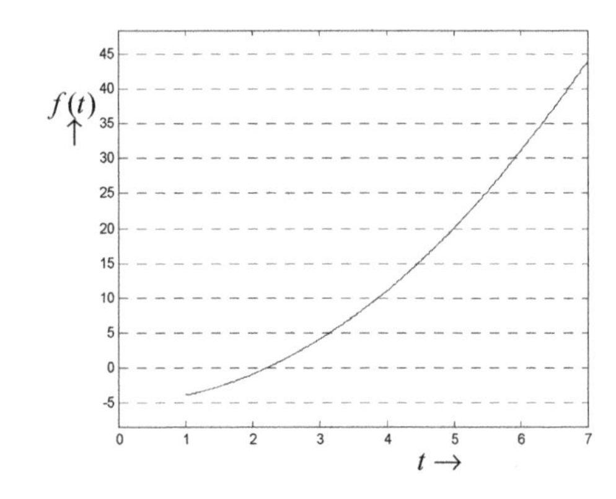

Figure 11.7(c) Plot of the signal $f(t)=t^2-5$ over $1\le t\le 7$ sec (right side figure)

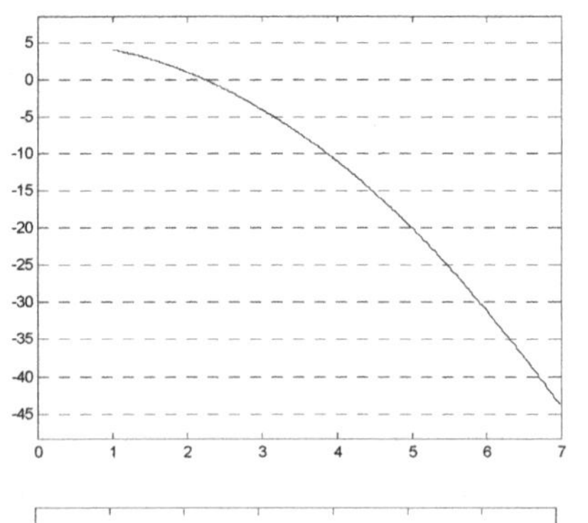

Figure 11.7(d) Flipping the signal of the figure 11.7(c) about the t axis (right side figure)

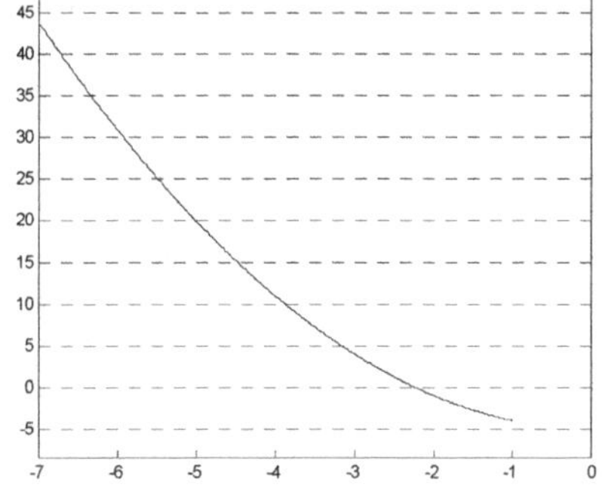

Figure 11.7(e) Flipping the signal of the figure 11.7(c) about the $f(t)$ axis (right side figure)

C. $Y(s) = \dfrac{s-5}{s^2(s+2)}$

Answers:

(a) $-\dfrac{5}{2}$ (b) 0

(c) does not exist (indicated by not a number or NaN)

Hint: section 10.8

Figure 11.7(f) Plot of the discrete signal $f[n] = n^2 - 5$ over $1 \le n \le 7$ sec (right side figure)

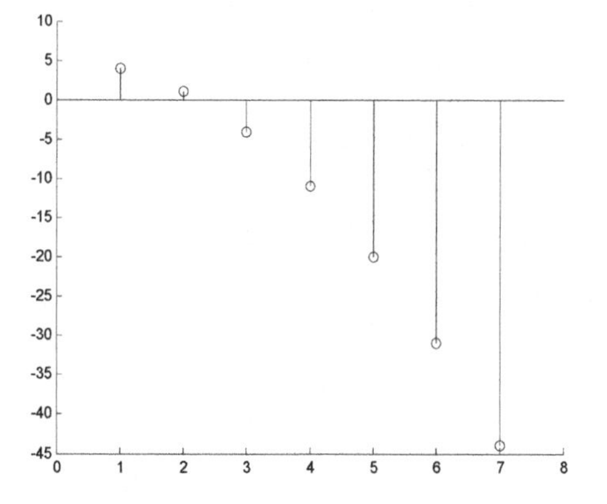

Figure 11.7(g) Flipping the discrete signal of the figure 11.7(f) about the n axis (right side figure)

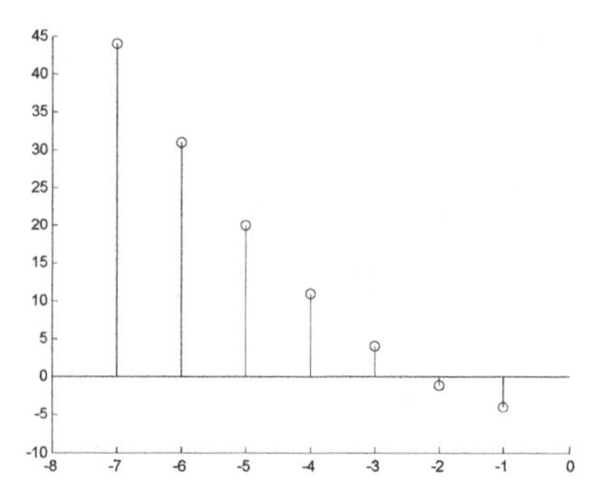

Figure 11.7(h) Flipping the signal of the figure 11.7(f) about the $f[n]$ axis (right side figure)

⬥⬥ Project 53 (on signal flipping)

(a) Figure 11.7(c) shows the plot of the parabolic signal $f(t)=t^2-5$ over $1 \le t \le 7$ sec. Verify that the figures 11.7(d) and 11.7(e) are the plots of the signal when the signal is flipped about the t and $f(t)$ axes respectively

(b) Figure 11.7(f) shows the plot of the discrete signal $f[n]=n^2-5$ over $1 \le n \le 7$ sec. Verify that the figures 11.7(g) and 11.7(h) are the plots of the signal when the signal is flipped about the n and $f[n]$ axes respectively

Hint: section 4.3

⬥⬥ Project 54 (on periodic signal sample generation)

Generate the continuous signal samples for each of the following waves and verify the wave shape after the sample generation:

(a) figure 11.2(g) shown triangular periodic wave with some off interval over $0 \le t \le 8m\sec$ (with the step 0.05 $m\sec$)

(b) figure 11.2(h) shown full rectified sine wave over $0 \le t \le \frac{1}{40}\sec$ (with the step 0.0001sec)

(c) figure 11.6(b) shown trapezoidal periodic wave over $0 \le t \le 12\sec$ (with the step 0.02sec)

(d) the rectangular wave with amplitude $\pm 100\,V$, frequency 60 Hz, and duty cycle 30% over $0 \le t \le \frac{1}{20}\sec$ (with the step 0.0001sec)

(e) the rectangular wave with amplitude swing $-30\,V$ to $90\,V$, frequency 20 Hz, and duty cycle 70% over $0 \le t \le 0.15\sec$ (with the step 0.0001sec)

Hint: sections 2.6 and 2.9

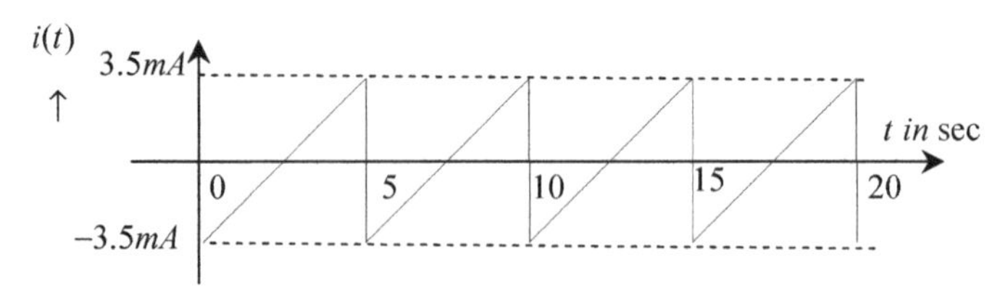

Figure 11.8(a) A sawtooth current wave

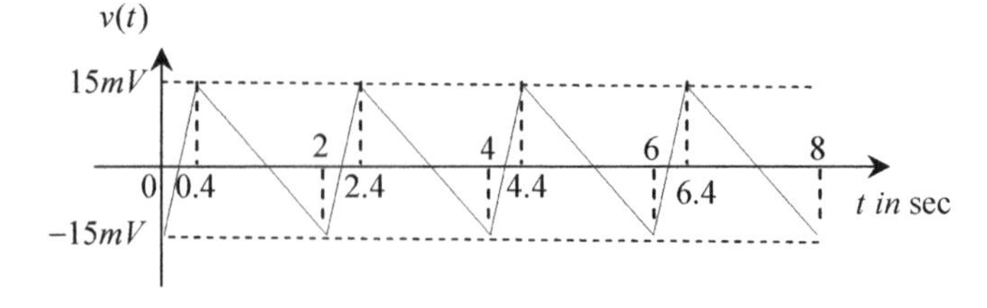

Figure 11.8(b) A triangular voltage wave

⬍⬍ Project 55 (on periodic signal sample generation)

Generate the continuous signal samples for each of the following waves and verify the wave shape after the sample generation:

(a) figure 11.8(a) shown sawtooth current wave (with step 0.05sec)

(b) figure 11.8(b) shown triangular voltage wave (with step 0.025sec)

(c) the power signal when the voltage and current expressions are $v(t)=200\times10^3\sin(2\pi ft-40^0)$ and $i(t)=70\times10^{-3}\cos(2\pi ft+28^0)$ respectively over $0\le t\le 0.08\text{sec}$ with f =25 Hz and step size 0.0001sec

Hint: sections 2.6, 2.7, and 2.9

⬍⬍ Project 56 (on periodic signal modeling)

Apply SIMULINK modeling to generate the periodic wave shapes of the projects 54 and 55 and verify each wave shape by sending it to the Scope.

Hint: sections 3.4, 3.5, and 3.6

⬍⬍ Project 57 (on sine wave derived signal modeling)

Apply SIMULINK modeling to generate each of the following sine derived signals and verify the wave shape by sending it to the Scope:

(a) a sine wave of frequency $3KHz$ and amplitude $\pm3\,V$ with wave existence $1\,m\text{sec}$

(b) a half rectified sine wave of frequency $3KHz$ and amplitude $\pm3\,V$ with wave existence $1\,m\text{sec}$

(c) two sinusoidal frequency wave $y(t)=30\sin 2\pi ft+2\sin 4\pi ft$ over $0\le t\le 1\mu\text{sec}$ where $f=2MHz$

(d) three sinusoidal frequency wave $y(t)=30\sin 2\pi ft+2\sin 4\pi ft-0.1\sin 5\pi ft$ over $0\le t\le 1\mu\text{sec}$ where $f=2MHz$

(e) two sinusoidal frequency with phase $y(t)=90\sin(2\pi ft+50^0)+7\sin(4\pi ft-30^0)$ over $0\le t\le 1\mu\text{sec}$ where $f=2MHz$

(f) the damped sine wave $y(t)=27e^{-0.3t}\sin(2\pi ft+20^0)$ over $0\le t\le 2\text{sec}$ where $f=2Hz$

(g) in example (c) the wave is clipped at ±20

(h) in example (a) the wave has a bias $0.5\,V$

Hint: section 3.4

⬍⬍ Project 58 (on quantization of a signal)

Suppose a continuous signal $f(t)=3t^2-7t+8$ exists over $0\le t\le 5$ secs. Digitize the signal by taking a sampling period 0.01sec and obtain it as a row matrix. Then the digitized signal is to be quantized in 16 levels. Obtain the signal as a row matrix after quantization. Graph the two digitized signals by using the plot of appendix F.

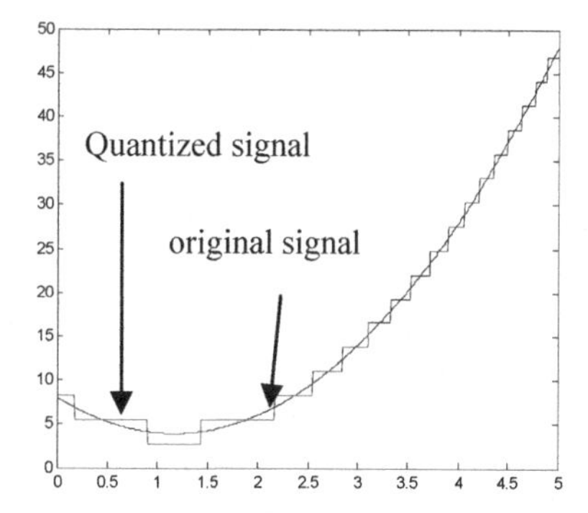

Figure 11.8(c) Plot of the original and quantized signal together

Answer: The final output should be like the figure 11.8(c)

Hint: section 4.7 and use the max and min of appendix C.4

♦♦ Project 59 (on quantization of a signal)

Suppose a continuous signal $f(t)=e^{-0.2t}\sin t$ exists over $0 \le t \le 2$ secs. Digitize the signal by taking a sampling frequency $100\ Hz$ and obtain it as a row matrix. Then the digitized signal is to be quantized by a 4-bit quantizer. Obtain the signal as a row matrix after quantization. From these two discrete signals, compute first the quantization error and then the signal to quantization noise power ratio.

Answer: SNR=33.301 dB

Hint: section 4.7

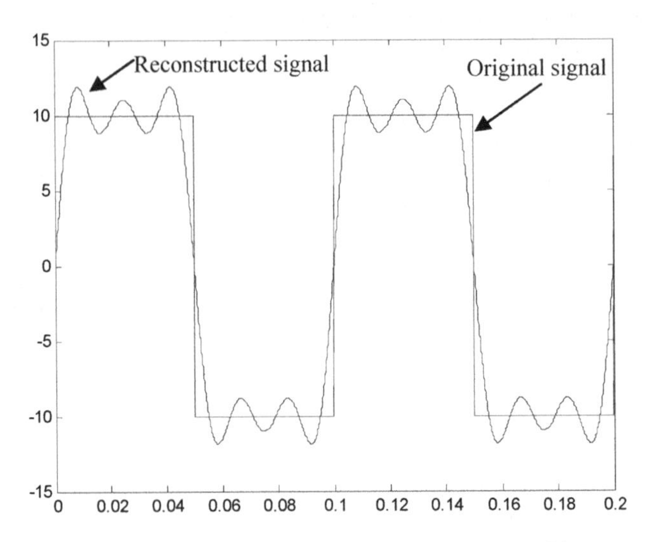

Figure 11.8(d) Plot of the original square wave and its Fourier reconstructed version together

♦♦ Project 60 (on reconstruction from Fourier series coefficients)

Consider a perfect square wave which has the amplitude swing ±10, frequency $10\ Hz$, and existence over $0 \le t \le 0.2$ secs. Verify that the square wave has the real form 1 Fourier series coefficients $A_n = \dfrac{-20\sin n\pi(\cos n\pi - 1)}{n\pi}$ and $B_n = \dfrac{20\cos n\pi(\cos n\pi - 1)}{n\pi}$ with 0 DC value. Based on the found coefficients, reconstruct the square wave for the first five harmonics and plot it with the original one by choosing a step 0.0001sec for both.

Answer: The final output should be like the figure 11.8(d)

Hint: section 2.6 and subsection 5.2.4

♦♦ Project 61 (on inverse discrete Fourier transform)

Consider the discrete signal $f[n]$=[0 1 3 9 −9]. Determine the forward discrete Fourier transform $F[k]$ of the signal. Verify that you are able to get the $f[n]$ from the found $F[k]$.

Do the same for the signal $f[n]=\sin 0.2n$ over $0 \le n \le 5$.

Again do the same for the signal $f[n]=n^3-3n+4$ over $0 \le n \le 7$.
Hint: section 5.4

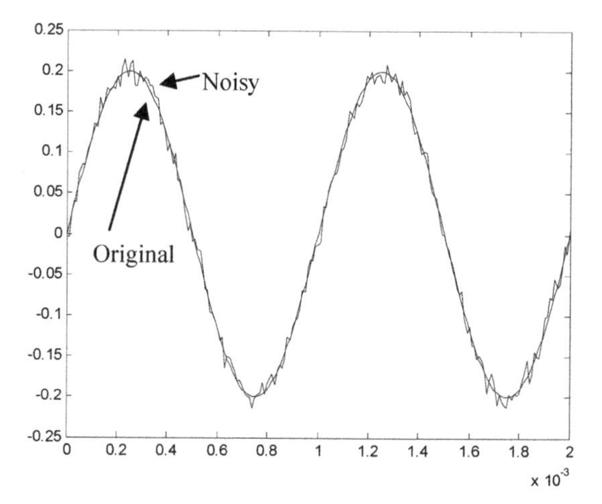

Figure 11.8(e) AWGN is added to the two cycle sine wave of the figure 2.3(a) – right side figure

♦♦ Project 62 (on noisy signal sample generation)

In communication problems we often generate the samples of a continuous signal by applying the technique of the chapter 2. After generating the samples, we intentionally add some noise specially additive white Gaussian (AWGN) to the signal for effect-on-noise analysis. Figure 11.8(f) demonstrates such noisy signal scheme.

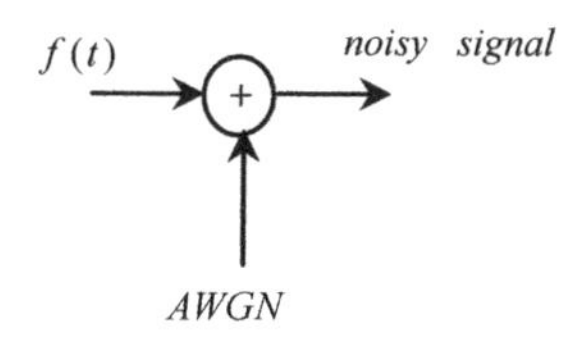

Figure 11.8(f) A continuous time signal is corrupted by adding AWGN

(a) Considering the two cycle sine wave signal of the figure 2.3(a), we wish to add an AWGN with 0 mean and 0.01 standard deviation. After adding the noise, the original and noisy signals are to be graphed together.
(b) In part (a) now add the AWGN with SNR 40.
(c) In part (a) now add the AWGN with SNR 20 dB.
(d) The figure 11.8(f) is for the additive noise. Obtain the noisy signal in part (a) for the multiplicative noise.

Hints:

(a) The final output should be very similar to the figure 11.8(e). The workspace v as regards to the example 1 of section 2.6 holds the samples of the $f(t)$ in figure 11.8(f). By the technique of the section 10.4, generate the AWGN samples for the same t as in v by the normrnd and keep the samples in some variable s. Then the noisy signal is simply s+v. For the graph, use plot(t,v,t,s+v) of the appendix F.

(b) Compute the power of the samples in v by using $P_{avg}=\frac{\sum_{n=0}^{N-1} f^2[n]T_s^2}{T}$ of section 10.4 then the noise power will be $\frac{P_{avg}}{40}$ where T =2 $m$sec and T_s= 0.01 $m$sec.

(c) The SNR in absolute scale will be $10^{\frac{dB}{10}}$ i.e. 100 – exactly like the part b but the ration 40 is replaced by 100.
(d) In part (a) the multiplicative noise is obtained by the command `v.*s` as regards to appendix B.

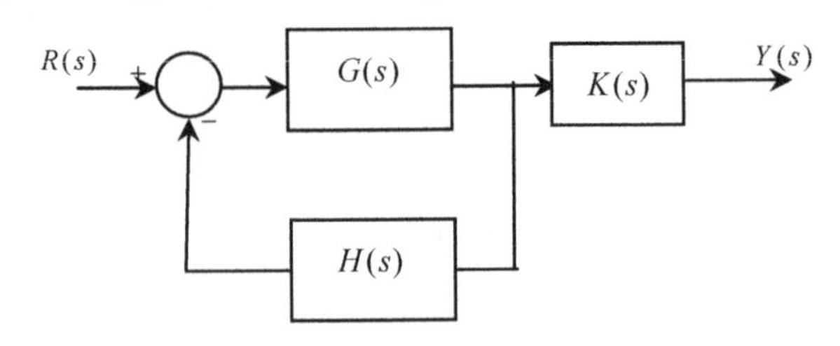

Figure 11.8(g) A negative feedback system in series with another system

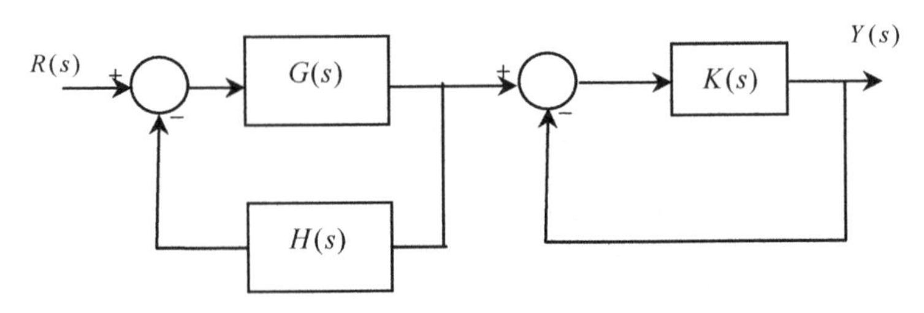

Figure 11.8(h) An interconnected system

◆◆ Project 63 (on inverse Fourier transform)

Apply the MATLAB built-in function `ifourier` to determine the inverse Fourier transform for each of the following frequency functions:

(a) $F(\omega)=\dfrac{-54}{1-j\omega-2\omega^2}$ (b) $F(\omega)=\dfrac{7e^{-j(3\omega+6)}}{2+(3+\omega)j}$

(c) $F(\omega)=e^{-j\omega}$ (d) $F(\omega)=\left\{\begin{array}{ll}2e^{-j\omega} & for\ \omega>0\\ 2e^{j\omega} & for\ \omega<0\end{array}\right\}$

(e) $F(\omega)=e^{-\frac{\omega^2}{9}}$ (f) $F(\omega)=e^{-j\omega}\dfrac{2\sin(6\omega-4)}{3\omega}$

Hint: subsection 5.3.2

Answers: (a) $f(t)=\dfrac{54\sin\frac{t\sqrt{7}}{4}e^{\frac{t}{4}}}{\sqrt{7}}[1-2u(t)]$ (b) $f(t)=7u(t-3)e^{6+3j-2t-3jt}$

(c) $f(t)=\delta(t-1)$ (d) $f(t)=\delta(t+1)+\delta(t-1)+\dfrac{2j}{\pi(t^2-1)}$

(e) $f(t)=\dfrac{3e^{-\frac{9}{4}t^2}}{2\sqrt{\pi}}$

(f) $f(t)=-\dfrac{1}{3}e^{4j}u(t-7)+j\dfrac{\sin 4}{3}+\dfrac{1}{3}e^{-4j}u(t+5)$

♦♦ Project 64 (on system modeling from block diagram)

By employing appropriate block and without input-output, construct a SIMULINK model to implement each of the following block diagrams:

(a) negative feedback system of the figure 8.2(e) with $G(s)=\dfrac{-7s}{s^2-2s-3}$ and $H(s)=\dfrac{3}{s}$

(b) positive feedback system of the figure 8.2(f) with $G(s)=\dfrac{-7(s+5)}{s(s+3)(s+2)}$ and $H(s)=\{A,B,C,D\}$ where $A=\begin{bmatrix}-2 & -1\\ 2 & 3\end{bmatrix}$, $B=\begin{bmatrix}-1\\ -2\end{bmatrix}$, $C=[-2 \quad 1]$, and $D=[-1]$

(c) the negative feedback system in series with another system as in figure 11.8(g) where $G(s)=\dfrac{-7s}{s+4}$, $H(s)=-3$, and $K(s)=s$

(d) the interconnected system of the figure 11.8(h) with $G(s)=\dfrac{-7s}{s+4}$, $H(s)=-3$, and $K(s)=s$

(e) the interconnected system of the figure 11.9(a) with $G_1(s)=\dfrac{-7s}{s+4}$, $G_2(s)=\{A,B,C,D\}$ where $A=\begin{bmatrix}-2 & -1\\ 2 & 3\end{bmatrix}$, $B=\begin{bmatrix}-1\\ -2\end{bmatrix}$, $C=[-2 \quad 1]$, and $D=[-1]$, $H(s)=-3$, and $G_3(s)=s$

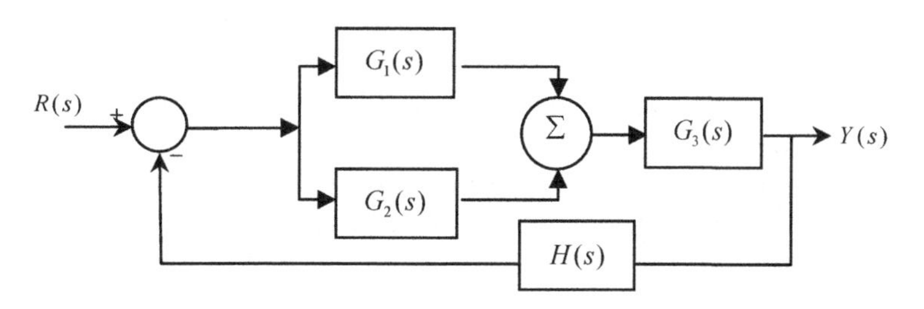

Figure 11.9(a) An interconnected system

♦♦ Project 65 (on frequency response of a system)

Verify the frequency response behavior for each of the following system functions by applying MATLAB built-in function **bode**:

(a) The system $G(s)=\dfrac{1}{s+100}$ shows a lowpass behavior

(b) The system $G(s)=\dfrac{1}{s+1000}$ increases the cutoff frequency compared to the one in part (a)

(c) The system $G(s)=\dfrac{s}{s+100}$ shows a highpass behavior

(d) The system $G(s)=\dfrac{s}{s+1000}$ increases the cutoff frequency compared to the one in part (c)

(e) The system $G(s)=\dfrac{s}{s^2+0.5s+100}$ shows a narrow band frequency response

(f) The system $G(s)=\dfrac{s}{s^2+0.5s+1000}$ increases the narrow band center frequency compared to the one in part (e)

Hint: section 9.4

♦♦ Project 66 (on modeling the step and the impulse responses of a system)

Apply SIMULINK modeling to verify the response in the Scope for each of the following systems:

(a) Step response for the project 26(a) mentioned system
(b) Step response for the project 26(b) mentioned system
(c) Impulse response for the project 28(a) mentioned system
(d) Impulse response for the project 28(b) mentioned system

Answers:

(a) figure 11.4(g) (b) figure 11.4(h)
(c) figure 11.5(a) (d) figure 11.5(b)

Hint: section 9.6

♦♦ Project 67 (on input-output on Laplace system)

(a) The Laplace system of the figure 10.3(e) has $H(s)=\dfrac{5}{2s+3}$. When an input $f(t)=3rect(t-0.5)$ is applied to the system in time domain, what should be the output $y(t)$ in s domain as well as in t domain?

(b) In part (a) now $f(t)=tri(t)$

(c) In circuit 10.3(b) an input $v_i=3rect(t-0.5)$ is applied with R=3Ω and L=2 H, what should be the output v_0 in s domain as well as in t domain?

(d) In circuit 10.3(b) an input $v_i=tri(t)$ is applied with R=3Ω and L= 2 H, what should be the output v_0 in s domain as well as in t domain?

Answers:

(a) $Y(s)=5\dfrac{3-3e^{-s}}{s(3+2s)}$ and $y(t)=10e^{-\frac{3t}{4}}\sinh\dfrac{3t}{4}-10u(t-1)e^{-\frac{3t}{4}+\frac{3}{4}}\sinh\left(\dfrac{3t}{4}-\dfrac{3}{4}\right)$

(b) $Y(s)=5\dfrac{s-1+e^{-s}}{s^2(3+2s)}$ and $y(t)=\dfrac{50}{9}e^{-\frac{3t}{4}}\sinh\dfrac{3t}{4}-\dfrac{5}{3}t+$

$$5u(t-1)\left\{\frac{t}{3}-\frac{1}{3}-\frac{4e^{-\frac{3}{4}t+\frac{3}{4}}}{9}\sinh\left(\frac{3t}{4}-\frac{3}{4}\right)\right\}$$

(c) both outputs of the part (a) just need a multiplication of $\dfrac{3}{5}$

(d) both outputs of the part (b) just need a multiplication of $\dfrac{3}{5}$

Hint: section 10.6

♦♦ Project 68 (on input-output on Fourier system)

(a) The Fourier system of the figure 10.3(a) has $H(\omega)=\dfrac{5}{2j\omega+3}$. When an input $f(t)=3rect(t-0.5)$ is applied to the system in time domain, what should be the output $y(t)$ in frequency domain as well as in t domain?

(b) In part (a) now $f(t)=tri(t)$

(c) In circuit 10.3(b) an input $v_i=3rect(t-0.5)$ is applied with $R=3\Omega$ and $L=2\,H$, what should be the output v_0 in frequency domain as well as in t domain?

(d) In circuit 10.3(b) an input $v_i=tri(t)$ is applied with $R=3\Omega$ and $L=2\,H$, what should be the output v_0 in frequency domain as well as in t domain?

Answers:

(a) $Y(\omega)=\dfrac{15j(e^{-j\omega}-1)}{\omega(3+j2\omega)}$ and $y(t)=5(1-e^{-\frac{3t}{2}})u(t)+5\left(-1+e^{-\frac{3t}{2}+\frac{3}{2}}\right)u(t-1)$

(b) $Y(\omega)=\dfrac{20\sin^2\frac{\omega}{2}}{\omega^2(3+j2\omega)}$ and $y(t)=$

$$\frac{5}{9}(3t+1+2e^{-\frac{3}{2}t-\frac{3}{2}})u(t+1)+\frac{5}{9}(2e^{-\frac{3}{2}t+\frac{3}{2}}+3t-5)u(t-1)+\frac{5}{9}(4-6t-4e^{-\frac{3}{2}t})u(t)$$

(c) both outputs of the part (a) just need a multiplication of $\frac{3}{5}$

(d) both outputs of the part (b) just need a multiplication of $\frac{3}{5}$

Hint: section 10.5

♦♦ Project 69 (on input-output on Z transform system)

(a) The Z transform system of the figure 10.4(a) has $H(z)=\dfrac{3-2z^{-1}}{3+2z^{-1}-z^{-2}}$. When an input $f[n]=2^{-n}$ is applied to the system in index domain, what should be the output $y[n]$ in z domain as well as in index domain?

(b) Graph the output response $y[n]$ versus n over $0\le n\le 7$ of the part (a)

Answers:

(a) $Y(z)=\dfrac{6z^3-4z^2}{6z^3+z^2-4z+1}$ and $y[n]=\dfrac{3^{-n}}{2}+\dfrac{5(-)^n}{6}-\dfrac{2^{-n}}{3}$

(b) suppose y holds the code of $y[n]$, then use the command stem(0:7, subs(y,0:7)) of appendix F, and the graph is not shown for the space reason

Hint: section 10.7

♦♦ Project 70 (on developing a circuit system)

(a) Referring to the circuit in figure 10.3(b), write down the time domain equations of the circuit

(b) Write down the s domain equations of the circuit in part (a)

(c) Develop the circuit system as a block diagram from input to output

(d) Model the system by using $R=3\Omega$ and $L=2\,H$ and by applying appropriate blocks of SIMULINK

Answers:

(a) $v_i = L\frac{di_0}{dt} + v_0$ and $v_0 = i_0 R$ where i_0 is the current flowing in the circuit of the figure 10.3(b) from top to down through the R

(b) $V_i = sLI_0 + V_0$ and $V_0 = I_0 R$

(c) figure 11.9(b)

(d) blocks needed: one Sum, one Transfer Fcn, and one Gain

Hint: sections 8.6 and 8.8

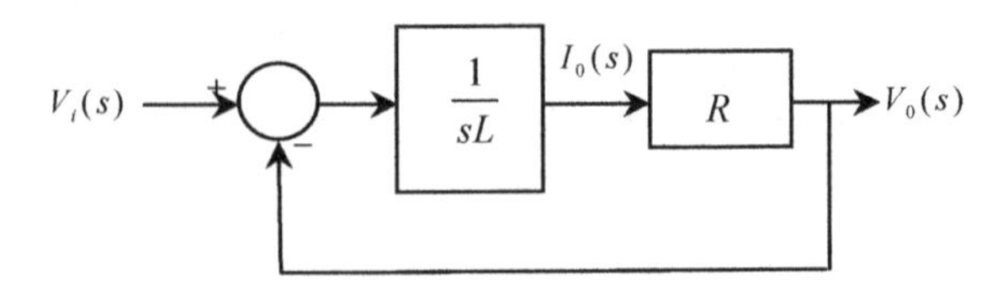

Figure 11.9(b) Modeling the RL circuit of figure 10.3(b)

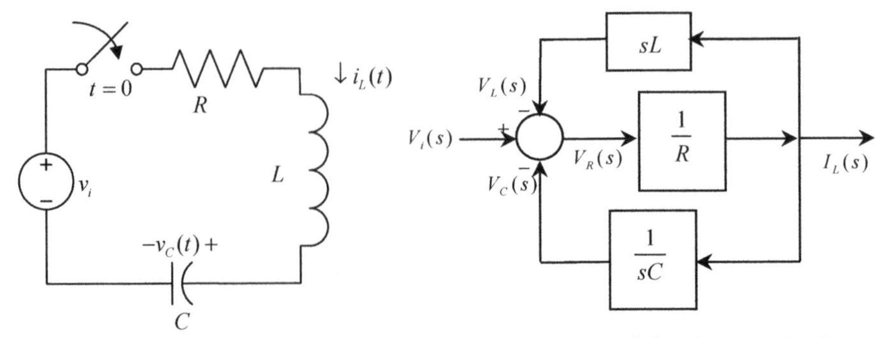

Figure 11.9(c) A second order circuit system

Figure 11.9(d) Modeling the second order RLC circuit of the figure 11.9(c)

♦♦ Project 71 (on developing a circuit system)

(a) Referring to the second order RLC circuit in figure 11.9(c), write down the time domain equations of the circuit

(b) Write down the s domain equations of the circuit in part (a)

(c) Develop the circuit system as a block diagram from input to output

(d) Model the system by using R=3Ω, L=2 H, and C=1 F and by applying appropriate blocks of SIMULINK

Answers:

(a) $v_i = v_R + v_L + v_C$, $v_R = i_L R$, $v_L = L\frac{di_L}{dt}$, and $v_C = \frac{\int i_L\, dt}{C}$ (v_R and v_L are the voltages across the R and L respectively and all unmentioned voltage polarity is in the direction of i_L)

(b) $V_i = V_R + V_L + V_C$, $V_R = I_L R$, $V_L = sLI_L$, and $V_C = \frac{I_L}{sC}$

(c) figure 11.9(d)

(d) blocks needed: one Sum, one Derivative and one Gain for sL, one Transfer Fcn for $\frac{1}{sC}$, and one Gain for $\frac{1}{R}$

Hint: sections 8.6 and 8.8

♦♦ Project 72 (on developing a circuit system)

(a) Referring to the series-parallel RC circuit in figure 11.9(e), write down the time domain equations of the circuit
(b) Write down the s domain equations of the circuit in part (a)
(c) Develop the circuit system as a block diagram from input to output
(d) Model the system by using R_1=3 $K\Omega$, R_2=4 $K\Omega$, and C=1 mF and by applying appropriate block of SIMULINK

Answers:

(a) $v_i - v_0 = R_1 i$, $v_0 = i_0 R_2$, $i_C = C\dfrac{dv_0}{dt}$, and $i_C + i_0 = i$ (where the currents i and i_C are in the $\rightarrow$ and $\downarrow$ directions through the R_1 and C respectively)

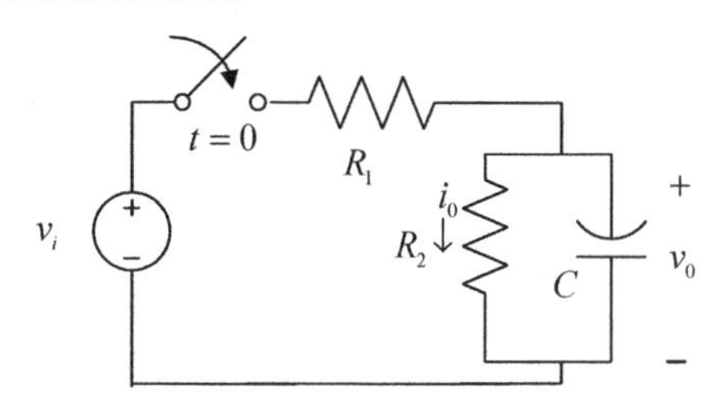

Figure 11.9(e) A series-parallel RC circuit

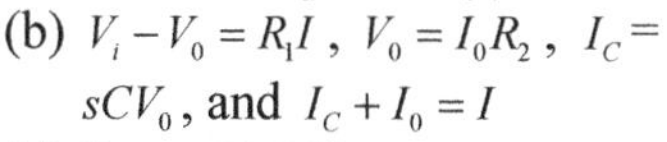

(b) $V_i - V_0 = R_1 I$, $V_0 = I_0 R_2$, $I_C = sCV_0$, and $I_C + I_0 = I$

(c) figure 11.9(f)

(d) blocks needed: two Sums, one Derivative, and three Gains

Hint: sections 8.6 and 8.8

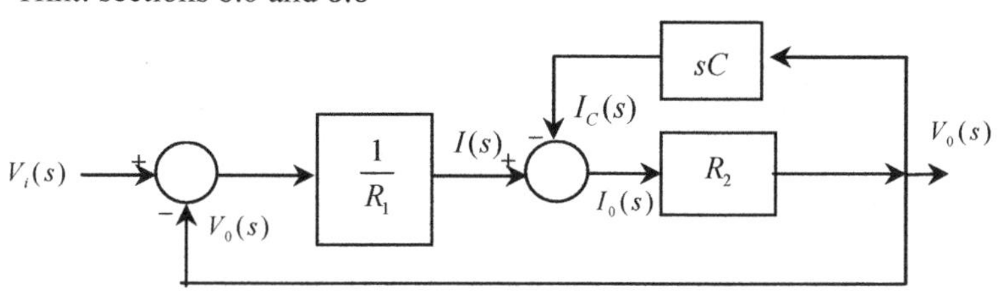

Figure 11.9(f) Modeling the circuit of figure 11.9(e)

♦♦ Project 73 (on developing a circuit system)

(a) By writing the time and transform domain equations, verify that the figure 11.9(h) is the block diagram representation of the parallel circuit in figure 11.9(g)
(b) Model the circuit system in SIMULINK by using R_1=3 $K\Omega$, R_2=4.1 $K\Omega$, L=3 mH, and C=1 mF and by applying appropriate blocks

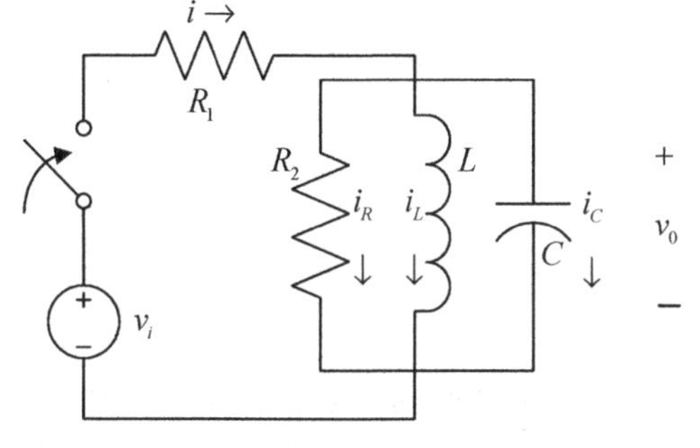

Figure 11.9(g) A parallel RLC circuit

Hint: sections 8.6 and 8.8 and blocks needed: two Sums, one Derivative, one Transfer Fcn, and three Gains

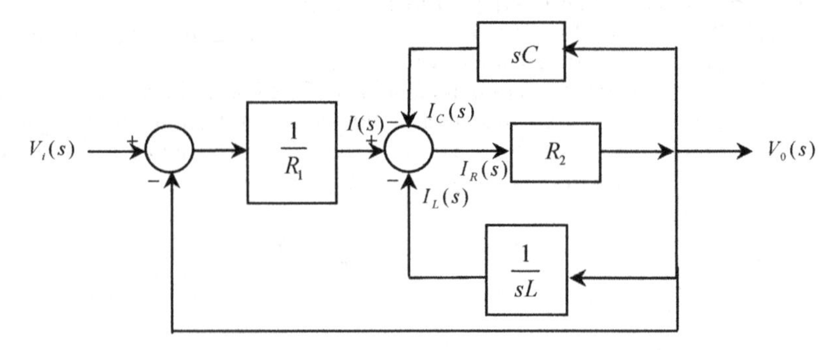

Figure 11.9(h) Modeling the parallel RLC circuit of the figure 11.9(g)

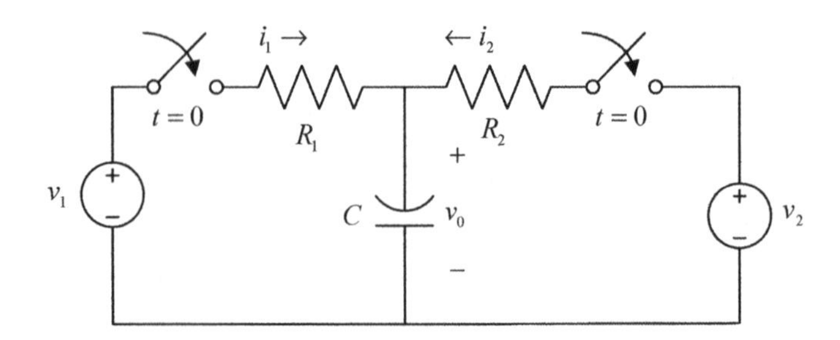

Figure 11.10(a) An RC circuit with two sources

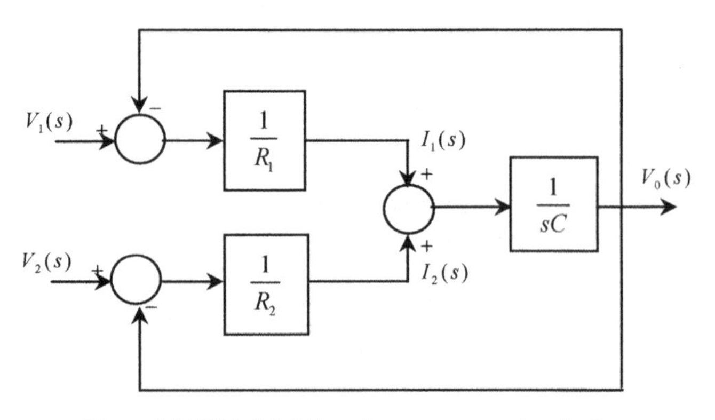

Figure 11.10(b) Modeling the two source circuit of the figure 11.10(a)

♦♦ Project 74 (on developing a circuit system)

(a) By writing the time and transform domain equations, verify that the figure 11.10(b) is the block diagram representation of the two source RC circuit of figure 11.10(a)

(b) Model the circuit system in SIMULINK by using R_1=3 $K\Omega$, R_2 =4.1 $K\Omega$, and C=1 μF and by applying appropriate blocks

Hint: sections 8.6 and 8.8 and blocks needed: three Sums, one Transfer Fcn, and two Gains

⬍⬍ Project 75 (on system function from a circuit system)

(a) In project 72 circuit system, verify the following transfer functions: $\frac{V_0}{V_i}=\frac{4}{12s+7}$ and $\frac{I_0}{V_i}=\frac{1}{1000(12s+7)}$

(b) In project 71 circuit system, verify the following transfer functions: $\frac{V_C}{V_i}=\frac{1}{2s^2+3s+1}$ and $\frac{V_i}{I}=\frac{2s^2+3s+1}{s}$

(c) In two stage filter circuit of the figure 11.10(c), verify the following transfer functions: $\frac{V_0}{V_i}=\frac{1080s^2}{1080s^2+106s+1}$ and $\frac{V_i}{I}=\frac{1080s^2+106s+1}{3s(76s+1)}$

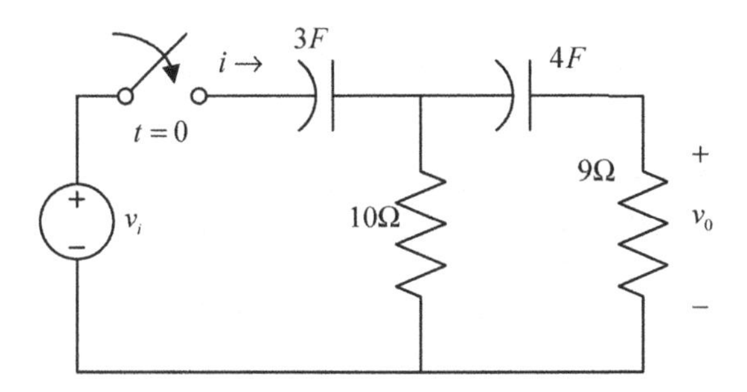

Figure 11.10(c) A two stage RC filter

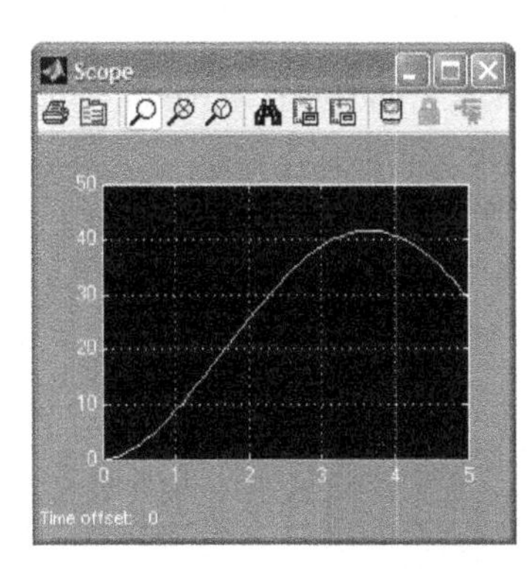

Figure 11.10(d) The v_0 output from the circuit in figure 11.9(e)

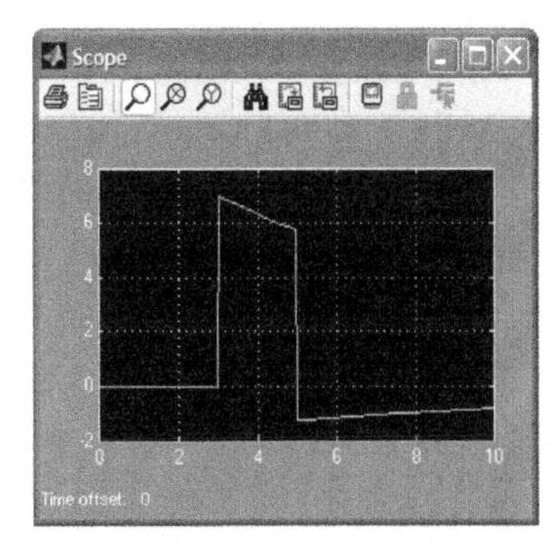

Figure 11.10(e) Response from the two stage filtering circuit

⬍⬍ Project 76 (on circuit system response)

In the following we apply some time domain input to some circuit and wish to see the circuit response in a Scope by modeling approach of SIMULINK:

(a) The circuit of the figure 11.9(e) is subjected to the input of the figure 10.3(c) and circuit parameter values are taken from the project 72. The output voltage v_0 versus t is to be displayed over $0 \le t \le 5$ secs

(b) The finite pulse of the figure 3.1(h) is applied as the v_i to the two stage filter circuit of the figure 11.10(c). The output voltage v_0 versus t is sought over $0 \le t \le 10$ secs

Answers:

(a) section 3.4 for source modeling, project 72 for the circuit system modeling, section 9.7 for similar example, and the Scope output for the v_0 should be like the figure 11.10(d)

(b) output should be like the figure 11.10(e), section 3.2 for the source modeling, and section 9.7 for similar example

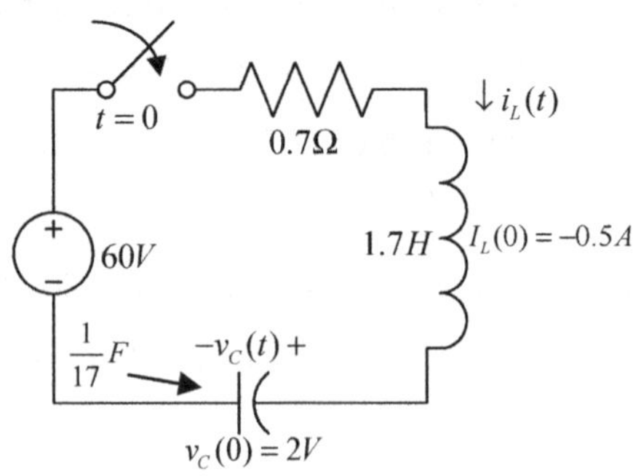

Figure 11.10(f) A second order transient circuit with initial conditions

♦♦ Project 77 (on time domain expression solution of circuits)

Performing time domain analysis on the second order circuit of the figure 11.10(f), show that the inductor current has the solution:

$$i_L(t) = -\frac{1}{578} e^{-\frac{7t}{34}} \left(-\frac{672503\sqrt{1279}}{3837} \sin\frac{3t\sqrt{1279}}{34} + 289\cos\frac{3t\sqrt{1279}}{34} \right) \quad \text{for } t \ge 0$$

and the capacitor voltage has the solution:

$$v_C(t) = 60 + e^{-\frac{7t}{34}} \left(-\frac{695\sqrt{1279}}{3837} \sin\frac{3t\sqrt{1279}}{34} - 58\cos\frac{3t\sqrt{1279}}{34} \right) \quad \text{for } t \ge 0$$

Also verify that the wave shapes of the $i_L(t)$ versus t and $v_C(t)$ versus t follow the oscillatorily decaying nature over $0 \le t \le 9$ secs respectively.

Hint: section 9.8 and appendix F cited ezplot for plotting

End of the chapter

Appendices

Appendix A

Block links for modeling signals and systems

When you open a new or work in a previously saved SIMULINK model file, the very next step is to know the exact location or link of a block which will be employed in modeling for different signal and system problems. Table A accumulates icon appearance, brief function, and location or link of SIMULINK blocks exercised in the text.

Table A. SIMULINK blocks and their links for modeling signals and systems (not arranged in the alphabetical order)

Block name	Icon Outlook	Function of the block or operation	Link or location
Display	0 Display	It shows the instantaneous value of the signal at the end of the simulation which it is connected to	SIMULINK → Sinks → Display
Constant	1 Constant	It generates user-defined constant values	SIMULINK → Sources → Constant
Product	× Product	It multiplies two or more signal values connected to its input ports	SIMULINK → Math Operations → Product
Fcn	f(u) Fcn	It performs user-defined mathematical operations on the signal to its input port assuming that the input signal name is u	SIMULINK → User-Defined Functions → Fcn
Sum	+ +	It adds two or more signal values connected to its input ports	SIMULINK → Math Operations → Sum
Ramp	Ramp	It generates straight line functions of various characteristics	SIMULINK → Sources → Ramp
Signal Builder	Signal 1 Signal Builder	It provides window interface for designing signal of user's choice	SIMULINK → Sources → Signal Builder
Sine Wave	Sine Wave	It generates sine waves of various characteristics	SIMULINK → Sources → Sine Wave
Saturation	Saturation	It clips positive and negative portions of a wave on user-definition	SIMULINK → Discontinuities → Saturation
Pulse Generator	Pulse Generator	It generates rectangular pulse waves of various characteristics	SIMULINK → Sources → Pulse Generator

Continuation of the last table:

Block name	Icon Outlook	Function of the block or operation	Link or location
Step	Step	It generates unit step function and its derived signals	SIMULINK → Sources → Step
Abs	|u| Abs	It returns the absolute value of the function to its input port	SIMULINK → Math Operations → Abs
Transport Delay	Transport Delay	It translates a function $f(t)$ to t_0 that is the input and output of the block are $f(t)$ and $f(t-t_0)$ respectively where t_0 is user-defined	SIMULINK → Continuous → Transport Delay
Gain	1 Gain	It multiplies the signal to its input port according to user-supplied gain	SIMULINK → Math Operations → Gain
Integrator	1/s Integrator	It performs the numerical integration of the signal to its input port in continuous sense	SIMULINK → Continuous → Integrator
Mux		It multiplexes two or more signals from single vector to group vector form	SIMULINK → Commonly Used Blocks → Mux
To Workspace	simout To Workspace	It exports signal data from SIMULINK to MATLAB workspace which it is connected to	SIMULINK → Sinks → To Workspace
Scope	Scope	It shows the functional variation (s) of some signal (s) which it is connected to	SIMULINK → Sinks → Scope
Derivative	du/dt Derivative	It performs the numerical differentiation of the signal to its input port in continuous sense	SIMULINK → Continuous → Derivative
Real-Imag to Complex	Re Im Real-Imag to Complex	It converts a real and imaginary form data or signal to complex or phasor form	SIMULINK → Math Operations → Real-Imag to Complex
Complex to Magnitude-Angle	|u| ∠u Complex to Magnitude-Angle	It converts any functional data from complex form to magnitude – phase angle (in radians) form	SIMULINK → Math Operations → Complex to Magnitude-Angle
Transfer Fcn	1/(s+1) Transfer Fcn	It implements Laplace transform system function in rational form when numerator and denominator in polynomial form	SIMULINK → Continuous → Transfer Fcn
Zero-Pole	(s-1)/(s(s+1)) Zero-Pole	It implements Laplace transform system function in rational form when numerator and denominator in factored form	SIMULINK → Continuous → Zero-Pole
State-Space	x' = Ax+Bu y = Cx+Du State-Space	It implements continuous system which follows the dynamic equations $\dot{x} = Ax + Bu$ and $y = Cx + Du$	SIMULINK → Continuous → State-Space

Appendix B

Coding in MATLAB or SIMULINK

MATLAB/SIMULINK executes the code of a signal or its expression in terms of string which is the set of keyboard characters placed consecutively. One distinguishing feature of MATLAB is that the workspace variable itself is a matrix. The strings adopted for computation are divided into two classes – scalar and vector. The scalar computation results the order of the output matrix same as that of the variable matrix. On the contrary, the order for the vector computation is determined in accordance with the matrix algebra rules. Some symbolic functions and their MATLAB counterparts are presented in table B.1. The operators for arithmetic computations are as follows:

addition	+
subtraction	–
multiplication	*
division	/
power	^

The operation sequence of different operators in a scalar or vector string observes the following order:

enclosing braces	( )	first,
power operator	^	then,
division operator	/	next,
multiplication operator	*	after that,
addition operator	+	then, and
subtraction operator	–	finally.

The syntax of the scalar computation urges to use .*, ./, and .^ in lieu of *, /, and ^ respectively. The operators *, /, or ^ are never preceded by . for the vector computation. The vector string is the MATLAB code of any symbolic expression or function often found in mathematics. Starting from the simplest one, we present some examples on writing the long expressions in MATLAB for the scalar form and the vector form as well.

♦♦ Write the MATLAB codes in scalar and vector forms for the following functions

$A.\ \sin^3 x \cos^5 x$ $\quad B.\ 2+\ln x$ $\quad C.\ x^4+3x-5$ $\quad D.\ \dfrac{x^3-5}{x^2-7x-7}$

$E.\ \sqrt{|x^3|+\sec^{-1} x}$ $\quad F.\ (1+e^{\sin x})^{x^2+3}$ $\quad G.\ \dfrac{\cosh x+3}{\sqrt{\dfrac{x+4}{\log_{10}(x^3-6)}}}$

$H.\ \dfrac{1}{(x-3)(x+4)(x-2)}$ $\quad I.\ \dfrac{1}{1+\dfrac{1}{1+\dfrac{1}{x}}}$ $\quad J.\ \dfrac{a}{x+a}+\dfrac{b}{y+b}+\dfrac{c}{z+c}$

$K.\ \dfrac{u^2v^3w^9}{x^4y^7z^6}$

In tabular form, they are coded as follows:

Example	String for scalar computation	String for vector computation
A	sin(x).^3.*cos(x).^5	sin(x)^3*cos(x)^5
B	2+log(x)	2+log(x)
C	x.^4+3*x-5	x^4+3*x-5
D	(x.^3-5)./(x.^2-7*x-7)	(x^3-5)/(x^2-7*x-7)
E	sqrt(abs(x.^3)+asec(x))	sqrt(abs(x^3)+asec(x))
F	(1+exp(sin(x))).^(x.^2+3)	(1+exp(sin(x)))^(x^2+3)
G	(cosh(x)+3)./sqrt((x+4)./log10(x.^3-6))	(cosh(x)+3)/sqrt((x+4)/log10(x^3-6))
H	1./(x-3)./(x+4)./(x-2)	1/(x-3)/(x+4)/(x-2)
I	1./(1+1./(1+1./x))	1/(1+1/(1+1/x))
J	a./(x+a)+b./(y+b)+c./(z+c)	a/(x+a)+b/(y+b)+c/(z+c)
K	u.^2.*v.^3.*w.^9./x.^4./y.^7./z.^6	u^2*v^3*w^9/x^4/y^7/z^6

Signal and system programming or simulation circumstance requires the type of code – whether scalar or vector.

Table B.1 Some mathematical functions and their MATLAB counterparts

Mathematical notation	MATLAB notation	Mathematical notation	MATLAB notation	Mathematical notation	MATLAB notation
$\sin x$	sin(x)	$\sin^{-1} x$	asin(x)	π	pi
$\cos x$	cos(x)	$\cos^{-1} x$	acos(x)	A+B	A+B
$\tan x$	tan(x)	$\tan^{-1} x$	atan(x)	A–B	A–B
$\cot x$	cot(x)	$\cot^{-1} x$	acot(x)	A×B	A*B
$\cos ecx$	csc(x)	$\sec^{-1} x$	asec(x)	e^x	exp(x)
$\sec x$	sec(x)	$\cos ec^{-1}x$	acsc(x)	A^B	A^B
$\sinh x$	sinh(x)	$\sinh^{-1} x$	asinh(x)	$\ln x$	log(x)
$\cosh x$	cosh(x)	$\cosh^{-1} x$	acosh(x)	$\log_{10} x$	log10(x)
$\sec hx$	sech(x)	$\sec h^{-1}x$	asech(x)	$\log_2 x$	log2(x)
$\cos echx$	csch(x)	$\cos ech^{-1}x$	acsch(x)	Σ	sum
$\tanh x$	tanh(x)	$\tanh^{-1} x$	atanh(x)	Π	prod
$\coth x$	coth(x)	$\coth^{-1} x$	acoth(x)	$\lvert x \rvert$	abs(x)
10^A	1e*A* e.g. 1e3	10^{-A}	1e-*A* e.g. 1e-3	$\sqrt{x}$	sqrt(x)

* In the six trigonometric functions for example sin(x), the x is in radian. If the x is in degree, we use sind(x). The other five functions also have the syntax cosd(d), tand(x), cotd(x), cscd(x), and secd(x) when the x is in degree. The default return from asin(x) is in radian, if you need the return to be in degree, use the command asind(x). Similar degree return is also possible from acosd(x), atand(x), acotd(x), asecd(x), and acscd(x).

We present some numerical example to quote the difference between the scalar and vector computations. Let us say we have the matrices $A=\begin{bmatrix}3 & 5\\ 7 & 8\end{bmatrix}$, $B=\begin{bmatrix}5 & 2 & 1\\ 0 & 1 & 7\end{bmatrix}$, and

$C=\begin{bmatrix}3 & 2 & 9\\4 & 0 & 2\end{bmatrix}$. The scalar computation is not possible between the matrices A and B because of their unequal order nor is between the matrices A and C for the same reason. On the contrary the scalar multiplication can be conducted between B and C for having the same order and which is $B.*C=\begin{bmatrix}15 & 4 & 9\\0 & 0 & 14\end{bmatrix}$ (element by element multiplication).

Matrix algebra rule says that any matrix A of order $M\times N$ can only be multiplied with another matrix B of order $N\times P$ so that the resulting matrix has the order $M\times P$. For the examples A and B, we have M=2, N=2, and P=3 and obtain the vector-multiplied matrix as $A\times B=\begin{bmatrix}3\times5+5\times0 & 3\times2+5\times1 & 3\times1+5\times7\\7\times5+8\times0 & 7\times2+1\times8 & 7\times1+8\times7\end{bmatrix}=$ $\begin{bmatrix}15 & 11 & 38\\35 & 22 & 63\end{bmatrix}$, and which has the MATLAB code A*B not A.*B. Similar interpretation follows for the operators * and /.

Whenever we write the scalar codes A.*B, A./B, and A.^B, we make it certain that both the A and B are identical matrix in size. The 3*A means all elements of matrix A are multiplied by 3 and we do not use 3.*A. Also do we not use A./3 but do A/3. The signs + and – are never preceded by the operator . in the scalar codes. The command 4./A means 4 is divided by all elements in A. The A.^4 means power on all elements of A is raised by 4 and so on.

♦♦ Scale factor in resistor, inductor, or capacitor units

The standard units of resistance, inductance, and capacitance are Ohm(Ω), Henry(H), and Farad (F) respectively. In electrical system problems quantity of interest is often given in units which are in the power of 10. Table B.2 presents the engineering scale factor units and their MATLAB equivalences. For example the resistor 10.7 $K\Omega$ is coded as 10.7e3. Again a capacitor of value 4.7 μF is entered by writing 4.7e-6 in standard unit.

Table B.2 Engineering unit scale factors and their MATLAB counterparts

Scale factor	Symbol	As power of 10	MATLAB code
giga	G	10^{9}	e9
mega	M	10^{6}	e6
kilo	K	10^{3}	e3
milli	m	10^{-3}	e-3
micro	μ	10^{-6}	e-6
nano	n	10^{-9}	e-9
pico	p	10^{-12}	e-12

♦♦ Scale factor in voltage, current, frequency, or power units

Just quoted scale factor is also practiced in voltage, current, frequency, or power units. For instance the voltage 2.3 mV, the power 1.11 μW, and the frequency 900 MHz are coded by 2.3e-3, 1.11e-6, and 900e6 respectively.

♦♦ Reactance coding

Reactance coding is extremely important in electrical systems because almost every circuit analysis requires the coding. The coding is related with two circuit elements – inductor and capacitor.

Any inductance L has the reactance $X_L = \omega L$ suppose L=5.7 mH and ω=45.7 $rad/\sec$ then the reactance coding for this inductor is 5.7e-3*45.7 (i.e. 45.7 $rad/\sec$ ×5.7 mH). When the frequency is given in Hz rather in $rad/\sec$ for example $f = 2.6KHz$, the reactance coding for the same inductor is 2*pi*2.6e3*5.7e-3(i.e. $2\pi \times 2.6KHz \times 5.7mH$) because $\omega = 2\pi f$.

Again any capacitance C has the reactance $X_C = \frac{1}{\omega C}$. Considering C= 5.7 pF and ω=5545.7 $rad/\sec$ the reactance coding for the capacitor is 1/5.7e-12/5545.7 (i.e. $\frac{1}{5545.7 rad/\sec \times 5.7 pF}$). If you write the code as 1/(5.7e-12*5545.7), that is also operational but involving unnecessary first brace in the coding. Machine is smart enough to realize the code if you write the former one. When Hertz frequency is given for instance $f = 2.6GHz$, the coding for the reactance is 1/2/pi/2.6e9/5.7e-12 (i.e. $\frac{1}{2\pi \times 2.6GHz \times 5.7 pF}$) for the same reason.

♦♦ Impedance coding

In impedance coding we just bring the unit imaginary number j (appendix C.12). Just mentioned two inductance codings (i.e. $Z_L = j\omega L$ and $Z_L = j2\pi f L$) in impedance form become j*5.7e-3*45.7 and j*2*pi*2.6e3*5.7e-3 for the angular and Hertz frequencies respectively.

Again the impedance codings ($Z_C = \frac{1}{j\omega C}$ and $Z_C = \frac{1}{j2\pi f C}$) for just mentioned two capacitances are 1/j/5.7e-12/5545.7 and 1/j/2/pi/2.6e9/5.7e-12 for the angular and Hertz frequencies respectively.

♦♦ s domain coding for inductor and capacitor

The s domain coding for inductance L and capacitance C are sL and $\frac{1}{sC}$ and which are s*5.7e-3 and 1/s/5.7e-12 for the reactance coding mentioned inductor and capacitor respectively. In symbolic form we write that as s*L and 1/s/C respectively.

♦♦ Rational form coding

In symbolic or expression form computation it is recommended that the reader use the rational form instead of fractional form otherwise machine finds the best digital representation for the data. For example the inductance 5.7 mH is coded by 57/10000 instead of 5.7e-3.

Appendix C

MATLAB functions useful for signals and systems

While working on signal and system problems in MATLAB, we come across lots of built-in MATLAB functions. In order to employ these functions for the analysis, we need to understand their input and output argument types and purpose of the functions. Functions exercised in the text with brief descriptions are in the following.

C.1 Matrix of ones, zeroes, and constants

MATLAB commands ones and zeros implement user-defined matrix of ones and zeroes respectively. Each function conceives two input arguments, the first and second of which are the required numbers of rows and columns respectively. Let us say we intend to form the matrices $A=\begin{bmatrix}1&1&1\\1&1&1\\1&1&1\\1&1&1\end{bmatrix}$, $B=\begin{bmatrix}1&1&1\\1&1&1\\1&1&1\end{bmatrix}$, and $C=\begin{bmatrix}1&1&1&1\\1&1&1&1\end{bmatrix}$.

Their orders are 4×3, 3×3, and 2×4 respectively and the implementations are as follows:

for A,
```
>>A=ones(4,3) ↵

A =
     1     1     1
     1     1     1
     1     1     1
     1     1     1
```

for B,
```
>>B=ones(3) ↵

B =
     1     1     1
     1     1     1
     1     1     1
```

for C,
```
>>C=ones(2,4) ↵

C =
     1     1     1     1
     1     1     1     1
```

Either the number of rows or columns will do if the matrix is a square. For the row and column matrices of ones for example of length 6, the commands would be ones(1,6) and ones(6,1) respectively.

Formation of the matrix of zeroes is quite similar to that of the matrix of ones. Replacing the function ones by zeros does the formation. Matrix of zeroes like $A=\begin{bmatrix}0&0&0\\0&0&0\\0&0&0\\0&0&0\end{bmatrix}$, $B=\begin{bmatrix}0&0&0\\0&0&0\\0&0&0\end{bmatrix}$, and $C=\begin{bmatrix}0&0&0&0\\0&0&0&0\end{bmatrix}$ (whose orders are 4×3, 3×3, and 2×4) we form by the commands A=zeros(4,3), B=zeros(3), and C=zeros(2,4) respectively. A row and a column matrices of 6 zeroes are formed by the commands zeros(1,6) and zeros(6,1) respectively.

A matrix of constants is obtained by first creating a matrix of ones of the required size and then multiplying by the constant number. For example the matrix $\begin{bmatrix}0.2&0.2&0.2\\0.2&0.2&0.2\\0.2&0.2&0.2\\0.2&0.2&0.2\end{bmatrix}$ can be generated by the command 0.2*ones(4,3).

C.2 Comparative and logical operators

Comparative operators are used for the comparison of two scalar elements, one scalar and one matrix elements, or two identical size matrix elements. There are six comparative operators as presented in the table C.1.

The output of the expression pertaining to the comparative operators is logical – either true (indicated by 1) or false (indicated by 0). For example when A=3 and B=4, the comparisons $A=B$, $A \neq B$, $A>B$, $A \geq B$, $A<B$, and $A \leq B$ should be false (0), true(1), false(0), false(0), true(1), and true(1) respectively. We implement these comparative operations as presented in the table C.2.

Table C.1 Equivalence of comparative operators

Comparative operation	Mathematical notation	MATLAB notation
equal to	$=$	==
not equal to	$\neq$	~=
greater than	$>$	>
greater than or equal to	$\geq$	>=
less than	$<$	<
less than or equal to	$\leq$	<=

There are two operands A and B in table C.2, each of which is a single scalar. Each of the operands can be a matrix in general. In that case the logical decision takes place element by element on all elements in the matrix. For instance if $A=\begin{bmatrix} 5 & 8 \\ 5 & 7 \end{bmatrix}$ and $B=\begin{bmatrix} 2 & 1 \\ -2 & 9 \end{bmatrix}$, A>B should be $\begin{bmatrix} 5>2 & 8>1 \\ 5>-2 & 7>9 \end{bmatrix}=\begin{bmatrix} 1 & 1 \\ 1 & 0 \end{bmatrix}$. Again if A happens to be a scalar (say A=4), the single scalar is compared to all elements in the B therefore $A \leq B$ should be $\begin{bmatrix} 4 \leq 2 & 4 \leq 1 \\ 4 \leq -2 & 4 \leq 9 \end{bmatrix}=\begin{bmatrix} 0 & 0 \\ 0 & 1 \end{bmatrix}$. In a similar fashion B also operates on A however the scalar and matrix related comparative implementation is presented in the table C.3.

Table C.2 Scalar comparative operation

>>A=3; B=4; ↵ >>A==B ↵ ans = 0 >>A~=B ↵ ans = 1	>>A>B ↵ ans = 0 >>A>=B ↵ ans = 0	>>A<B ↵ ans = 1 >>A<=B ↵ ans = 1

Table C.3 Scalar and matrix comparative operation

when A and B are matrices,	when A is scalar and B is matrix,
>>A=[5 8;5 7]; ↵ >>B=[2 1;-2 9]; ↵ >>A>B ↵ ans = 1 1 1 0	>>A=4; ↵ >>B=[2 1;-2 9]; ↵ >>A<=B ↵ ans = 0 0 0 1

Some basic logical operations are NOT, OR, and AND. The characters ~, |, and & of the keyboard are adopted for the logical NOT, OR, and AND respectively. In all logical outputs the 1 and 0 stand for true and false respectively. All logical operators apply to the matrices in general. For the matrix $A=\begin{bmatrix} 0 & 0 \\ 0 & 1 \end{bmatrix}$, NOT(A) operation should provide $\begin{bmatrix} 1 & 1 \\ 1 & 0 \end{bmatrix}$ (see table C.4).

Table C.4 Basic logical operations on matrix elements

for NOT(A) operation, >>A=[0 0;0 1]; ↵ >>~A ↵ ans = 1 1 1 0	for A OR B, >>A=[1 1;0 1]; ↵ >>B=[0 1;1 1]; ↵ >>A|B ↵ ans = 1 1 1 1	for A AND B, >>A&B ↵ ans = 0 1 0 1	for A XOR B, >>xor(A,B) ↵ ans = 1 0 1 0

The logical OR and AND operations on the like positional elements of the two matrices A=$\begin{bmatrix}1 & 1\\0 & 1\end{bmatrix}$ and B=$\begin{bmatrix}0 & 1\\1 & 1\end{bmatrix}$ must return $\begin{bmatrix}1 & 1\\1 & 1\end{bmatrix}$ and $\begin{bmatrix}0 & 1\\0 & 1\end{bmatrix}$ respectively. Table C.4 shows both implementations.

If A or B is a single 1 or 0, it operates on all elements of the other.

Sometimes we need to check the interval of the independent variable of some function for instance $-6 \le x \le 8$. The interval is split in two parts $-6 \le x$ and $x \le 8$. In terms of the logical statement one expresses the $-6 \le x \le 8$ as (-6<=x)&(x <=8).

There is no operator for the XOR logical operation instead the MATLAB function xor syntaxed by xor(A,B) implements the operation as presented in the table C.4.

C.3 Position indexes of matrix elements with conditions

MATLAB function find looks for the position indexes of matrix elements subject to some logical condition whose general format is [R C]=find(condition) where the indexes returned to R and C are meant to be for the row and column directions respectively. The R and C are some user-chosen workspace variable names. Let us consider $A=\begin{bmatrix}11 & 10 & 11 & 10\\12 & 10 & -2 & 0\\-7 & 17 & 1 & -1\end{bmatrix}$ which we enter by the following:

>>A=[11 10 11 10;12 10 -2 0;-7 17 1 -1]; ↵ ← A is assigned to A

We would like to know what the position indexes of A where the elements are greater than 10 are. In matrix A the left-upper most element has the position index (1,1). The elements of A being greater than 10 have the position indexes (1,1), (2,1), (3,2), and (1,3). MATLAB finds the required index in accordance with columns. Placing the row and column indexes vertically, we have $\begin{bmatrix}1\\2\\3\\1\end{bmatrix}$ and $\begin{bmatrix}1\\1\\2\\3\end{bmatrix}$ respectively. The output arguments R and C of the find receive these two column matrices respectively. The input argument of the find must be a logical statement, any element in A greater than 10 is written as A>10. The position indexes are found as shown on the right side attached text box.

where elements of A are greater than 10,

```
>>[R C]=find(A>10) ↵

R =

     1
     2
     3
     1

C =

     1
     1
     2
     3
```

where elements of A =10,

```
>>[R C]=find(A==10) ↵

R =

     1
     2
     1

C =

     2
     2
     4
```

where elements of $A \le 0$,

```
>>[R C]=find(A<=0) ↵

R =

     3
     2
     2
     3

C =

     1
     3
     4
     4
```

for the row matrix D,

```
>>D=[-10 34 1 2 8 4]; ↵
>>R=find(D>=8) ↵

R =

     2     5
```

for the column matrix E,

```
>>E=[-2 8 -2 7]'; ↵
>>C=find(E~=-2) ↵

C =

     2
     4
```

To exercise more conditions, what are the position indexes in the matrix A where the elements are equal to 10? The answer is (1,2), (2,2), and (1,4). Again the position indexes where the elements are less than or equal to zero are (3,1), (2,3), (2,4), and (3,4).

The comparative operators $>$, $<$, $\geq$, $\leq$, and $\neq$ have the MATLAB counterparts >, <, >=, <=, and ~= respectively.

So far we considered a rectangular matrix for demonstration of position index finding. Let us see how the find works for a row or column matrix. Let us take $D=[-10 \quad 34 \quad 1 \quad 2 \quad 8 \quad 4]$ from which we find the position indexes of the elements where they are greater than or equal to 8. Obviously they are the 2nd and 5th elements. Here we do not need to place two output arguments to the find.

Again let us find the position indexes of the elements of the column matrix $E=\begin{bmatrix}-2\\8\\-2\\7\end{bmatrix}$ where the elements are not equal to –2. The 2nd and 4th elements are not equal to –2. The output of the function find is a row one for the row matrix input and a column one for the column matrix input.

Presented in the last paragraph of the last page are the executions for all these conditional findings.

C.4 Finding the maximum/minimum numerically

Given a matrix, one finds the maximum element from the matrix by using the command max (min for the minimum). Let us say we have three matrices $R=[1 \quad -2 \quad 3 \quad 9]$, $C=\begin{bmatrix}23\\-20\\30\\8\end{bmatrix}$, and $A=\begin{bmatrix}2 & 4 & 7\\-2 & 7 & 9\\3 & 8 & -8\end{bmatrix}$ whose maxima are 9, 30, and 9 (from all elements in the matrix) and minima are –2, –20, and –8 respectively. We find the maxima first entering (subsection 1.1.2) the respective matrices as follows:

for the row matrix,	for the column matrix,	for the rectangular matrix,
>>R=[1 -2 3 9]; ↵	>>C=[23;-20;30;8]; ↵	>>A=[2 4 7;-2 7 9;3 8 -8]; ↵
>>max(R) ↵	>>max(C) ↵	>>max(max(A)) ↵
ans =	ans =	ans =
9	30	9
>>min(R) ↵	>>min(C) ↵	>>min(min(A)) ↵
ans =	ans =	ans =
-2	-20	-8

for index finding in R,

```
>>[M,I]=max(R) ↵

M =
      9
I =
      4
```

Font equivalence is maintained by using the same letter for example A⇔A in above implementation. If the matrix is a row or column one, we apply one max or min. For a rectangular matrix, the max or min operates on each column individually that is why two max or min functions are required. The functions are equally applicable for decimal number elements.

In the row matrix R, the maximum 9 is

occurring as the fourth element in the matrix. Suppose we also intend to find the position index (that is 4) of the maximum element in the R. Now we need two output arguments – one for the maximum and the other for its index. Its implementation is shown on the right side attached text box of the last paragraph in the last page in which the two output arguments M and I correspond to the maximum and its integer index respectively.

The function min also keeps this type of integer index returning option in a similar fashion.

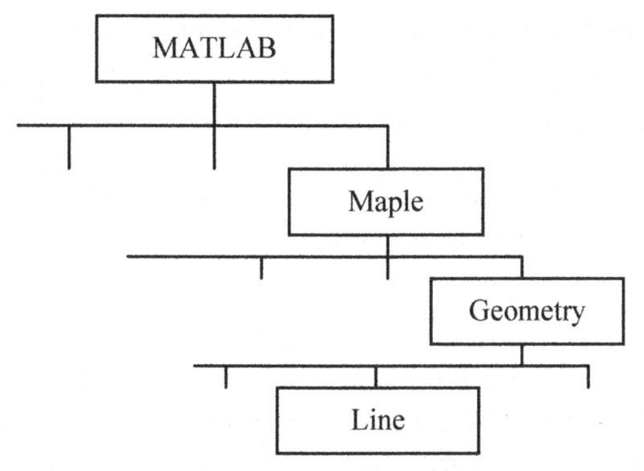

Figure C.1(a) Hierarchy of the Line function in MATLAB

C.5 Difference between MATLAB and maple functions

MATLAB is full of library and library functions, one specific library is called maple which is very convenient for symbolic computation. Any maple statement is executed in MATLAB with the syntax maple(statement under quote). In maple there is a function by the name Line in a family called Geometry so hierarchically we organize the Line function in maple as shown in figure C.1(a).

Recall the function fourier of chapter 5 which exists both in maple and MATLAB. Their calling differences on some F are shown below:

```
>>fourier(F) ↵               ← when called in MATLAB
>>maple('fourier(F)') ↵      ← when called in maple
```

C.6 Equation of a straight line passing through two points

A two dimensional line can be defined from two given points like the figure C.1(b). Let us say we have two points whose coordinates are $A(0,4)$ and $B(4,-3)$. The straight line AB passing through A and B has the equation $7x+4y-16=0$ which we wish to obtain.

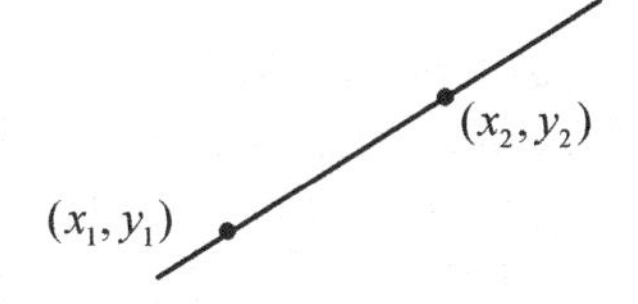

Figure C.1(b) A straight line passing through two points

The maple (appendix C.5) library geometry keeps many two dimensional geometry computational functions (reference 4). Any two dimensional point coordinates are described by the built-in function point with the syntax point(user-given point name, x coordinate, y coordinate) but within the maple. For example the point $A(0,4)$ is written as point(A,0,4). The maple program statements are separated by a comma and we activate the geometry package by writing with(geometry) therefore we enter the two points as follows:

```
>>maple('with(geometry),point(A,0,4),point(B,4,-3)'); ↵
```

Again the built-in function line within the maple constructs a two dimensional line object from two given points with the syntax line(user-given line name, [first point name, second point name], [horizontal axis variable name of the line, vertical axis variable name of the line]) so the line object we form by exercising:

```
>>maple('line(L,[A,B],[x,y])'); ↵
```

Just conducted statement indicates that we defined the line object as L (our-chosen name). The second input argument A,B must be under the third brace and holds earlier entered point names. The third input argument x,y must be under the third brace and indicates the chosen horizontal and vertical axes variables as x and y respectively. If we wish to choose the axis variables as t - $f(t)$, the command would be [t,f]. However within the maple the line object is held in the variable L.

Now if we intend to see the equation of the line, we apply the maple built-in function Equation on the L as follows:

```
>>maple('Equation(L)') ↵

ans =

-16+7*x+4*y = 0              ← The equation 7x+4y−16=0
```

You could have implemented all these commands by executing one line statement maple('with(geometry):point(A,0,4):point(B,4,-3):line(L,[A,B],[x,y]):Equation(L)') at the command prompt. Note that the suppression operators in MATLAB and maple are ; and : respectively. The use of : between two line statements in maple does not display what is going on within maple.

Sometimes it is desirable that we express the y in terms of x from the equation $7x+4y-16=0$ and which should be $y=\frac{16-7x}{4}$. In that case we assign the return of the maple('Equation(L)') to some user-chosen variable but in MATLAB not in maple for example

```
>>e=maple('Equation(L)') ↵        ← The e is our-chosen name

e =

-16+7*x+4*y = 0                   ← The e retains the equation 7x+4y−16=0
```

MATLAB built-in function solve mainly solves algebraic equations (appendix D). We apply the solve on the e for the solution of y as follows:

```
>>solve(e,'y') ↵   ← The return can be assigned again to some variable for further
                                                                   calculation
ans =

4-7/4*x              ← Meaning (16−7x)/4
```

C.7 Simple if/if-else/nested if syntax

Conditional commands are exercised by the if-else statements (reserve words). Also comparisons and checkings need if-else statements. We can have different if-else structures namely simple-if, if-else, or nested-if depending on the programming circumstances, some of which we discuss in the following.

Simple if

The program syntax of the simple-if is as follows:

if *logical expression*

Executable MATLAB command(s)

end

Logical expression usually requires the use of comparative operators whose reference is in appendix C.2. If the logical expression beside the if is true, the command between the if and end is executed otherwise not. In tabular form the simple-if implementation is as follows:

Example: If $x \geq 1$, we compute $y = \sin x$. When $x = 2$, we should see $y = \sin 2 =$ 0.9093.	Executable M-file:
	x=2;
	if x>=1
	  y=sin(x);
	end
		Steps: Save the statements in a new M-file (subsection 1.1.2) by the name test and execute the following:
		>>test ↵
			Check from the command window after running the M-file:
			>>y ↵
			
			y =
			0.9093

If-else

The general program syntax for the if-else structure is as follows:

if *logical expression*
Executable MATLAB command(s)
else
Executable MATLAB command(s)
end

If the logical expression beside the if is true, the command between the if and else is executed else the command between else and end is executed. In tabular form, the if-else-end implementation is the following:

Example: When $x = 1$, we compute $y = \sin\frac{x\pi}{2} = 1$ otherwise $y = \cos\frac{x\pi}{2} = 0$.	Executable M-file:
	x=1;
	if x==1
	  y=sin(x*pi/2);
	else
	  y=cos(x*pi/2);
	end
		Steps: Save the statements in a new M-file by the name test and execute the following:
		>>test ↵
			Check from the command window after running the M-file:
			>>y ↵
			
			y =
			1

If we had x=2; in the first line of the M-file, we would see y= $\cos\pi = -1$.

Nested-if

The third type of the if structure is the nested-if whose program syntax is as follows:

if *logical expression*
Executable MATLAB command(s)
elseif *logical expression*
Executable MATLAB command(s)
⋮
elseif *logical expression*
Executable MATLAB command(s)
else
Executable MATLAB command(s)
end

Clearly the syntax takes care of multiple logical expressions which we demonstrate by one example as shown in the following.

Example: The best example can be taking the decision of grades out of 100 based on the achieved number of a student. The grading policy is stated as if the achieved number of a student is greater than or equal 90, greater than or equal to 80 but less than 90, greater than or equal to 70 but less than 80, greater than or equal to 60 but less than 70, greater than or equal to 50 but less than 60, and less than 50, then the grade is decided as A, B, C, D, E, and F respectively.	Executable M-file: N=77; if N>=90 g='A'; elseif (N<90)&(N>=80) g='B'; elseif (N<80)&(N>=70) g='C'; elseif (N<70)&(N>=60) g='D'; elseif (N<60)&(N>=50) g='E'; else g='F'; end	In the executable M-file, the **N** and **g** refer to the number achieved and the grade respectively. If the number **N** is 77, the grade **g** should be **C**. Any character is argumented under the single inverted comma. Steps: Save the left statements in a new M-file by the name **test** and execute the following: >>test ↵	Check from the command window after running the M-file: >>g ↵ g = C

C.8 Data accumulation

Sometimes it is necessary that we perform the appending operation on an existing matrix in MATLAB workspace.

◆ Appending rows

Assume that $A=\begin{bmatrix}1 & 3 & 5\\ 2 & 6 & 8\\ 9 & 5 & 0\\ 4 & 7 & 8\end{bmatrix}$ is formed by appending two row matrices [9 5 0] and [4 7 8] with the matrix $B=\begin{bmatrix}1 & 3 & 5\\ 2 & 6 & 8\end{bmatrix}$. We first enter matrix B (subsection 1.1.2) into MATLAB and append one row after another by using the command presented below:

for entering B, >>B=[1 3 5;2 6 8] ↵ B = 1 3 5 2 6 8	for appending the first row, >>B=[B;[9 5 0]] ↵ B = 1 3 5 2 6 8 9 5 0	for appending the second row, >>A=[B;[4 7 8]] ↵ A = 1 3 5 2 6 8 9 5 0 4 7 8

The command B=[B;[9 5 0]] tells that the row [9 5 0] is to be appended with the existing B (inside the third bracket) and that the result is again assigned to B. You can append as many rows as you want. The important point is the number of elements in each row that is to be appended must be equal to the number of columns in the matrix B.

◆ Appending columns

Suppose $C=\begin{bmatrix}1 & 3 & 5 & 9 & 3\\ 2 & 6 & 8 & 0 & 1\\ 9 & 5 & 0 & 1 & 9\end{bmatrix}$ is formed by appending two column matrices $\begin{bmatrix} 9 \\ 0 \\ 1 \end{bmatrix}$ and $\begin{bmatrix} 3 \\ 1 \\ 9 \end{bmatrix}$ with matrix $D = \begin{bmatrix} 1 & 3 & 5 \\ 2 & 6 & 8 \\ 9 & 5 & 0 \end{bmatrix}$. We get the matrix D into MATLAB and append one column after another as follows:

for entering D,	for appending the first column,	for appending the second column,
>>D=[1 3 5;2 6 8;9 5 0] ↵ D = 1 3 5 2 6 8 9 5 0	>>D=[D [9 0 1]'] ↵ D = 1 3 5 9 2 6 8 0 9 5 0 1	>>C=[D [3 1 9]'] ↵ C = 1 3 5 9 3 2 6 8 0 1 9 5 0 1 9

The column matrix [9 0 1]' and D has one space gap within the third brace. In the second of above implementation, the resultant matrix is again assigned to D. Append as many columns as you want just remember that the number of elements in each column that is to be appended must be equal to the number of rows in the matrix D.

◆ **Data accumulation by using these two techniques**

Suppose initially there is nothing in the f matrix, which in MATLAB we write by the statement f=[ ]; (an empty matrix is assigned to f). An empty matrix does not have any size and completely empty, it follows the null symbol ∅ of the matrix algebra. Let us say k=2 and perform the assignment as follows:

```
>>f=[ ]; k=2; ↵
```

Now if we execute f=[f k] time and again first f=[f k] returns 2, second f=[f k] returns [2 2], third f=[f k] returns 2 2 2, and so on. This is called row-wise data accumulation. Column-wise data accumulation occurs by executing f=[f;k] each time.

The demonstrated k is just a scalar but it can be return from some function, row matrix, or column matrix.

C.9 For-loop syntax

A for-loop performs similar operations for a specific number of times and must be started with the for and terminated by an end statements. Following the for there must be a counter. The counter of the for-loop can be any variable that counts integer or fractional values depending on the increment or decrement. If the MATLAB command statements between the for and end of a for-loop are few words lengthy, one can even write the whole for-loop in one line. The programming syntax and some examples on the for-loop are as follows:

◆ **Program syntax**

for *counter*=starting value:increment or decrement of the counter value:final value

Executable MATLAB command(s)

end

◆ **Example 1**

Our problem statement is to compute $y = \cos x$ for $x = 10^0$ to 70^0 with the increment 10^0. Let us assign the computed values to some variable y, where y should be [$\cos 10^0$ $\cos 20^0$ $\cos 30^0$ $\cos 40^0$ $\cos 50^0$ $\cos 60^0$ $\cos 70^0$]=[0.9848 0.9397 0.866 0.766 0.6428 0.5 0.342].

In the programming context, y(1) means the first element in the row matrix y, y(2) means the second element in the row matrix y, and so on. The MATLAB code for the $\cos x$ is cosd(x) where x is in degree. The for-loop counter expression should be k=1:1:7 or k=1:7 to have the control on the position index in the row matrix y (because there are 7 elements or indexes in y). Since the computation needs 10 to 70, one generates that writing k*10. Following is the implementation:

Executable M-file:
```
for k=1:1:7
        y(k)=cosd(k*10);
end
```

Or, as a one line:
```
for k=1:1:7 y(k)=cosd(k*10); end
```

Steps we need:

Open a new M-file (subsection 1.1.2), type the executable M-file statements in the M-file editor, save the editor contents by the name test in your working path, and call the test as shown below.

```
>>test ↵
>>y ↵

y =
     0.9848 0.9397 0.8660 0.7660  0.6428 0.5000  0.3420
```

♦ Example 2

For-loop helps us accumulate data (appendix C.8) consecutively controlled by the loop index. In this example we accumulate some data row-wise according to the for-loop counter index.

For $k=1$, 2, and 3, we intend to accumulate the k^2 side by side. At the end we should have [1 4 9] assigned to some variable f – this is our problem statement.

for the right shifting,
```
>>f=[ ]; for k=1:3 f=[f k^2]; end ↵
>>f ↵

f =
     1   4   9
```

for the left shifting,
```
>>f=[ ]; for k=1:3 f=[k^2 f]; end ↵
>>f ↵

f =
     9   4   1
```

The for-loop for the accumulation is presented above (corresponding to the right shifting). The vector code (appendix B) for k^2 is k^2. The statement f=[ ]; means that an empty matrix is assigned to f outside the loop but at the beginning. An empty matrix does not have any size and completely empty, it follows the null symbol ∅ of the matrix algebra. The k variation in our problem is put as the for-loop counter. How the for-loop accumulates is shown below.

When k=1, f=[f k^2]; returns f=[[ ] 1^2]; ⇒ f=1;
When k=2, f=[f k^2]; returns f=[1 2^2]; ⇒ f=[1 4];
When k=3, f=[f k^2]; returns f=[1 4 3^2]; ⇒ f=[1 4 9];

The accumulation is happening from the left to the right. A single change provides the shifting from the right to the left which is f=[k^2 f];. The complete code and its execution result are also shown beside the previous execution (corresponding to the left shifting) of the last page.

♦ **Example 3**

Another accumulation can be columnwise that is we wish to see the output like $\begin{bmatrix}1\\4\\9\end{bmatrix}$ in example 2.

We just insert the row separator of a rectangular matrix (done by the operator ;) in the command f=[f k^2];. Again the shifting can happen either from the up to down or from the down to up. Both implementations are shown below.

for the down shifting,

```
>>f=[ ]; for k=1:3 f=[f;k^2]; end ↵
>>f ↵

f =
         1
         4
         9
```

for the up shifting,

```
>>f=[ ]; for k=1:3 f=[k^2;f]; end ↵
>>f ↵

f =
         9
         4
         1
```

♦ **Example 4**

Many scientific problems need writing multiple for-loops. Usually one loop is for one dimensional function, two loops are for two dimensional function, and so on. One dimensional functional data takes the form of a row or column matrix.

Suppose we have the one dimensional data as y=[9 6 7 4 6]. We wish to access to every data in y. A single for-loop helps us do that as shown below.

```
>>y=[9 6 7 4 6]; for k=1:length(y) v=y(k); end ↵
```

First we assigned the data to workspace y as a row matrix. The command length finds the number of elements in the row matrix y. The y(k) means the k-th element in the y which we assigned to workspace v.(any user-chosen name). Every single data in y is sequentially available in v. The contents of y can also be a column matrix.

C.10 Mathematics readable form by pretty

There is a function called pretty which returns any functional expression as close as mathematical form provided that its input argument is in vector string or code (appendix B) form. For example in vector string form, the x^3 is coded as x^3. When x^3 is the input argument of the pretty, the return of the pretty is x^3 provided that the independent variable x is defined by using the syms before. The pretty is applied only for the display reason, no computation is conducted on the expression. If a function has the code 2*x, you will see it as 2 x. It does not work on numeric values but use of the command sym on the numeric values shows the rational form.

For example pretty(sym(3.3)) displays $\frac{33}{10}$. Just to show by another example, 1/(x+1) is $\frac{1}{x+1}$ which can be executed as

>>syms x ↵ ← Declaring the related x of $\frac{1}{x+1}$ by the syms

>>y=1/(x+1); ↵ ← Assigning the code of $\frac{1}{x+1}$ to y where y is a user-chosen variable

>>pretty(y) ↵ ← Applying the pretty on the codes stored in y

```
    1
  ------
  x + 1
```

C.11 Summing matrix elements

MATLAB function sum adds all elements in a row, column, or rectangular matrix when the matrix is its input argument. Example matrices are R=[1 –2 3 9], $C=\begin{bmatrix} 23 \\ -20 \\ 30 \\ 8 \end{bmatrix}$, and $A=\begin{bmatrix} 2 & 4 & 7 \\ -2 & 7 & 9 \\ 3 & 8 & -8 \end{bmatrix}$ whose all element sums are 11, 41, and 30 for the R, C, and A respectively. We execute the summations as follows (font equivalence is maintained by using the same letter for example A⇔ A):

Sum for the row matrix,
```
>>R=[1 -2 3 9]; ↵
>>sum(R) ↵

ans =
       11
```

Sum for the column matrix,
```
>>C=[23 -20 30 8]'; ↵
>>sum(C) ↵

ans =
       41
```

Sum for the rectangular matrix,
```
>>A=[2 4 7;-2 7 9;3 8 -8]; ↵
>>sum(sum(A)) ↵
ans =
       30
```

For a rectangular matrix, two functions are required because the inner sum performs the summing over each column and the result is a row matrix. The outer sum provides the sum over the resulting row matrix.

The function is operational for real, complex, even for symbolic variables like x or y.

C.12 Complex number basics

Symbolically the imaginary unit of a complex number is denoted by i, j, or $\sqrt{-1}$ whose MATLAB representation is i, j, or sqrt(-1). As an example the complex number $4+j5$ is entered into MATLAB by any of the following expressions 4+5i, 4+5*i, 4+i*5, 4+5*j, or 4+5*sqrt(-1).

Matrix of complex numbers follows similar entering style to that of the integer or real number with little difference in conjugateness (section 1.1.2). Let us enter the complex number matrices $R=[3-j \;\; 4j \;\; -4]$, $C=\begin{bmatrix} 7j \\ -4+5j \\ 8j \end{bmatrix}$, and $A=\begin{bmatrix} 2 & 5-j & 9j \\ 7j & 2+j & 11j \end{bmatrix}$ into MATLAB as conducted in the following:

for R,
```
>>R=[3-i 4i -4] ↵

R =
      3.0000 - 1.0000i        0 + 4.0000i   -4.0000
```
for C,
```
>>C=[7i -4+5i 8i].' ↵

C =
        0 + 7.0000i
  -4.0000 + 5.0000i
        0 + 8.0000i
```
for A,
```
>>A=[2 5-i 9i;7i 2+i 11i] ↵

A =
     2.0000                  5.0000 - 1.0000i        0 + 9.0000i
          0 + 7.0000i        2.0000 + 1.0000i        0 +11.0000i
```
The operators .' and ' mean transpose without and with conjugate respectively. In the column matrix case if we use the operator ' at the end, we would assign $\begin{bmatrix} -7j \\ -4-5j \\ -8j \end{bmatrix}$ to C.

Modulus or absolute value of a complex number $A+jB$ is given by $\sqrt{A^2+B^2}$. To take the modulus of a complex number, we use the command abs (abbreviation for absolute value) with the syntax abs(complex scalar or matrix). For example the modulus of $4+j3$ and elements in $R=[12+j5 \quad -4-j3 \quad -8+j6]$ are 5 and [13 5 10] respectively which we compute by using the right side attached command. In both cases we assigned the return to workspace A which can be any user-given name.

modulus for the single complex number,
```
>>A=abs(4+3i) ↵

A =
      5
```
modulus for the complex row matrix elements in R,
```
>>R=[12+5i -4-3i -8+6i]; ↵
>>A=abs(R) ↵

A =
      13    5   10
```

Argument of a complex number $A+jB$ is given by $\tan^{-1}\frac{B}{A}$. To find the argument, we use the function angle with the syntax angle (complex scalar or matrix name). The function returns any value from –π to π. For instance the arguments of $4+j3$ and each element in $R=[12+j5 \quad -4-j3 \quad -8+j6]$ are $\tan^{-1}\frac{3}{4}$=0.6435^c and [0.3948^c –2.4981^c 2.4981^c] respectively which we implement by the right side attached commands. In both cases we assigned the return to the workspace P which can be any user-given name. For degree to radian and radian to degree conversions we use the commands deg2rad and rad2deg respectively for example on P as rad2deg(P) for the degree.

argument for the single complex number:
```
>>angle(4+3i) ↵

P =
      0.6435
```
argument for the complex row matrix elements in R,
```
>>P=angle(R) ↵

P =
     0.3948  -2.4981   2.4981
```

The conjugate of a complex number $A+jB$ is given by $A-jB$. To find the conjugate of a complex number, we apply the function conj with the syntax conj(complex scalar or matrix name). As an example the conjugate of $4+j3$ and all elements in R=[$12+j5$ $-4-j3$ $-8+j6$] are $4-j3$ and [$12-j5$ $-4+j3$ $-8-j6$] respectively. Both implementations are shown below and assigned to the workspace C (user-given name).

conjugate for the single complex number,

```
>>C=conj(4+3i) ↵

C =
        4.0000 - 3.0000i
```

conjugate of the elements in the row matrix R,

```
>>R=[12+5i -4-3i -8+6i]; ↵
>>C=conj(R) ↵

C =
        12.0000-5.0000i -4.0000+3.0000i -8.0000- 6.0000i
```

```
>>real(R) ↵         ← for the real elements in R

ans =
        12   -4   -8
>>imag(R) ↵         ← for the imaginary elements in R

ans =
        5   -3   6
```

A complex number $A+jB$ has the real part A and the imaginary part B. To find the real and imaginary parts from complex number(s), we apply the functions real and imag with the syntax real (complex scalar or matrix name) and imag (complex scalar or matrix name) respectively. The real and imaginary parts for the elements in R=[$12+j5$ $-4-j3$ $-8+j6$] are [12 −4 −8] and [5 −3 6] respectively whose findings are attached above on the right side in this paragraph. The returns could have been assigned to some user-supplied variables.

from rectangular to polar conversion,

```
>>[t r]=cart2pol(5,4) ↵

t =
        0.6747
r =
        6.4031
```

from polar to rectangular conversion,

```
>>[A B]=pol2cart(0.6747,6.4031) ↵

A =
        5.0001
B =
        3.9998
```

Rectangular to polar conversion is widely used in electrical engineering which we address now. Given a complex number in rectangular form $A+jB$, the polar or exponential form of the number is $re^{j\theta}$ where $r=\sqrt{A^2+B^2}$ and $\theta=\tan^{-1}\frac{B}{A}$. Its MATLAB counterpart is cart2pol (abbreviation for the Cartesian to (2) polar) and we use the syntax [θ, r]=cart2pol(A, B) where θ is in radian. Again given the polar form of a complex number $re^{j\theta}$, the reverse conversion is $A=r\cos\theta$ and $B=r\sin\theta$ and the resembling MATLAB function is pol2cart (abbreviation for the polar to (2) Cartesian) with the syntax [A, B]=pol2cart(θ in radian, r). Unavailability of θ makes us write t instead of θ. The rectangular form number $5+j4$ has the polar form (r, θ)=(6.4031,0.6747^c) which we intend to obtain and vice versa. We see both implementations above on the right side box in this paragraph. Slight discrepancy is seen in the implementation, instead of 5 we are getting 5.0001. When we write

0.6747 as θ, we ignore the fifth digit and whatsoafter. The use of [A B]=pol2cart(t,r) returns the perfect result because t holds the complete data from the computation. Anyhow the input arguments of both functions can be matrix as well and output arguments are so.

Now we address how we turn some real data to complex number. Suppose we have two identical size row matrices x=[5 6 7] and y=[8 9 –9] and we wish to form the complex matrix $A=[5+j8 \quad 6+j9 \quad 7-j9]$ which needs us to exercise the commands x=[5 6 7]; y=[8 9 -9]; A=x+i*y;. Again say we have some polar or exponential form data like r=[6 4 3] and $\theta=[\frac{\pi}{3} \quad -\frac{\pi}{7} \quad \frac{\pi}{2}]$ and we intend to form $A=[6e^{i\frac{\pi}{3}} \quad 4e^{-i\frac{\pi}{7}} \quad 3e^{i\frac{\pi}{2}}]$ which requires us to execute the following: r=[6 4 3]; t=[pi/3 -pi/7 pi/2]; A=r.*exp(i*t); where t for θ and we use the scalar code .* for the multiplication (appendix B). If t were in degrees, the command would be A= r.*exp(i*t*pi/180);.

Appendix D

Algebraic equation solver

By virtue of the built-in master function solve, we find the solution of a single or multiple algebraic equations when the equations are its input arguments. The notion of the solution is symbolic and a substantial number of simultaneous linear, algebraic, or trigonometric equations can be solved by using the solve. The common syntax of the implementation is solve (equation-1, equation-2,......so on in vector string form – appendix B, unknowns of the equations separated by a comma but put under quote).

The return from the function solve is in general a structure array which is beyond the discussion of the text (see reference [4]). Very briefly a structure array is composed of several members. In order to view the solution from the solve, one needs to call individual member of the array. If s is a structure array and u is one of its members, we call the member by using the command s.u. One can assign the s.u to some other workspace variable if it is necessary. Following points must be considered while using the solve for the signal and system analysis:

(a) In basic signal and system study mainly we solve the simultaneous linear equations or the equations in which all related variables have power 1. For example the one variable equation $2x-7=5x$ is written as 2*x-7=5*x in code form, the two variable equations $7x+y-7=0$ and $2y+4=5x$ are written as 7*x+y-7=0 and 2*y+4=5*x respectively, and so on.

(b) Since the solution approach is completely symbolic, we enter rational value of the equation coefficient in case of decimal data for example the equation $2.4x-7.5=5x$ is written as 24*x/10-75/10=5*x.

(c) In signal and system course customarily we use i for the current. But the i or j is the unit imaginary number in MATLAB (appendix C.12). Such use sometimes turns the function solve non-executable. If any variable i is in the system equations, we use a dummy variable like c for the current.

(d) The return from the solve is usually in rational form for instance 24/10 instead of 2.4. If we need the decimal value, we employ the command double on the return.

(e) The equations are assigned under quote while entering as the input arguments to the solve or assigning to some variables.

Let us go through the solution finding for following three examples.

⬥⬥ Example 1

The solution of the equation $2x+7=9x$ is $x=1$ which we wish to find. We execute the following for the solution at the command prompt:

```
>>s=solve('2*x+7=9*x','x') ↵

s =

1
```

The s in the last execution is any user-chosen name. The s=1 return indicates the $x=1$ solution. The independent variable in the given equation is x that is why the second input argument of the solve is 'x'.

Example 2

The equation set $\begin{Bmatrix} 6x-y=-8 \\ 9x=8y+5 \end{Bmatrix}$ has the solution $x=-\frac{23}{13}$=–1.7692 and $y=-\frac{34}{13}$=–2.6154 and our objective is to obtain the solution.

The given two equations have the codes 6*x-y=-8 and 9*x=8*y+5 respectively. The related variables in the two equations are x and y therefore we carry out the following at the command prompt:

```
>>s=solve('6*x-y=-8','9*x=8*y+5','x','y') ↵

s =

          x: [1x1 sym]
          y: [1x1 sym]
```

The s in above execution is also any user-chosen name. The solve returns the solution to the s. As we mentioned earlier that the return from the solve is a structure array and its members are the x and y (related variables in the given equations). The return is an object (called symbolic object, indicated by the sym) rather than data. If we intend to pick up the solution of x and y from s, we need to exercise the commands s.x and s.y respectively. Let us see what we obtain as the solution:

For the x value:

```
>>s.x ↵

ans =

-23/13
>>double(s.x) ↵

ans =
          -1.7692
```

For the y value:

```
>>s.y ↵

ans =

-34/13
>>double(s.y) ↵

ans =
          -2.6154
```

The result is as expected. We could have assigned the return to some variable for example s.x or double(s.x) to a by writing a=s.x or a=double(s.x).

Example 3

For the multiple equations it is not feasible that we enter all equations as one line in the solve. Instead we first assign the given equations to some user-chosen names and then call the solve with these names as the input arguments. The equation set $\{x-y-3.2z+2u=-8,\ 8.5y-7z+u=5,\ x-4y+2z=76,\ \ -3.4x+6z+7u=-12\}$ has the solution $u=\frac{29598}{1387}$=21.3396, $x=\frac{348525}{2774}$=125.6399, $y=\frac{95869}{2774}$=34.5598, and $z=\frac{245775}{5548}$=44.2997 which we find by exercising ongoing function and terminology as follows:

```
>>e1='x-y-32*z/10+2*u=-8'; ↵ ← Assigning the first equation to e1
>>e2='85*y/10-7*z+u=5'; ↵     ← Assigning the second equation to e2
```

```
>>e3='x-4*y+2*z=76'; ↵                  ← Assigning the third equation to e3
>>e4='-34*x/10+6*z+7*u=-12'; ↵          ← Assigning the fourth equation to e4
>>s=solve(e1,e2,e3,e4,'x','y','z','u') ↵ ← calling the solve on e1, e2, e3, and e4

s =                            ← s holds the solution as a structure array
        u: [1x1 sym]           ← u is a member of s
        x: [1x1 sym]           ← x is a member of s
        y: [1x1 sym]           ← y is a member of s
        z: [1x1 sym]           ← z is a member of s
```

The e1, e2, e3, and e4 are all user chosen names in above implementation. The next step is to see the values return by the solve:

for the rational value of x :

```
>>s.x ↵

ans =

348525/2774
```

for the rational value of y :

```
>>s.y ↵

ans =

95869/2774
```

for the rational value of z :

```
>>s.z ↵

ans =

245775/5548
```

for the rational value of u :

```
>>s.u ↵

ans =

29598/1387
```

for the decimal value of x :

```
>>double(s.x) ↵

ans =
          125.6399
```

for the decimal value of y :

```
>>double(s.y) ↵

ans =
          34.5598
```

for the decimal value of z :

```
>>double(s.z) ↵

ans =
          44.2997
```

for the decimal value of u :

```
>>double(s.u) ↵

ans =
          21.3396
```

Note that when we assign the equations we use the quote but inside the solve the assignees do not have the quote for example e1 not 'e1'. If we wish to see all four values as a four element row matrix, we exercise the following:

```
>>[s.x s.y s.z s.u] ↵

ans =

[ 348525/2774,  95869/2774, 245775/5548,  29598/1387]
```

All four values in decimal form as a row matrix are seen as follows:

```
>>double([s.x s.y s.z s.u]) ↵

ans =
          125.6399   34.5598   44.2997   21.3396
```

Obviously the return is in the order we placed respectively.

Appendix E

Creating a function file

A function file is a special type of M-file (subsection 1.1.2) which has some user-defined input and output arguments. Both arguments can be single or multiple. The first line in a function file always starts with the reserve word function. The function file must be in your working path or its path must be defined in MATLAB. Depending on the problem, a function file is written by the user and can be called from the MATLAB command prompt or from another M-file. For convenience, long and clumsy programs are split into smaller modules and these modules are written in a function file. However the basic structure of a function file is as follows:

MATLAB Command Prompt → function file

$>> g$ =call f ⟹ $\underbrace{g(y_1, y_2, \ldots y_m)}_{\text{output arguments}} = f\underbrace{(x_1, x_2, x_3, \ldots x_n)}_{\text{input arguments}}$

We present following examples for the illustration of function files keeping in mind that the arguments' order and type of the caller and the function file are identical.

Example 1

Let us say computation of $f(x) = x^2 - x - 8$ is to be implemented as a function file. When $x = -3$ and $x = 5$, we should be having 4 and 12 respectively. The vector code (appendix B) for the function is x^2-x-8 assuming the x is a scalar and obviously x is for x. We have one input (which is x) and one output (which is $f(x)$). Open a new M-file editor, type the codes of the figure E.1(a) exactly as they appear in the M-file, and save the file by the name f. The assignee y and independent variable x can be any name of your choice and which are the output and input arguments of the function respectively. Again the file and function name f can be any user-chosen name only the point is the chosen function or file name should not exist in MATLAB. Let us call the function f(x) to verify the programming as shown above in right side text box of this paragraph. You can write dozens of MATLAB executable statements in the file but whatever is assigned to the last y returns the function f(x) to g. Writing the = sign between the y and f(x) in the function file is compulsory.

Calling for example 1: for $x = -3$,

```
>>g=f(-3) ↵      ← call f(x) for x = -3

g =
      4
```

for $x = 5$,

```
>>g=f(5) ↵       ← call f(x) for x = 5

g =
     12
```

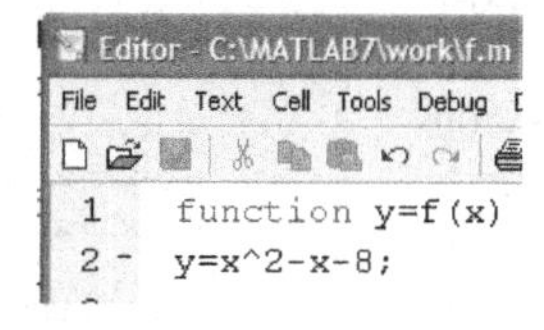

Figure E.1(a) Single input – single output function file

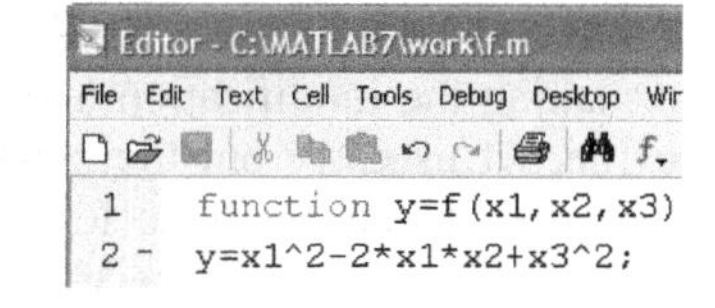

Figure E.1(b) Multiple inputs – single output function file

Example 2

Example 1 presents one input-one output function how if we handle multiple inputs and one output? The input argument names are separated by commas in a function file. A three variable function $f(x_1, x_2, x_3) = x_1^2 - 2x_1x_2 + x_3^2$ is to be computed from a function file. The

Calling for example 2: when input arguments are all scalar:

```
>>g=f(3,4,5) ↵     ← calling f(x1, x2, x3) for x1=3, x2=4, and x3=5

g =
      10
```

Calling for the example 2: when input arguments are all column matrix:

```
>>x1=[2 3 4]'; ↵   ← x1 values are assigned to x1 as a column matrix
>>x2=[-2 2 5]'; ↵  ← x2 values are assigned to x2 as a column matrix
>>x3=[1 0 3]'; ↵   ← x3 values are assigned to x3 as a column matrix
>>f(x1,x2,x3) ↵    ← calling f(x1, x2, x3) using column matrix input arguments

ans =
        13
        -3
       -15
```

input arguments (assuming all scalar) are x_1, x_2, and x_3 and the output argument is the functional value of the function. The x_1 is written as x1, and so is the others. Follow the procedure of the example 1 but the code should be as shown in the figure E.1(b). Let us inspect the function (with the specific x_1=3, x_2=4, and x_3=5, the output value of the three variable function must be $f(3,4,5) = 3^2 - 2\times3\times4 + 5^2 = 10$) as presented in above text box.

The function not only works for the scalar inputs but also does for matrices in general for example a set of input argument values are $x_1 = \begin{bmatrix} 2 \\ 3 \\ 4 \end{bmatrix}$, $x_2 = \begin{bmatrix} -2 \\ 2 \\ 5 \end{bmatrix}$, and $x_3 = \begin{bmatrix} 1 \\ 0 \\ 3 \end{bmatrix}$ for which the $f(x_1, x_2, x_3)$ values should be $\begin{bmatrix} 13 \\ -3 \\ -15 \end{bmatrix}$ respectively. The computation needs the scalar code (appendix B) of $f(x_1, x_2, x_3)$ regarding x_1, x_2, and x_3. The modified second line statement of the figure E.1(b) now should be y=x1.^2-2*x1.*x2+x3.^2;. On making the modification and saving the file, let us carry out the commands which are placed in above text box of this page too. If it is necessary, the output can be assigned to user-given workspace variable v by writing v=f(x1,x2,x3) at the command prompt. The return from the function file also follows the same input matrix order. If the input arguments of f(x1,x2,x3) are rectangular

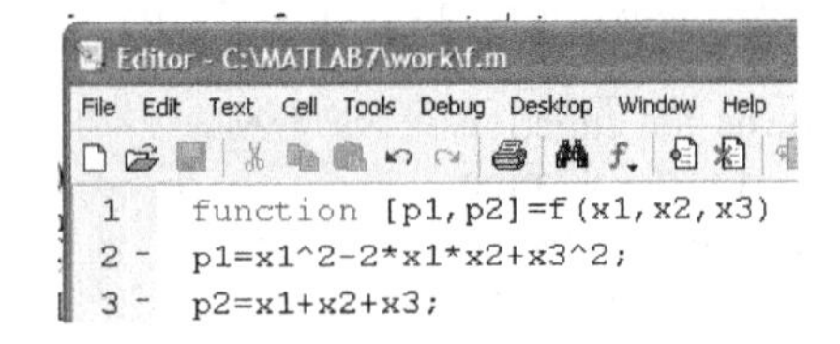

Figure E.1(c) Function file for three input and two output arguments

matrix, so is the output. The input arguments of the function file do not have to be the mathematics symbol. Suppose x_1=ID, x_2=Value, and x_3=Data, one could have written the first and second lines of the function file in the figure E.1(b) as function y=f(ID,Value,Data) and y=ID.^2-2*ID.*Value +Data.^2; respectively.

Example 3

To illustrate the multi-input and multi-output function file, let us consider that p_1 and p_2 are to be found from three variables x_1, x_2, and x_3 (all are scalars) by employing the expressions $p_1 = x_1^2 - 2x_1 x_2 + x_3^2$ and $p_2 = x_1 + x_2 + x_3$ whose function file (type the codes in a new M-file editor and save the file by the name f) is presented in the figure E.1(c).

Choosing x_1=4, x_2=5, and x_3=6, one should get p_1=12 and p_2=15 for which right side text box commands are conducted at the command prompt. More than one output arguments (which are here p_1 are p_2 and represented by p1 and p2 respectively) are separated by commas and placed inside the third brace following the word function of the figure E.1(c).

Function file calling for the example 3:

```
>>[p1,p2]=f(4,5,6) ↵ ← calling the function file f for p1 and
                        p2 using x1=4, x2=5, and x3=6

p1 =

    12

p2 =

    15
```

When we call the function from the command prompt, the output argument writing is similar to that of the function file (that is why we write [p1,p2] as output arguments at the command prompt). The output argument variable names do not have to be p1 and p2 and can be any name of user's choice. If there were three output arguments p_1, p_2, and p_3, the output arguments in the function file would be written as [p1,p2,p3] and their calling would happen in a like manner.

Note: We saved different function files by the same name f just for simplicity and maintaining unifying approach. By this action any previously saved file by the name f disappears. What we suggest is save the function file by other name like f1 and call accordingly for instance the first line of figure E.1(c) would be function [p1,p2]=f1(x1,x2,x3) and calling would take place [p1,p2]=f1(4,5,6) for the last illustration.

Appendix F

Graphing functions in MATLAB

One of the nicest features of MATLAB is you can have your graphics drawn while you are programming the signal and system problems. There are so many easy accessible built-in graphics functions that one finds it very interesting when the input-output argumentation style of these functions is understood. Some graphing functions which we frequently applied in previous chapters are addressed here for the syntax details.

Functions of the form $y = f(x)$

If any function is of the form $y = f(x)$ and the $f(x)$ versus x is to be graphed, the built-in function ezplot is the best option which uses a syntax ezplot(function vector code under quote according to appendix B, interval bounds as a two element row matrix) where the first and second elements in the row matrix are the beginning and ending bounds of the interval respectively. The ezplot graphs $y = f(x)$ in the default interval $-2\pi \le x \le 2\pi$ when no interval description is argumented.

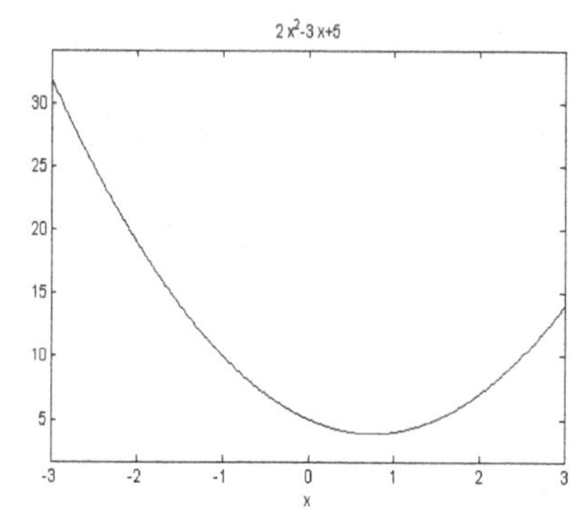

Figure F.1(a) Plot of $y = 2x^2 - 3x + 5$ versus x over $-3 \le x \le 3$

Let us say we intend to plot the function $y = 2x^2 - 3x + 5$ over the interval $-3 \le x \le 3$. We first give $2x^2 - 3x + 5$ MATLAB vector code and then assign that to y as follows:

```
>>y='2*x^2-3*x+5'; ↵
```

In above implementation the y is any user-chosen name. The interval $-3 \le x \le 3$ is entered by [-3 3]. To obtain the plot of y in the given interval, we execute the following at the command prompt:

```
>>ezplot(y,[-3,3]) ↵
```

Above command results in the figure F.1(a).

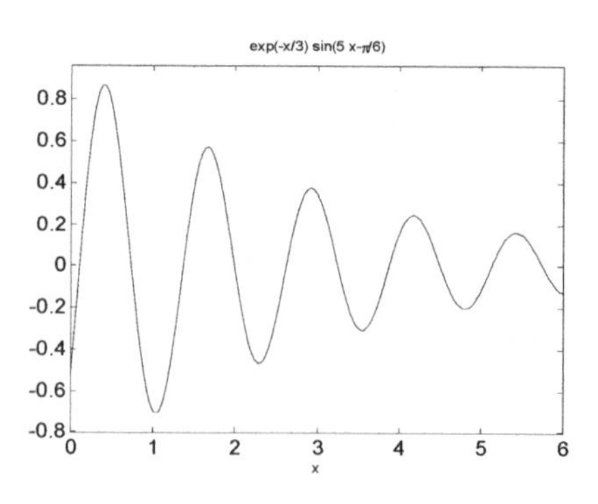

Figure F.1(b) Plot of $y = e^{-\frac{x}{3}} \sin(5x - \frac{\pi}{6})$ versus x over $0 \le x \le 6$

As another example, let us plot the damped sine wave $y = e^{-\frac{x}{3}} \sin\left(5x - \frac{\pi}{6}\right)$ over the interval $0 \le x \le 6$ whose execution is the following:

```
>>y='exp(-x/3)*sin(5*x-pi/6)'; ↵
```

```
>>ezplot(y,[0,6]) ↵
```

Figure F.1(b) is the outcome after the execution.

◆◆ y versus x data

The function plot graphs y versus x data. Let us say we have the attached (on the right side) tabular data. We intend to plot these data as y versus x graph. Commands to plot the data are also presented above on the rightmost side in this paragraph. First we assign the x and y data to the workspace variables x and y (any user-given name) respectively and then call the function plot to see the figure F.1(c). The plot has two input arguments, the first and second of which are the x and y data both as a row or column matrix of identical size respectively.

Tabular data							Command to plot the y vs x data:
x	-6	-4	0	4	5	7	>>x=[-6 -4 0 4 5 7]; ↵ >>y=[9 3 -3 -5 2 0]; ↵ >>plot(x,y) ↵
y	9	3	-3	-5	2	0	

In order to plot the mathematical expression using the plot, one first needs to calculate the functional values using the scalar code (appendix B) and then applies the function. During the calculation, computational step selection is mandatory which is completely user-defined.

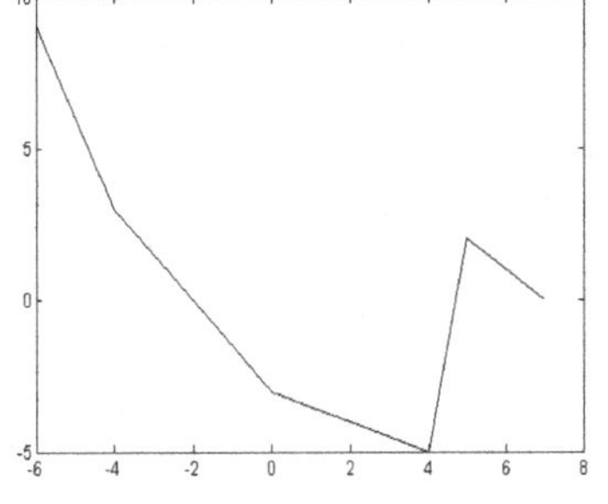

Figure F.1(c) y vs x plot of the tabular data

For instance we wish to plot the function $f(x)=x^2-x+2$ over $-2\leq x\leq 3$. Let us choose some x step size say 0.1. The x vector as a row matrix is generated by x=-2:0.1:3; (subsection 1.1.2). At every element in x vector, the functional value is computed and assigned to workspace f by f=x.^2-x+2;. The f is any user-given name. Now we call the function by writing plot(x,f) to see the graph (not shown for space reason). The function plot just draws the graph, no graphical features such as x axis label or title are added to the graph. It is the user who is supposed to add these graphical features.

◆◆ Multiple y data versus common x data

The plot keeps many options, one of which is just discussed. We graph several y data versus common x data with the help of the same plot but with different number of input arguments. Let us choose the right side attached table for the graphing. We intend to plot the y_1, y_2, and y_3 data on common x data. To do so,

Tabular data for multiple y versus common x:						
x	-6	-4	0	4	5	7
y_1	9	3	-3	-5	2	0
y_2	0	-2	1	0	5	7.7
y_3	-1	2	8	1	0	-3

```
>>x=[-6 -4 0 4 5 7]; ↵        ← Assigning the x data as a row matrix to x
>>y1=[9 3 -3 -5 2 0]; ↵       ← Assigning the y1 data as a row matrix to y1
>>y2=[0 -2 1 0 5 7.7]; ↵      ← Assigning the y2 data as a row matrix to y2
>>y3=[-1 2 8 1 0 -3]; ↵       ← Assigning the y3 data as a row matrix to y3
```

>>plot(x,y1,x,y2,x,y3) ↵ ← Applying the function plot

The plot now has six input arguments – two for each graph, the first and second of which are the common x data and y data to be plotted respectively. If there were four y data, the command would be plot(x,y1,x,y2,x,y3,x,y4). Once the data is plotted for several y, identifying the y traces is obvious and which is carried out by the command legend. The command legend('y1','y2','y3') puts identifying marks/colors among various graphs. The input argument of the legend is any user-given name but under quote and separated by a comma. The number of y traces must be equal to the number of input arguments of the legend. We gave the names y1, y2, and y3 for the three y traces respectively. In doing so, we end up with the figure F.1(d). You can even move the legend on the plot area by using the mouse. You see all graphics throughout the text as black and white because we did not include color graphics in the text (for expense reason). But MATLAB displays figures in color plots, which you can easily identify.

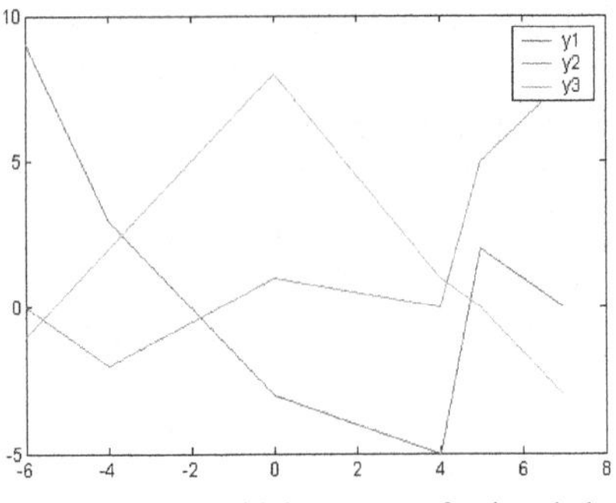

Figure F.1(d) Multiple y vs x for the tabular data

Another situation can be we have several functions and we intend to plot them on common x variation. For instance we wish to graph $y_1 = x^3 - x^2 + 4$ and $y_2 = x^2 - 7x - 5$ on common $-1 \le x \le 3$.

Under these circumstances, the step selection of the x data is compulsory. Without calculating the functional values of the given y curves, we can not graph the functions and for which we use the scalar code. Let us choose the x step size as 0.1. We first generate the common x vector as a row matrix by writing x=-1:0.1:3; and then calculate the y_1 and y_2 (y1⇔y_1 and y2⇔y_2) data by writing y1=x.^3-x.^2+4; y2=x.^2-7*x-5; and eventually the plot appears by executing plot(x,y1,x,y2), graph not shown for space reason. Thus you can plot three or more functions.

Commands for the figure F.1(e):

>>subplot(121) ↵ ← It handles the first graph
>>ezplot('x') ↵ ← Plotting $y = x$
>>subplot(122) ↵ ← It handles the second graph
>>ezplot('exp(-x)') ↵ ← Plotting $y = e^{-x}$

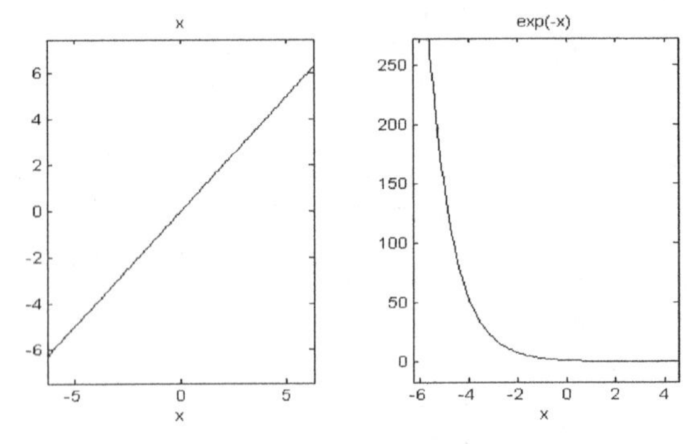

Figure F.1(e) Plots of $y = x$ and $y = e^{-x}$ side by side in the same window

Multiple graphs in the same window

The function subplot splits a figure window in subwindows based on the user definition. It accepts three positive integer numbers as the input arguments, the first and second of which indicate the number of subwindows in the horizontal and the number of subwindows in the vertical directions respectively. For example 22 means two subwindows horizontally and two subwindows vertically, 32 means three subwindows horizontally and two subwindows vertically, ... and so on. The third integer in the input argument numbered consecutively just offers the control on the subwindows so generated. If the first two digits are 32, there should be 6 subwindows and they are numbered and controlled by using 1 through 6. When you plot some graph in a subwindow, as if you are handling an independent figure window.

Let us say we intend to graph $y=x$ and $y=e^{-x}$ side by side as two different plots by using earlier mentioned ezplot but in the same window. If we imagine the subfigures as matrix elements, we have a figure matrix of size 1×2 (one row and two columns). That is why the first two integers of the input argument of the subplot should be 12. Attached commands on the right side of the last paragraph in the last page show the figure F.1(e). The third integers 1 and 2 in the subplot give

Commands for the figure F.1(f):

>>subplot(221) ↵	← Subfigure selection for $y=x$
>>ezplot('x') ↵	← Plotting $y=x$
>>subplot(222) ↵	← Subfigure selection for $y=e^{-x}$
>>ezplot('exp(-x)') ↵	← Plotting $y=e^{-x}$
>>subplot(212) ↵	← Subfigure selection for $y=(1-e^{-x})$
>>ezplot('1-exp(-x)') ↵	← Plotting $y=(1-e^{-x})$

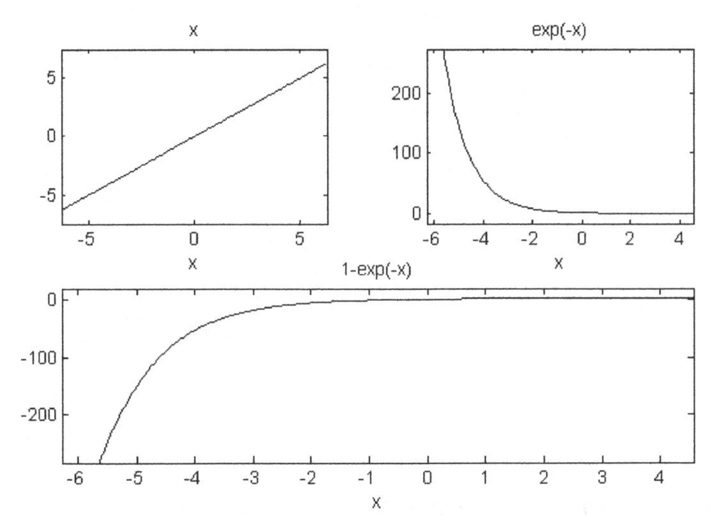

Figure F.1(f) Plots of $y=x$ and $y=e^{-x}$ in the upper row and $y=(1-e^{-x})$ in the lower row in the same window

the control on the first and second subfigures respectively.

As another example we wish to plot $y = x$ and $y = e^{-x}$ in the upper row and only $y = (1 - e^{-x})$ in the lower row subfigures in the same window whose implementation needs second paragraph attached text box commands of the last page and whose final output is the figure F.1(f). We are supposed to have four figures when the integer input argument of the `subplot` is 22 (two for rows and two for columns). The arguments 221, 222, 223, and 224 provide handle on the four figures consecutively. But the figures could have been plotted on 223 and 224 are absent so we ignore them. The argument 21 creates two subfigures (two rows and one column) handled by 211 and 212, but 211 is absent so we ignore that. Let us see the input arguments of the `subplot` for different subfigures (each third brace set [ ] is one subfigure in the following tabular representation) as follows:

Subfigures needed	First two input integers of subplot	Third input integer of subplot	Commands we need
[ ] [ ] [ ] [ ]	22	[1] [2] [3] [4]	subplot(221) subplot(222) subplot(223) subplot(224)
[ ] [ ] [ ]	22 for upper two (lower two remain empty) 21 for the lower one (upper one remains empty)	[1] [2] [2]	subplot(221) subplot(222) subplot(212)
[ ] [ ] [ ]	21 for the upper one (lower one remains empty) 22 for the lower two (upper two remain empty)	[1] [3] [4]	subplot(211) subplot(223) subplot(224)
[ ] (tall [ ]) [ ]	22 for the left two (right two remain empty) 12 for the right one (left one remains empty)	[1] (tall [2]) [3]	subplot(221) subplot(223) subplot(122)
(tall [ ]) [ ] [ ]	22 for right two (left two remain empty) 12 for the left one (right one remains empty)	(tall [1]) [2] [4]	subplot(222) subplot(224) subplot(121)

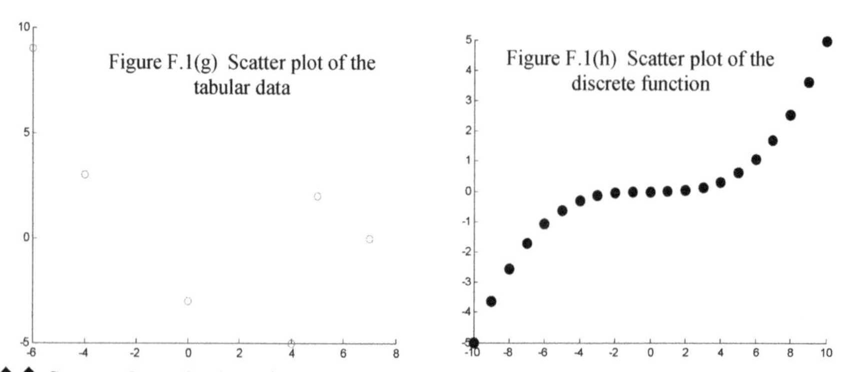

Figure F.1(g) Scatter plot of the tabular data

Figure F.1(h) Scatter plot of the discrete function

❖ Scatter data plot by using small circles for discrete signal

Instead of having a graph as continuous line, it is possible to have the graph in terms of bold dots or round circles like the figure F.1(g). The function `scatter` executes this sort of graph for which the common syntax is scatter(x data as a row matrix, y data as a row matrix, size of the circle, color of the circle). The function also accepts the first two input arguments. The size of the circle is any user-given integer number. The larger is the number, the bigger is the size for example 75, 100, etc.

Let us graph the table F.1 data as the scatter plot. The command we need is placed on the right side attached text box. Upon execution of the command, we see the figure F.1(g). The color of the circle is blue by default but any three element row matrix sets the user-defined color. The three element row matrix refers to red, green, and blue components respectively each one within 0 and 1. Black color means all zero, white means all 1, red means other two components zero, and so on. The circle displayed in the figure F.1(g) is all empty but one fills the circle by using the reserve word filled under quote and included as another input argument to the scatter. Let us say we intend to scatter graph with circle size 100 and the circles should be filled with black color. The necessary command is scatter(x,y,100,[0 0 0],'filled'), graph is not shown for space reason.

Table F.1: Some x - y data

x	-6	-4	0	4	5	7
y	9	3	-3	-5	2	0

Scatter plot for the table F.1 data:

```
>>x=[-6 -4 0 4 5 7]; ↵
>>y=[9 3 -3 -5 2 0]; ↵
>>scatter(x,y) ↵
```

This type of graph is suitable for representation of the function which is discrete in nature. For instance the discrete function $y[n]=\frac{n^3}{200}$ over $-10 \le n \le 10$ is to be plotted with black circles of size 100 where n is integer. As a procedure, we form a row matrix n to generate (subsection 1.1.2) the interval with start value -10, increment 1, and end value 10 writing n=-10:10;. The scalar code of appendix B computes the $y[n]$ values and assigns those to workspace y. However the complete code is placed on the right side text box which brings about the graph of the figure F.1(h).

Scatter plot for discrete function:

```
>>n=-10:10; y=n.^3/200; ↵
>>scatter(n,y,100,[0 0 0],'filled') ↵
```

♦♦ Discrete function or data plotting using vertical lines

Just now we discussed how one graphs the discrete functional data by using bold dots. There is another option which graphs any discrete data by using vertical lines proportionate to the discrete functional values. A discrete function may exist in two forms – data and expression based. The built-in function that graphs a discrete function is stem which has a syntax stem(x data as a row matrix, y data

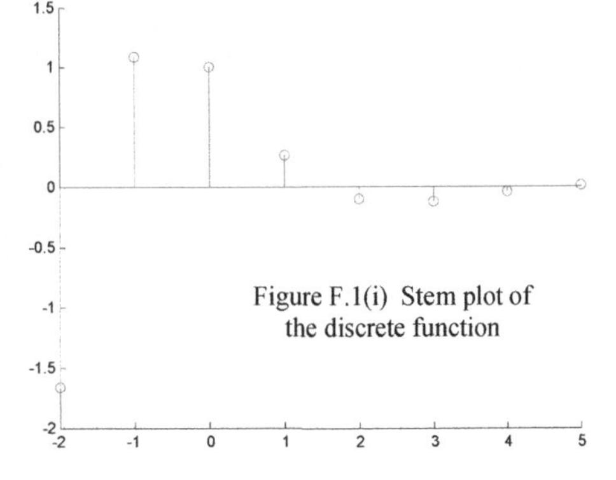

Figure F.1(i) Stem plot of the discrete function

as a row matrix). If we have some expression based discrete function, first the sample values of the discrete function need to be calculated through the scalar code (appendix B) and after that the graphing is performed.

Let us plot the discrete function $f[n]=2^{-n}\cos n$ over the integer interval $-2\le n\le 5$. Attached on the right side text box is the implementation of the command which results in the figure F.1(i). In the implementation, the workspace n and f hold the eight integers from –2 to 5 and sample values of $f[n]$ both as a row matrix.

> Discrete function plot using vertical lines:
> ```
> >>n=-2:5; ↵
> >>f=2.^(-n).*cos(n); ↵
> >>stem(n,f) ↵
> ```

By default the vertical line color of the stem plot is blue. User-defined color of the vertical lines is obtainable adding one more input argument to the stem mentioning the color type but under quote (r for red, g for green, b for blue, c for cyan, m for magenta, y for yellow, k for black, and w for white). If we wish to set the vertical line color as green for the graph of the figure F.1(i), we exercise the command stem(n,f,'g'). The vertical line head circles can be filled by using the command stem(n,f,'g','filled') where filled is a reserve word placed under quote.

If the reader wants to plot the table F.1 data considering discrete signals $x[n]$ and $y[n]$ as stem plot, following needs to be exercised:

```
>>x=[-6 -4 0 4 5 7]; ↵          ← x holds x[n] data
>>y=[9 3 -3 -5 2 0]; ↵          ←y holds y[n] data
>>stem(x,y,'k','filled') ↵      ← calling the stem for plot with black line and filled circles
```

Figure F.1(j) is output from above executions in which the horizontal and vertical axes of the figure correspond to $x[n]$ and $y[n]$ data respectively.

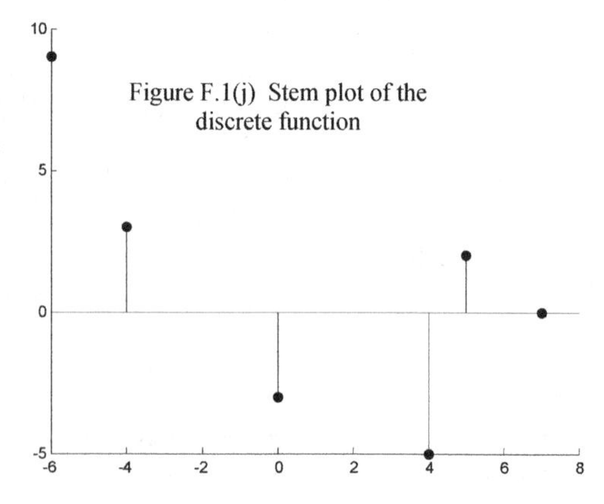

Figure F.1(j) Stem plot of the discrete function

Appendix G

Symbolic integration of functions

For symbolic integration of functions we utilize the function int (abbreviation for the integration). One can say that the int of MATLAB is equivalent to the integration symbol $\int \ldots\, dx$ or $\int \ldots\, dt$ of calculus. Indefinite integration of a function is accompanied by a constant which is not returned by the int. The independent variable of the integrand is declared by using the syms before any integration is conducted. The syntax we use for the integration is int(function or integrand as a vector string – appendix B). All through in this appendix we are going to maintain the font equivalence like x$\Leftrightarrow x$. Known is $\int \cos x dx = \sin x$ and we wish to compute it. The $\cos x$ has the independent variable x and code cos(x). We carry out the integration as follows:

```
>>syms x, I=int(cos(x)) ↵   ← one space gap between syms and x

I =

sin(x)
```

We use a comma for the separation between two MATLAB statements. The integration is conducted and the return from the int is assigned to the workspace I which is any user-given name. The return is also the vector code of the resulting computation. Once you have the integration result in I, use the command pretty to see the nice readable form of the integration. Let us see more examples on the symbolic integration.

Single indefinite integration

In the following we include some single indefinite integration alongside its final integration result which we wish to compute.

$$A.\ \int \frac{(\ln x)^4}{x} dx = \tfrac{1}{5}(\ln x)^5$$

$$B.\ \int \frac{1}{x(x^2+1)^2} dx = \ln x - \frac{1}{2}\ln(x^2+1) + \frac{1}{2(x^2+1)}$$

$$C.\ \int \sin^3 x \cos^3 x dx = -\frac{1}{6}\sin^2 x \cos^4 x - \frac{1}{12}\cos^4 x$$

The MATLAB executions maintaining earlier mentioned symbology and function are the following:

For example A:

```
>>syms x, I=int((log(x)^4)/x); pretty(I) ↵
                  5
        1/5  log(x)
```

For example B:

```
>>syms x, I=int(1/x/(x^2+1)^2); pretty(I) ↵

                          2                  1
      log(x)  -  1/2 log(x   + 1) +  1/2  ---------
                                              2
                                             x  + 1
```

For example C:

```
>>syms x, I=int(sin(x)^3*cos(x)^3); pretty(I) ↵

                   2       4              4
       - 1/6 sin(x)  cos(x)  - 1/12 cos(x)
```

Single definite integration

Aforementioned int is also applicable here but with different number of input arguments which accounts for the lower and upper limits of the integration. To evaluate a single definite integral of the form $\int_{x=a}^{x=b} f(x)\,dx$, we use the syntax int(vector code of $f(x)$, x, a, b). Writing the independent variable inside the int is optional. Let us consider following definite integrals whose computed values are also given and which we intend to compute.

A. $\int_{x=1}^{x=4} \frac{(\ln x)^4}{x} dx = \frac{32}{5}(\ln 2)^5$

B. $\int_{x=3}^{x=9} \frac{1}{x(x^2+1)^2} dx = \ln 3 - \frac{1}{2}\ln\frac{41}{5} - \frac{9}{205}$

C. $\int_{x=-\frac{\pi}{2}}^{x=\frac{\pi}{2}} \sin^3 x \cos^3 x dx = 0$

Following indicates the single definite integration at MATLAB command prompt:

For example A:

```
>>syms x, I=int(log(x)^4/x,1,4); pretty(I) ↵

                     5
       32/5  log(2)
```

For example B:

```
>>syms x, I=int(1/x/(x^2+1)^2,3,9); ↵
>>pretty(I) ↵

       log(3) - 1/2 log(41) - 9/205 + 1/2 log(5)
```

For example C:

```
>>syms x, I=int(sin(x)^3*cos(x)^3,-pi/2,pi/2); ↵
>>pretty(I) ↵

       0
```

In all examples we assigned the computed values to the workspace I which holds the vector code of the computation and the pretty just shows the readable form.

One might be interested about the decimal value of the integral following the computation for example $\frac{32}{5}(\ln 2)^5 = 1.024$ in the example A for that we use the command double(I).

Double and triple integrations

In a similar fashion one conducts the double and triple definite or indefinite integrations by employing two and three int functions respectively (reference 4).

Appendix H

Symbolic differential equation solver of MATLAB

Frequently derivatives constitute the circuit dynamics of an electrical system. Algebraic equations involving the derivatives are called differential equations. When we have one dependent and one independent variables, the differential equation is termed as the ordinary differential equation (ODE). For instance $\frac{dy}{dx}+y=9$ and $\frac{d^4y}{dx^4}+y=9$ are the ordinary differential equations, where y is the dependent and x is the independent variables. Order of a differential equation is the highest order of derivatives present in the equation. A first order ordinary differential equation has the form $f\left(x,y,\frac{dy}{dx}\right)$=0, a second order ordinary differential equation has the form $f\left(x,y,\frac{dy}{dx},\frac{d^2y}{dx^2}\right)$=0…so forth. The solution of an ordinary differential equation is the function that satisfies the differential equation. Method of finding the symbolic solution of an ordinary differential equation in MATLAB is addressed in this appendix.

In the case of electrical circuits, the differential equation dependent variables usually are inductor current $i_L(t)$ and capacitor voltage $v_C(t)$ and the independent variable is t. The differential equations that have to be solved in most electrical systems are related to the $\frac{di_L(t)}{dt}$ and/or $\frac{dv_C(t)}{dt}$ or their higher order derivatives.

The master function dsolve (abbreviation for differential equation solve) finds almost all widely practiced ordinary differential equation's analytical solution in close mathematical form if it exists. The input to or output from the dsolve is assumed to be symbolic. The vector code (appendix B) of the given differential equation is written for finding the

First order ODE: for example A,

```
>>S=dsolve('Dy=4*x^4*y^2','x'); ↵
>>pretty(S) ↵
                 5
     - --------------
              5
         4 x  - 5 C1
```

First order ODE: for example B,

```
>>S=dsolve('Dy=4*x^4*y^2','y(0)=2','x'); ↵
>>pretty(S) ↵
                 5
         - -----------
                5
           4 x  - 5/2
```

First order ODE: for example C,

```
>>S=dsolve('2*x*Dy=y^2/x+4*y','x'); ↵
>>pretty(S) ↵
                2
               x
         -2 ----------
            x - 2 C1
```

First order ODE: for example D,

```
>>S=dsolve('x^3*Dy+x*y=y^2','x'); ↵
>>pretty(S) ↵
                 x
     ----------------------------
     1 - x + exp(- 1/x) C1 x
```

First order ODE: for example E,

```
>>S=dsolve('Dy=y^2/t+y/t-2/t,y(1)=2'); ↵
>>pretty(S) ↵
                3
        - 1/2 t  - 1
        -------------
                    3
          -1 + 1/4 t
```

solution. Unless description of the independent variable is inserted, t is understood as the independent variable to the dsolve. The code of the equation, initial condition, and independent variable all are written under quote but separated by a comma. We maintain font equivalence in all computations onwards (like x⇔ x). The return from the dsolve can be assigned to some user-given name say S which essentially holds the vector code of the solution. The built-in function pretty (appendix C.10) is applied on the codes stored in S to view the readable form.

⬧⬧ First order ordinary differential equations

We address some examples of the first order differential equations with and without the initial conditions in the following. In order to feed the first order derivative, the operator D is used which indicates the $\frac{d}{dt}$ or $\frac{d}{dx}$ of calculus on which the $\frac{dy}{dt}$ is written as Dy. If the dependent variable is u , we write Du for the derivative. Regardless of the independent variable, the Dy or Du represents the first order derivative. Some first order differential equations accompanying their final solutions are provided in the following which we intend to compute in MATLAB.

A. $\frac{dy}{dx}=4x^4y^2$ has the solution $y=\frac{-5}{4x^5+5C}$

B. $\frac{dy}{dx}=4x^4y^2$ with the initial condition $y(0)=2$ has the solution $y=\frac{-10}{8x^5-5}$

C. $2x\frac{dy}{dx}=\frac{y^2}{x}+4y$ has the solution $y=\frac{2x^2}{C-x}$

D. Bernoulli equation $x^3y'+xy=y^2$ has the solution $y=\frac{x}{1-x+Cxe^{-\frac{1}{x}}}$

E. Riccati equation $\frac{dy}{dt}=\frac{y^2}{t}+\frac{y}{t}-\frac{2}{t}$ with the initial condition $y(1)=2$ has the solution $y=\frac{2t^3+4}{4-t^3}$

Attached commands in the right side text box of the last page are the implementation of all examples. Concerning the attached text box example A of the last page, the equation has the vector code Dy=4*x^4*y^2 and independent variable x which become the input arguments to the dsolve respectively. The - C1 means the arbitrary constant C. In example B, there is one initial condition $y(0)=2$ which is written as y(0)=2. When the initial condition is present, the dsolve has three input arguments, the first, second, and third of which are the equation code, initial condition, and independent variable respectively. The return might need rearranging to be identical with the given expression (as in examples B and E). The arbitrary constant C in the examples C and D are equivalent to 2 C1 and C1 respectively. In the example E, the independent variable is t which is the default one so there is no necessity to write about it. The workspace S holds the code for the solution of y in all five examples.

For some differential equation the output might seem bizarre. Let us try to find the solution of the differential equation $\frac{dy}{dx}+y\sin x=x$ by applying the same function and symbology as follows:

```
>>S=dsolve('Dy+sin(x)*y=x','x'); pretty(S) ↵
                               /
                               |
          exp(cos(x))          | x exp(-cos(x)) dx + exp(cos(x)) C1
                               |
                               /
```

As the return says, the solution of the equation is $y = e^{\cos x}\int xe^{-\cos x}dx + Ce^{\cos x}$ that means the solution is possible through some integration.

◆◆ Second order ordinary differential equations

A second order ordinary differential equation has the form $f\left(x, y, \frac{dy}{dx}, \frac{d^2y}{dx^2}\right)$=0. The second order derivative like $\frac{d^2y}{dt^2}$ or $\frac{d^2u}{dt^2}$ is represented by D2y or D2u respectively where D2 is the second order derivative operator irrespective of the independent variable. There must be two initial values in the second order equation if the initial values are given. We follow the same function and workspace variable as conducted for the first order examples. The same vector code is used to write the ODE. We are going to verify the following second order differential equation examples in MATLAB whose solutions are given beside.

A. $\frac{d^2y}{dt^2} - 2\frac{dy}{dt} - 15y = 0$ has the solution $y(t) = C_1e^{5t} + C_2e^{-3t}$ where C_1 and C_2 are the two arbitrary constants

B. $u'' + u' + u = 0$ with $u(0) = 1$ and $u'(0) = \sqrt{3}$ has the solution $u(t) =$

$$e^{-\frac{t}{2}}\cos\frac{\sqrt{3}}{2}t + e^{-\frac{t}{2}}\left(2 + \frac{1}{\sqrt{3}}\right)\sin\frac{\sqrt{3}}{2}t$$

C. $y'' + 4y' + 4y = e^x + e^{-2x}$ has the solution $y = e^{-2x}(C_1 + C_2x) + \frac{e^x}{9} + \frac{x^2e^{-2x}}{2}$

D. Euler equation $2t^2\frac{d^2y}{dt^2} - 3t\frac{dy}{dt} - 3y = 0$ has the solution $y(t) = \frac{C_1 + C_2t^{\frac{7}{2}}}{t^{\frac{1}{2}}}$

```
Second order ODE: for example A,
>>S=dsolve('D2y-2*Dy-15*y=0'); pretty(S) ↵
       C1 exp(5 t) + C2 exp(-3 t)
Second order ODE: for example B,
>>S=dsolve('D2u+Du+u=0','u(0)=1,Du(0)=sqrt(3)'); pretty(S) ↵
           1/2                        1/2                         1/2
    (1/3 3    + 2) exp(- 1/2 t) sin(1/2 3    t) + exp(- 1/2 t) cos(1/2 3    t)
Second order ODE: for example C,
>>S=dsolve('D2y+4*Dy+4*y=exp(x)+exp(-2*x)','x'); ↵
>>pretty(S) ↵
                                                          2
  exp(-2 x) C2 + exp(-2 x) x C1 + 1/18 (2 exp(3 x) + 9 x   ) exp(-2 x)
Second order ODE: for example D,
>>S=dsolve('2*t^2*D2y-3*t*Dy-3*y=0'); ↵
>>pretty(S) ↵
          C1          3
         ------ + C2 t
          1/2
         t
```

Shown inside the text box of the last page is the implementation of all four second order ODE examples. In example A, the equation has the code D2y-2*Dy-15*y=0 whose default variable is t. The C1 and C2 indicate the arbitrary constants C_1 and C_2 respectively. In example B, the dependent variable is u on that D2u⇔$\frac{d^2u}{dt^2}=u''$ and Du⇔$\frac{du}{dt}=u'$. The initial conditions $u(0)=1$ and $u'(0)=\sqrt{3}$ are passed by writing u(0)=1 and Du(0)=sqrt(3) respectively. The dsolve has two input arguments, the first and second of which are the equation code and initial condition set respectively. All initial conditions as a set are put under the quote but each one is separated by a comma. In the example C, the independent variable is x so the default t does not work any more. Information of the independent variable has to be passed on through the second input argument of the dsolve. The solution of the example D needs rearrangement to have given identical expression. In all examples workspace S holds the code of the solution.

♦♦ Higher order ordinary differential equations

The methodology so described for the first and second order ODE is extended to handle the higher order ordinary differential equations. Applying the symbol, function, and input argument style of foregoing examples, we carry out the computation of the following higher order differential equations in MATLAB whose final solutions are also provided beside.

A. $y''' + y' - 2y = 0$ has the solution $y(x) = C_1 e^x + e^{-\frac{x}{2}}\left[C_2 \cos\frac{\sqrt{7}x}{2} + C_3 \sin\frac{\sqrt{7}x}{2}\right]$

where C_1, C_2, and C_3 are the arbitrary constants

B. $y''' - y'' = 0$ with the initial conditions $y(1)=1$, $y'(0)=3$, and $y''(-1)=-1$ and the independent variable u has the solution $y(u)=e^2-e-2+(3+e)u-e^{u+1}$

Solution of both examples is presented in the right side attached text box. Regarding the implementation of example A, the third order derivative y''' has the code D3y where D3 is the derivative operator regardless of the independent variable. The C1, C2, and C3 indicate the three arbitrary constants C_1, C_2, and C_3 respectively. The three initial conditions $y(1)=1$, $y'(0)=3$, and $y''(-1)=-1$ of example B are passed on writing y(1)=1, Dy(0)=3, and D2y(-1)=-1 respectively. In both examples, the return

```
Higher order ODE: for example A,
>>S=dsolve('D3y+Dy-2*y=0','x'); pretty(S) ↵
                                            1/2
  C1 exp(x) + C2 exp(- 1/2 x) sin(1/2 7    x)

                                   1/2
        + C3 exp(- 1/2 x) cos(1/2 7    x)
Higher order ODE: for example B,
>>S=dsolve('D3y-D2y=0','Dy(0)=3,D2y(-1)=-1,y(1)=1','u'); ↵
>>pretty(simplify(S)) ↵
        -exp(1) - 2 + exp(2) + u exp(1) + 3 u - exp(1 + u)
for example B: entering input arguments separately,
>>E='D3y-D2y=0'; ↵
>>I='Dy(0)=3,D2y(-1)=-1,y(1)=1'; ↵
>>S=dsolve(E,I,'u'); pretty(simplify(S)) ↵
      -exp(1) - 2 + exp(2) + u exp(1) + 3 u - exp(1 + u)
```

is assigned to the workspace S which holds the code for $y(x)$ or $y(u)$. In the example B, we utilized the command simplify to simplify the return stored in S.

If a differential equation is lengthy, we first enter the equation or the initial conditions and call the dsolve afterwards. We again presented the implementation for the example B in just mentioned textbox in which the first and second lines are to assign the differential equation and the initial conditions to the workspace variables E and I (any user-given name) respectively, both of which are under the quote. In the dsolve, we do not use the quote again but maintain the same input argument order as done previously.

♦♦ System of differential equations

A system of differential equations is given by $\begin{Bmatrix} \frac{dy_1}{dt} = f_1(t, y_1, y_2, \ldots\ldots, y_n) \\ \frac{dy_2}{dt} = f_2(t, y_1, y_2, \ldots\ldots, y_n) \\ \vdots \\ \frac{dy_n}{dt} = f_n(t, y_1, y_2, \ldots\ldots, y_n) \end{Bmatrix}$

where y_1, y_2, ..., and y_n are the n dependent variables and t is the independent variable. Our intention is to find the functions y_1, y_2, ..., and y_n satisfying the differential equations of the system simultaneously.

Ongoing function dsolve is so adaptable that it finds the solution of the system of differential equations as well. The common syntax for the solution to a system of differential equation is dsolve(equation 1, equation 2, etc, initial condition set, independent variable). The coding of the equations or initial conditions follows the same style as conducted previously. There is some exception regarding the return from the dsolve which is now a structure array (beyond the scope of the text, reference 4). Let us see the following examples on the system of ODEs.

Example 1:

It is given that $\begin{bmatrix} y_1 \\ y_2 \end{bmatrix} = \begin{bmatrix} -\frac{3}{5}e^t + \frac{8}{5}e^{6t} \\ \frac{6}{5}e^t + \frac{4}{5}e^{6t} \end{bmatrix}$ is the solution of the system $\begin{Bmatrix} \frac{dy_1}{dt} = 5y_1 + 2y_2 \\ \frac{dy_2}{dt} = 2y_1 + 2y_2 \end{Bmatrix}$ subject to the initial conditions $\begin{Bmatrix} y_1(0) = 1 \\ y_2(0) = 2 \end{Bmatrix}$ which we wish to compute in MATLAB.

The dependent variables y_1 and y_2, the derivatives $\frac{dy_1}{dt}$ and $\frac{dy_2}{dt}$, and the initial conditions $y_1(0) = 1$ and $y_2(0) = 2$ are written as y1 and y2, Dy1 and Dy2, and y1(0)=1 and y2(0)=2 respectively where the y1 and y2 are user-chosen names.

The text box on the right side of the next page shows the computation of the system solution in which the first two lines are to assign the vector codes of the equations to the workspace e1 and e2 (can be any user-given names) respectively. The third line is to assign the given initial conditions to the workspace I (can any name of your choice). Note that each equation is under a quote but the initial condition set is under the quote. The fourth line calls the dsolve for the solution and assigns the outcome to the user-given S. The dsolve has three input arguments, the first, second, and third of which are the given first equation, given second equation, and the given initial condition set respectively.

The contents of S say that the solution of the system is returned as a structure array which has two members y1 and y2, obviously they are the dependent variables of the given system. In order to pick up the member y1 from the structure S, we use the command S.y1. Similar explanation also goes for the y2. We assigned the members of the array to the workspace p1 and p2 (can be any other user-given names) which essentially hold the code for the solution of y_1 and y_2 respectively from what the pretty just displays the nice form.

Solving system of ODEs for the example 1:

```
>>e1='Dy1=5*y1+2*y2'; ↵
>>e2='Dy2=2*y1+2*y2'; ↵
>>I='y1(0)=1,y2(0)=2'; ↵
>>S=dsolve(e1,e2,I) ↵

S =
      y1: [1x1 sym]
      y2: [1x1 sym]
>>p1=S.y1; p2=S.y2; ↵
>>pretty(p1) ↵
      8/5 exp(6 t) - 3/5 exp(t)
>>pretty(p2) ↵
      4/5 exp(6 t) + 6/5 exp(t)
```

If the independent variable were x instead of the default t, the command would be S=dsolve(e1,e2,I,'x') in which the fourth input argument is the independent variable indicatory letter under a quote.

Without the presence of the initial conditions, the command would be S=dsolve(e1,e2,'x') and we would see the result in terms of the arbitrary constants.

Example 2:

Let us compute the solution of the nonhomogeneous system $\begin{Bmatrix} \frac{dx}{dt} = 31x - 21y + 9z - e^{-3t} \\ \frac{dy}{dt} = 44x - 30y + 12z + 2t \\ \frac{dz}{dt} = -22x + 14y - 8z + \sin t \end{Bmatrix}$ satisfying the initial conditions $\begin{Bmatrix} x(0) = -2 \\ y(0) = 1 \\ z(0) = 0 \end{Bmatrix}$ in MATLAB, which is given by $\begin{bmatrix} x \\ y \\ z \end{bmatrix} =$

$$\begin{bmatrix} (33t + \frac{3743}{30})e^{-3t} - \frac{1317}{10}e^{-2t} - 7t + \frac{35}{6} - \frac{9}{10}\cos t + \frac{9}{10}\sin t \\ (44t + \frac{7426}{45})e^{-3t} - \frac{1701}{10}e^{-2t} - \frac{25}{3}t + \frac{131}{18} - \frac{6}{5}\cos t + \frac{6}{5}\sin t \\ (-22t - \frac{3713}{45})e^{-3t} + 86e^{-2t} + \frac{2}{5}\cos t - \frac{1}{5}\sin t + \frac{14}{3}t - \frac{35}{9} \end{bmatrix}.$$

Following the symbology and functions from ongoing discussion, we find the solution of the system as attached in the text box of next page, in which the first two lines are to assign the three differential equations and the initial condition set to the workspace variables e1, e2, e3, and I respectively. The third line calls the function dsolve for the solution which returns a structure holding three members x, y, and z representing $x(t)$, $y(t)$, and $z(t)$ respectively. We picked up the $x(t)$, $y(t)$, and $z(t)$ from the S in the fourth line command and assigned them to the workspace p1, p2, and p3 respectively. So to say, the last three variables hold the codes for the $x(t)$, $y(t)$, and $z(t)$ we are looking for respectively. Just to view the solution for $x(t)$, we execute the command pretty(p1). In a similar fashion you can verify the symbolic solution for the $y(t)$ and $z(t)$ by executing pretty(p2) and pretty(p3) respectively.

Solving the nonhomogeneous system of ODEs for the example 2:

```
>>e1='Dx=31*x-21*y+9*z-exp(-3*t)'; e2='Dy=44*x-30*y+12*z+2*t'; ↵
>>e3='Dz=-22*x+14*y-8*z+sin(t)'; I='x(0)=-2,y(0)=1,z(0)=0'; ↵
>>S=dsolve(e1,e2,e3,I) ↵

S =
        x: [1x1 sym]
        y: [1x1 sym]
        z: [1x1 sym]
>>p1=S.x; p2=S.y; p3=S.z; ↵     ← p1⇔ x(t), p2⇔ y(t), p3⇔ z(t)
>>pretty(p1) ↵
                    3743
  -7 t + 35/6 + ------ exp(-3 t) - 9/10 cos(t) + 9/10 sin(t) + 33 t exp(-3 t)
                     30

        1317
      - ------ exp(-2 t)
         10
```

References

>> >> Signal and System Fundamentals >> >>

[1] Robert L. Boylestad, "*Introductory Circuit Analysis*", 9th Edition, 2000, Prentice Hall, Inc., Pearson Education, Upper Saddle River, New Jersey.

[2] Leonard S. Bobrow, "*Elementary Linear Circuit Analysis*", 1987, Second Edition, Harcourt Brace Jovanovich College Publishers, New York.

[3] Raymond A. DeCarlo and Pen-Min Lin, "*Linear Circuit Analysis – Time Domain, Phasor, and Laplace Transform Approaches*", Second Edition, 2001, Oxford University Press, New York.

[4] James A. Cadzow and Hugh F. Van Landingham, "*Signals, Systems, and Transforms*", 1985, Prentice Hall, Inc., Englewood Cliffs, New Jersey.

[5] Rulph Chassaing and Darrell W. Horning, "*Digital Signal Processing with the TMS320C25*", 1990, John Wiley & Sons, New York.

[6] Maurice Bellanger, "*Digital Processing of Signals – Theory and Practice*", Second Edition, 1989, John Wiley & Sons Ltd, New York.

[7] C. K. Chui and G. Chen, "*Signal Processing and Systems Theory – Selected Topics*", 1992, Springer-Verlag, New York.

[8] K. G. Beauchamp, "*Transforms for Engineers – A Guide to Signal Processing*", 1987, Oxford University Press, New York.

[9] David K. Cheng, "*Analysis of Linear Systems*", 1990, Narosa Publishing House, New Delhi.

[10] Mourad Barkat, "*Signal – Detection and Estimation*", 1991, Artech House, Inc., Norwood, MA.

[11] H. Baher, "*Analog & Digital Signal Processing*", 1990, John Wiley & Sons, New York.

[12] D. Brook and R. J. Wynne, "*Signal Processing – Principles and Applications*", 1988, Edward Arnold, London.

[13] Richard E. Blahut, "*Algebraic Methods for Signal Processing and Communications Coding*", 1992, Springer-Verlag, New York.

[14] Douglas F. Elliott, "*Handbook of Digital Signal Processing: Engineering Applications*", 1987, Academic Press.

[15] Sanjit K. Mitra and James F. Kaiser, "*Handbook for Digital Signal Processing*", 1993, John Wiley & Sons.

[16] Monson H. Hayes, "*Schaum's Outline of Theory and Problems of Digital Signal Processing*", 1999, McGraw-Hill Companies, Inc.

>> >> MATLAB and SIMULINK >> >>

[1] Mohammad Nuruzzaman, "*Tutorials on Mathematics to MATLAB*", 2003, AuthorHouse, Bloomington, Indiana.

[2] Mohammad Nuruzzaman, "*Modeling and Simulation in SIMULINK for Engineers and Scientists*", 2005, AuthorHouse, Bloomington, Indiana.

[3] Mohammad Nuruzzaman, "*Digital Image Fundamentals in MATLAB*", 2005, AuthorHouse, Bloomington, Indiana.

[4] Mohammad Nuruzzaman, "*Technical Computation and Visualization in MATLAB for Engineers and Scientists*", February, 2007, AuthorHouse, Bloomington, Indiana.

[5] Mohammad Nuruzzaman, "*Electric Circuit Fundamentals in MATLAB and SIMULINK*", October, 2007, BookSurge Publishing, Charleston, South Carolina.

[6] Duffy, Dean G., *"Advanced Engineering Mathematics with MATLAB"*, Second Edition, 2003, Chapman & Hall, CRC, Boca Raton.

[7] Hanselman, Duane C. and Littlefield, Bruce R., "*Mastering MATLAB 5: A Comprehensive Tutorial*", 1998, Prentice Hall, Upper Saddle River, New Jersey.

[8] Shampine, Lawrence F. and Reichelt, Mark W., *"The MATLAB ODE Suite"*, 1996, The Math-Works, Inc., Natick, MA.

[9] Marcus, Marvin, "*Matrices and MATLAB - A Tutorial*", 1993, Prentice Hall, Englewood Cliffs, N. J.

[10] Ogata, Katsuhiko, "*Solving Control Engineering Problems with MATLAB*", 1994, Englewood Cliffs, N. J. Prentice Hall.

[11] Part-Enander, Eva, "*The MATLAB Handbook*", 1998, Harlow: Addisson Wesley.

[12] Prentice Hall, Inc., "*The Student Edition of MATLAB for MS-DOS Personal Computers*", 1992, Prentice Hall, Englewood Cliffs, N. J.

[13] Saadat, Hadi., "*Computational Aids in Control Systems Using MATLAB*", 1993, McGraw-Hill, New York.

[14] Gander, Walter. and Hrebicek, Jiri., "*Solving Problems in Scientific Computing Using MAPLE and MATLAB*", 1997, Third Edition, Springer Verlag, New York.

[15] Biran, Adrian B and Breiner, Moshe, "*MATLAB for Engineers*", 1997, Addison Wesley, Harlow, Eng.

[16] D. M. Etter, "*Engineering Problem Solving with MATLAB*", 1993, Prentice Hall, Englewood Cliffs, N. J.

[17] Shahian, Bahram. and Hassul, Michael., "*Control System Design Using MATLAB*", 1993, Prentice Hall, Englewood Cliffs, N. J.

[18] Prentice Hall, Inc., "*The Student Edition of MATLAB for Macintosh Computers*", 1992, Prentice Hall, Englewood Cliffs, N. J.

[19] Ogata, Katshuiko, "*Designing Linear Control Systems with MATLAB*", 1994, Prentice Hall, Englewood Cliffs, N. J.

[20] Bishop, Robert H., "*Modern Control Systems Analysis and Design Using MATLAB*", 1993, Addsison Wesley, Reading, MA.

[21] Moscinski, Jerzy and Ogonowski, Zbigniew., "*Advanced Control with MATLAB and Simulink*", 1995, E. Horwood, Chichester, Eng.

[22] Alberto Cavallo, Roberto Setola, and Francesco Vasca, "*Using MATLAB Simulink and Control Systems Toolbox - A Practical Approach*", 1996, Prentice Hall, London.

[23] Jackson, Leland B., "*Digital Filters and Signal Processing with MATLAB Exercises*", Third Edition, 1996, Kluwer Academic Publishers, Boston.

[24] Kuo, Benjamin C. and Hanselman, Duanec., "*MATLAB Tools for Control System Analysis and Design*", 1994, Prentice Hall, Englewood Cliffs, N. J.

[25] Chipperfield, A. J. and Fleming, P. J., "*MATLAB Toolboxes and Applications for Control*", 1993, London, New York: Peter Peregrinus on Behalf of the Institute of Electrical Engineers.

[26] Math Works Inc., "*MATLAB Reference Guide*", Math Works Inc., 1993, Natick, Massachusetts.

[27] Cleve Moler and Peter J. Costa, "*MATLAB Symbolic Math Toolbox*", User's Guide, Version 2.0, May 1997, Natick, Massachusetts.

[28] James W. Nilsson and Susan A. Riedel, "*Using Computer Tools for Electric Circuits*", 1996, Fifth Edition, Addison Wesley Publishing Company, Natick, Massachusetts.

[29] Theodore F. Bogart, "*Computer Simulation of Linear Circuits and Systems*", 1983, John Wiley and Sons, Inc., New York.

Subject Index

R

S

ISBN 978-1-4196-9934-4

51899 >

9 781419 699344

Printed in the USA

Made in the USA
Middletown, DE
07 October 2023

40417685R00176